# Montana Stirrups, Sage and Shenanigans
## Western Ranch Life in a Forgotten Era

Francie Brink Berg
Anne Brink Sallgren Krickel
Jeanie Brink Thiessen

# Acknowledgments

*We are grateful to the following who have contributed and assisted with this book: Jean Nielson, Research Librarian, Miles City Public Library; Carol Watts, Custer County Conservation District; Montana Historical Society Research Center, Helena; Montana Fish, Wildlife and Parks, Helena, and Scotty Denson, Miles City regional office; Wm. (Butch) Krutzfeldt, current owner of the ranch; Edwin Hill, Tusler school friend; family members and friends, especially Diane Sallgren Covington and Julie West for their consistent help. Our appreciation to Dylan Edwards, Graphic Designer for the Theodore Roosevelt Medora Foundation, whose modern cover captures the spirit of the old west. A special thanks to Ronda I. Fink for her skill and dedication in the book production. In addition we want to acknowledge our debt to the pioneers and neighbors we knew growing up who influenced our lives and taught us, along with our parents, the ways of the west.*

*Montana Stirrups, Sage and Shenanigans:*
*Western Ranch Life in a Forgotten Era*
Francie Brink Berg
Anne Brink Sallgren Krickel
Jeanie Brink Thiessen

First printing
Copyright ©2013
by Frances M. Berg
All rights reserved. Reproduction in
whole or part prohibited without
publisher's written permission.

ISBN: 978-0-918532-75-6 Hardcover
ISBN: 978-0-918532-76-3 Softcover

Cover design by Dylan Edwards
Layout and production by Ronda Turbiville Fink
Illustrations by Anne Brink Krickel
Photos are from the Brink family collections

Published by
Flying Diamond Books
402 South 14th Street
Hettinger, ND 58639

info@montanastirrupsandsage.com
www.montanastirrupsandsage.com

# Introduction

We trailed our cattle out that first time north of Miles City across the Yellowstone River, finding our way through lovely, but rugged, unknown country. Just the two of us, a hundred head of cows, a borrowed horse and our trusty Eagle, blind in one eye. In four days of hard riding we reached our summer pasture, making ten or fifteen miles a day.

But we weren't finished yet. The fifth morning we saddled up and rode our horses those many miles home again. Coming out of the Depression, we didn't own a stock rack for our truck and had never seen a horse trailer.

It was a time of independence and self-reliance, of ranchers who lived with a sense of freedom, love of the land and joy of life, and knew the closeness of family and community. There were no telephones or all-terrain vehicles, and few ranches had electricity.

We write of ranch life as lived in Montana and other western states through a legacy of pioneer values and traditions. In many ways it is the story of young ranching families all across the west during a time that demanded resourcefulness, hard work and courage.

In our stories of branding and trailing cattle, breaking horses, hunting and Nazi Prisoners of War in our fields, we hope the reader will recognize the traditional values we lived by, our challenges and the satisfaction of jobs well done. The transformation of a wild range colt into a savvy and affectionate cow horse, our delight in the boss cow of the trail herd lifting her head at first scent of home, overcoming emergencies of accident and fire, helping a needy stranger and learning to love nature and the wildlife that surrounded us. Even though our work could be difficult, it was also threaded through by pranks, jokes and the fun we had creating them.

Our unique location, as the last place in the valley, where the river swings in close against high, steeply eroded bluffs, was the natural route of entry for travelers up the Yellowstone River—even today a scenic river, flowing free without dams for its entire length. Alongside it, the easiest way through the valley cuts right across our land, avoiding river crossings. Thus most people who came west as far as Miles City, visited Yellowstone Park, or crossed the continent at this latitude, almost certainly traversed the land that became the Brink ranch, or rode the rails or river along its edge. It was where early travelers naturally came across and where modern road builders find the easiest course between valley and high bluffs.

Circling rimrocked hills overlooked our ranch buildings. Who knew what fascinating bits of history these rims had seen in the past several hundred years? It didn't take much to visualize Crow and Cheyenne hunters running buffalo over the steep sandstone cliffs or early pioneers wending horse-drawn wagons through gaps in the towering rocks. All this only a quarter-mile from our house. We spent many hours scouring these hills, exploring, stalking deer, tracing the routes of five major trails and roads built across our ranch.

A good place for a cattle ranch, sheltered and with good water sources, ours was established in the 1880s as headquarters for a big cattle operation. An early owner, Henry Tusler, once ran thousands of cattle on both sides of the Yellowstone River.

By the time we came, the era of cattle kingdoms and free range had ended, along with the rush of homestead years. Advances in ranching were yet to come. This is a period when ranch activities are seldom noted by western historians. Largely unrecorded, unheralded, in a sense it is almost forgotten.

Our family and our neighbors struggled to overcome the depression—a desperate time to raise four small children—in which many lost everything, as we did. Then war came. Hitler goose-stepped his armies through Europe with his goal of conquering the world, while Japan captured every island one after the other on its way south through the Pacific. All this created growing anxiety among Americans and we did all we could to help on the home front.

Western ranching changed dramatically during those years. It was a time of transition—almost as if one foot remained in the old west, while the other reached out ahead toward an easier but more complex future. New technologies and a new social culture were still ahead.

In many ways our parents did things in the old ways, in others they eagerly adopted emerging science.

We lived their heritage intertwined with ours. Both their families came west near the turn of century: Mom in 1906 at age five, travelling by covered wagon; Dad from Minnesota three years later at age thirteen. They respected and lived the exuberant spirit of the west and kept its sense of humor alive through stories told around the supper table. We felt a strong connection to those earlier days as we explored the remnants of rangeland homesteads, abandoned for decades. Riding through these ruins, we wondered about the mother who watered those few stunted trees and the child who clung to a faded, lumpy teddy bear we found grown through by thorny wild roses.

Our parents lived each day with courage, humor, enthusiasm, a positive attitude and hard work. We learned to laugh and enjoy what life dealt us.

We accepted the unwritten, unspoken code of the west with its solid values of character, tolerance and respect for the privacy of others. Ethics of honesty and honor that once impelled Grandma to turn down at the land office, a much needed free 160 acres of good pasture, because she refused to sign a paper that said they would use that 'stone quarter' for its building stone, as the federal government demanded.

We were taught to appreciate and respect the uncommon people of the west, no matter their outward appearance, whether a widow crocheting in her rocking chair, living alone on a hilly gravel-poor claim, scrawny and opinionated, soft-hearted toward kids, ready with a ginger cookie and story. How did she live? Neighbors stopped by. We girls cringed at the visit of the bachelor neighbor, nose dripping in winter, yet we set another plate at the table.

We learned the connectedness of all living things—a soaring bald eagle above and dinosaur bones beneath our feet, in the famed Hell Creek Formation. Often we picked up fossil remnants, petrified wood, and tiny seashells from anthills on a gravel ridge, denizens of a time our drought-prone land teemed with swamps, jungles and prehistoric monsters.

It is our hope that through these pages others will experience these connections and a sense of stewardship of the land. We lifted our heads to the song of a meadowlark and coo of the mourning dove, dusted off buffalo skulls, glimpsed the skunk, raccoon and porcupine, and startled a pair of shy whitetail deer from a cedar draw. Living close to nature, we valued the land and its gifts. And yes, even coyotes and rattlesnakes, old adversaries, ever deserving our respect.

It's been said those were simpler times. In some ways, they were. Yet in complex ways, unknown today, our family planned ahead. We learned survival skills and used them daily. Before bringing up a jar of canned meat from a basement shelf and grinding it with dill pickles for sandwiches, we had to butcher the beef, raise the pig or shoot the deer, then cut and can the meat in a pressure cooker. To bake bread, begun hours before, we slipped on a coat to chop wood, added coal and built up a hot fire—but not too hot—then put a hand in the oven to gauge its baking temperature. A block of ice cooling milk in our ice box in August emerged from the Yellowstone River by way of a January neighborhood ice-cutting party. In a sense, we were the ants in the old story. Grasshoppers never made it.

Yes, it could be cold. Crisp and frosty, so our nostrils sucked together with each breath. We heeded Mom's warning against lying down and giving in to drowsiness: "People can freeze to death that way." Then suddenly, came a January thaw in the night. We woke to a Chinook wind—that warm, spring-like breeze swishing through the cottonwoods and melting snow from the hillsides. Deep winter disappeared and the air smelled delightfully damp and warm.

We could have hail, pounding rain or summer heat—or gorgeous weather in every season. We remember crisp sunny mornings and the fresh, clean essence of spring, amid the scents of cedar and sage carried on a soft breeze, while a meadowlark sang out his piercingly lovely melody from the top of the tallest pine and our cows strung out eagerly down the trail. Summer days were never more beautiful than after a refreshing rain, when sun flooded the cliffs with golden light and each color glistened—rose scoria, black coal veins, layers of slick gumbo dusted with thin slices of sparkling mica, and the vivid, varied greens of grassy draws edged in silver juniper and sage—while a double rainbow arched across the eastern sky.

Little did we know in September 2001, as we three sisters camped on the bluffs overlooking our lower pasture, that our reminiscing would take book form. A death in the family had brought us together from distant homes and on a whim we set up a tent one evening on the ridge overlooking our ranch and the Yellowstone River below.

Next morning, hiking that ridge, we pointed out below a rock, a trail, a water hole, triggering one memory after another.

"Do you remember the time when…we found that coyote den with six pups…watched the Yellowstone flooding…hunted deer in the school section…pulled the prank that…"

The stories flowed and laughter erupted as we recalled one incident after the other, interweaving the memories, transported back a half century in time, telling each other, "You should write that down!"

The idea didn't go away. We did write some of it down. Then more, including a few shenanigans. As the stories took shape over months and even years, we knew again the pleasures of working together, depending on each other to get the job done. It's been a joy renewing the close friendships and love we once knew, this time putting the experiences on paper. Often we missed Beverley, older sister, award-winning journalist and charismatic leader of many adventures. We like to think her enthusiasm and laughter is alive in these pages.

Looking down on the blue flowing ribbon of the Yellowstone that day from the bluff reminded us of the long history of this land. How fleeting were the years we spent here. The ranching style of that era, now gone. We've endeavored to capture that time and those long ago challenges with integrity, a fresh sense of adventure and humor.

We want to bring to life those ranching years among the badlands, buttes and valley lands of eastern Montana. This, then, is our story.

Francie Brink Berg
Anne Brink Sallgren Krickel
Jeanie Brink Thiessen

# Contents

**Introduction**

1. **Stirrups, Sage and Shenanigans**
   Trailing into the Blizzard    11
   Powder River Romeo    15
   Lunch for Lawrence    19
   First Hunt    22
   Riding the Ditcher    25
   The Goose That Stayed    29
   Move to Brink Ranch    34

2. **Horses We Knew**
   Rider Over Backwards    39
   Facing the Fast Freight    42
   Buck—Mr. Personality    46
   Hung Up in the Stirrup    52
   Corralling Spooky Horses    57
   Loading Horses on the Range    60
   Horses We Knew and Loved
              Most of the Time    61
   Wild Horses    66

3. **Trailing Cattle**
   First Time Trailing    73
   Finding the Way    79
   Spring Creek Canyon    83
   String 'Em Out    86
   Trailing Troubles    91
   Third Horse High Jinks    97
   Sleeping in the Open    99
   Jerky and Beans    105

4. **Creating Ranch Fun**
   Fishing the Yellowstone    108
   My How Hot    113

Taxidermy Aromas   119
Prairie Dog Pets   121
Tight Fit   128
Feedlot Rodeo   130
Cutting the Christmas Tree   134
How We Celebrated Christmas   137

## 5. Riding Summer Ranges

Comet Runaway   141
Snake Bite   144
Chuckwagon Camp   147
Riding the Wide Open   151
Stallion Attack   157
Prairie Fire   162

## 6. Hunting—Rancher Style

Roadkill on the Rez   169
Elk Firing Line   172
Lucky Shots   178
Game Warden Dilemma   181
Rangeland Hunting Ethics   187
Crow Rock Antelope Flight   188
Lively Skunks   192
Two-Fifty Savage   195
Seven Deer in Seven Days   200

## 7. On the Home Front

Amazing Cherry Pie   203
For the War Effort   206
Garden Bountiful   208
Frog Egg Breakfast   212
Saved by a Whirling Dervish   215
Echoes in the Kitchen   218
Three-Legged Kitchen Help   224
The Gutsy Bug   226
Gathering 'Round the Piano   229
Rangeland Hospitality Code   230
Pioneer Storytelling   236
Running the Tumbleweeds   243

## 8. Working Livestock

Wild Heifer   246
Working Cattle   248
Coyote Attacks   250
Purple Coyote Carcasses   255
Bringing Home the Cows   258
Branding   260
Milking Antics   263

Cream Separator Do's and Don'ts  268
Patrolman Thomas, Always Vigilant  272
Feeding in Winter  274
Bloated Livestock  276
Sally Was No Lady  282
Mangled by the Greyhound Bus  285

9. **Rural School**

Long Walk Home  289
Playground Creativity  293
The School Trek  299
Horseback to School  304
Intensive Learning at Tusler  307
Three Students on Deadman Road  314
Nothing as Fearful  319

10. **Field Work**

Wheatfield Fire  323
Prisoners of War on our Ranch  326
Overturned John Deere  332
Child Truck Drivers  335
Hayfield Highs  337
Bee City  341
Rattlesnake Risks  344
Tractor Driving Beats Milking  351
Rat Trap Harvest  355
Blowing the Spring  357

11. **Neighbors**

Matchmaking Horseplay  362
Yellowstone Ice Jam  364
Red Ribboned Skunk  369
Neighborhood Fun  373

## Brink Ranch Legacy

A Good Place for Cattle Ranching  380
Our Ranch House  382
Ranch Ownership  385
Major Trails and Roads across our Ranch  386
Celebrating Travelers, Events and Routes through this Land  391
Historical Timeline  392

## Maps

## Tribute

## About the Authors

*To our parents and sister Beverley*

# 1 Stirrups, Sage and Shenanigans

## Trailing into the Blizzard

Snow hit our faces in sharp little pellets as the wind swirled around us. The horses turned their heads sideways into the storm and the cattle bunched up tightly, balking as we urged them forward. We'd been on the trail four days, and planned this one to be our last. We had expected to get the cattle home before nightfall, but we didn't know a blizzard was coming.

ANNE

Light snow started falling before noon and increased in intensity to heavy snow as the afternoon progressed. The temperature dropped and every so often Francie and I got off our horses and walked to warm our feet. We were stiff with cold, and the bottoms of our feet stung with a thousand needle pricks when we eased out of the stirrups and hit the ground.

The storm continued to worsen. The snow hit in icy flakes and the wind swirled it up with the ground snow and slammed it in our faces. We had to find somewhere to hold the cattle till the blizzard blew itself out, however long that took. Dad thought that in a couple of miles we'd reach a pasture where we might leave the herd for the night.

When our lead cow turned tail to the wind, we knew we were in trouble. Dad was following us in the truck and chasing the cattle on foot to help us keep them together and heading in the right direction.

"That ranch is up ahead," he called out. "I'm going up there and make arrangements for a place to hold up until this blows over."

According to the radio the blizzard could last a couple of days, or more. In this hard, fierce wind, we'd need a fenced pasture or corral, so the cattle wouldn't drift with the wind and scatter in all directions.

An especially hard gust of wind hit, and my horse whipped himself around, rump to the wind. It did feel better having the wind

*When our lead cow turned tail to the wind, we knew we were in trouble.*

*Western Ranch Life in a Forgotten Era*

*Our hands stung with cold, holding the reins into the wind and frosty air.*

Our cattle move slowly in arctic cold through the Pine Hills not wanting to face a harsh wind.

at my back, but that was no way to chase cattle.

I faced back into the wind, but the cows didn't want to go that direction any more than my horse did.

One good thing, we were wearing the flannel-lined hoods Mom made for us, with wolverine fur around the edge next to our faces. As we turned at an angle to the wind, the fur stuck out from the hoods three inches, deflecting the sharp wind and snow. Even better, with its own unique quality, when moisture from our breath caught on wolverine fur it never frosted up, as it did on other furs. Since there were no wolverines in eastern Montana, Dad had traded another pelt to Mr. Goldstein from the fur company in Miles City for it. It was so much better to be out in that cold weather without icicles rubbing your face from the fur, or getting wet from your breath.

We wore our warm mittens, too, that Grandma had knit for us from her heavy, maroon wool double yarn. Made with deep wrists, they kept out the snow.

Last spring when we trailed our cattle through the Pine Hills toward our summer range, Francie had lost one of these mittens somehow while dashing after a small wild calf that ran back down the fence a mile or so. When he finally turned in the right direction, still running erratically, there was no time to backtrack and hunt the mitten.

That was a cold, snowy day too, and I saw she was cold, switching her single mitten from hand to hand, stuffing the bare hand in her pocket as she tried to keep them both warm with only one covering. Hands can get so cold since they're always in the wind and weather holding the reins.

Far from any roads, we had been surprised when an old blue pickup came over a hill and jounced down the draw to meet us. A weathered-faced rancher stepped out holding Francie's red mitten in his hand.

"One of you girls lose this?" he asked. "I found it when I drove down to check on my heifers."

"Oh, thank you! How did you ever find it?" Francie broke into a smile and pulled on her mitten, mighty glad not to have to ride the whole day in the cold without it.

"Well, it's red! It showed up real good in the fresh snow."

He chuckled, climbed back in his pickup and disappeared over the hill.

That man had driven miles out of his way as he circled around Pine Hills roads to bring her that mitten. A kindness from unknown ranchers was always a marvel.

They went to great lengths to help a person in need. But then, Mom and Dad did the same—it's what country people do to help each other.

So, we were stunned during this blizzard when we were refused help from the rancher down the road.

Dad wasn't gone long, but when he came back we could see he was angry. He told us that while talking to the rancher, Mr. X was sitting beside his stove warming his stocking feet. He wasn't what we'd call an authentic rancher; he'd recently bought his small place in the Pine Hills and worked full-time in Miles City.

"No," he said, "you can't leave your cattle here. I don't want them. They might break down my fences."

Dad tried to persuade him, explaining that his daughters were driving cattle against the blizzard and needed a place to wait it out. He had a load of hay to feed them, so they wouldn't need care—just a place to wait out the storm—and if they damaged anything, he'd repair it.

To Dad's amazement, Mr. X shook his head.

"No, you can't hold your cattle on my place. I know it's storming and your daughters are out there, but you'll just have to keep going."

No wonder Dad was upset.

We were following a county road with the cattle on unfenced land, though we'd soon come to the fenced, paved highway. This area contained two hundred acres of triangle-shaped land, fenced on two sides.

"We'll hold the cattle up there," Dad told us, "I've got to get you girls out of the cold."

Good news! Francie and I looked at each other with relief.

"But what about the cattle?" we asked. "They'll move on with only two fences, and we'll never find them."

"There's a high hill on the north side that will give them some protection. When I get the load of hay spread from the truck, they'll eat and then bed down. I'm staying with them. Mom will come up in the car and pick up you girls." He continued, "You need to get out of this weather."

We sat in the truck to warm up while Dad took one of the horses and kept moving the cattle against the wind. Truck heaters in those days weren't very warm, but it was good getting out of the storm. It was hard to make ourselves get back out in the blizzard when Dad returned. He said it wasn't much farther, and once more we got on our horses to move the cows.

Mom met us at the highway in the car with a thermos of cocoa. It tasted sooo good! Swallowing the rich, hot liquid warmed us all the way through.

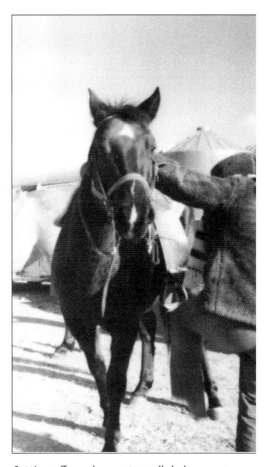

*Getting off our horses to walk helps warm us on frost-bitten days. Anne's icy feet sting when they hit the ground and it's not easy to mount again when stiff with cold.*

"I know it's storming and your daughters are out there, but you'll just have to keep going."

*It was hard making ourselves get back into the blizzard when Dad returned.*

We didn't have to go far to our new stopping point. Dad had our little chuckwagon hooked behind the truck, and he told Mom he'd keep a fire going all night and make sure the cattle bedded down.

In the morning Mom returned with Francie and me to watch the cattle when Dad went home for another load of hay. We kept the cattle there for two days while the blizzard blew itself out.

When the wind died next morning, we pushed the cattle on home, making it in half a day, despite the bitter cold that followed the blizzard. Our lead cow and her trusty cohorts knew we were almost home, making the trail easier. It was just a matter of one of us keeping ahead of them to open gates. From past experience, these cows knew that plenty of hay, water and shelter awaited them at home.

An amusing sequel to this adventure: A month or two later, our folks advertised his polled Hereford bull for sale. Dad, an innovator when it came to agriculture and intrigued by the benefits of cattle without horns, had bought Shorty as a yearling from a breeder in the upper Paradise Valley. He proved to be a good bull, but Dad felt it was time to sell Shorty, to avoid inbreeding young cows.

We listed our phone number in the advertisement but not the price of the bull. With Mom and Dad gone for the day, Jeanie—home from college for the Christmas holiday—was left with explicit directions on answering the phone, including the price for Shorty.

A couple of ranchers called and Jeanie did her best, but none of them wanted to pay our price.

Next day, the phone rang and Mom talked to a prospective bull buyer. Jeanie listened in amazement as Mom quoted Shorty's price, far more than she was supposed to ask. When Mom hung up she had a smug, satisfied look on her face.

"I sold the bull!" she exclaimed. "And I sold him for almost twice what we planned."

"Why did you do that?" Jeanie asked, shocked, knowing that Mom had the highest principles and never deviated from them.

"The man who bought the bull was Mr. X from the Pine Hills," Mom replied. "The extra amount he's paying is what it cost him to leave my girls out in that blizzard!"

## Powder River Romeo

"Am I going to have to drive out there on that slippery, muddy road?" asked Beverley. "Why can't we wait a couple days until it quits raining and dries up?"

"Can't do that," Dad said. "The bull has to get out there. He's gotta start breeding or we'll have late calves."

"But all those animals shifting around in the back of the truck?"

"They'll be okay. And you might need the horses."

Beverley accepted the inevitable with good cheer. "Come on Francie and Anne, let's get the stock rack on."

There was no choice: we had to go.

Made of heavy plank extensions which we struggled to handle and bolt into place, the stock rack converted our grain truck into a high-sided livestock truck. No easy task in the rain.

When ready, we loaded the bull with a cow and calf and a couple of horses. We cinched a saddle with its long coiled 'ketch' rope onto the top rail—just in case we needed to saddle a horse and chase the cattle into distant pastures.

We headed toward our summer range at Kemp's, about sixty miles out in the Powder River badlands. Beverley drove with great confidence even after we left Highway 12 and turned south onto the red scoria Mizpah road. I had a strong foreboding as the scoria turned to wet gumbo.

Then we hit a stretch of road under repair. The loads of earth moved in to build the road higher and wider had turned to deep, soft mud.

Beverley toughed it out, giving the truck more gas on the slides and into curves while letting up on the straight-away. I didn't say anything. Anne's eyes grew big in her scared face and she gripped the seat with white knuckles. Beverley kept driving on and on through the slippery mud, one stretch after the other, wrenching the wheel back and forth. The truck sashayed through wet gumbo, the animals shifting their weight, as we slid toward a drop-off above a washout ravine. It was still raining hard.

Beverley smiled, "Well, we aren't in the ditch yet!"

"What about that poor little calf? What if he gets trampled?" I worried. Dad always cautioned us to drive with extra care when hauling livestock.

We had partitioned off the cattle in front with two horses be-

*The truck sashayed through wet gumbo, the animals shifting their weight, as we slid toward a drop-off above the washout ravine.*

*Separating cattle and horses with a wood panel partition, we loaded bull, cow and calf and two horses for the muddy trek to our Powder River summer range.*

Western Ranch Life in a Forgotten Era

hind, but we could feel them moving back and forth. Beverley tried to slow down as they struggled to maintain their balance, but still drive fast enough to avoid getting stuck.

We were okay in the middle of the greasy, rain-slicked road for awhile, but then abruptly the truck swung out and back, in a wider and wider sweep, moving closer to the ditch.

Suddenly, down we slid over the edge as if in slow motion, off the soft shoulder into a creek bed.

Beverley, a very good driver, wrenched the wheel to hit the draw straight-on; otherwise we'd have tipped. The truck rocked back and forth, but to our surprise, stayed upright, coming to rest wedged crosswise in the steep-sided creek.

"Hey! That was quite a ride!" Beverley exclaimed. "We stayed on the road longer than I expected!"

I glanced at Anne. She still looked pale and scared. I felt the same way.

We climbed out, shaken, relieved to be in the bottom of the draw, our livestock okay. But no chance of driving out. The creek ran a small stream of rainwater beneath the truck. Luckily, as it turned out, the truck end gate lined up directly against the bank and, while somewhat jammed, could be pried opened.

We looked around for help. No sign of humans anywhere. No traffic—we hadn't met anyone on the road for miles, not since we turned onto that slick Mizpah road. No ranches. No cattle. No nothing, only badlands, buttes and grassy flats—as far as we could see. And dusk closing in.

"Look! There's a light!" Anne pointed into the distance.

Sure enough, far off down along Powder River glinted the faint light of a ranch house. A wisp of smoke rose off the horizon.

"Let's head down there," I said.

We hardly knew how to get there. Three of us and only one saddle: Our two saddle horses were Comet—half-wild, often uncontrollable, never before ridden bareback or double—and Buck, gentle but never ridden without saddle or double either. The bull, now snubbed to the stock rack with a rope halter, had never been tied or led before. Bumping up against him, an unfriendly cow worried over her wet and bedraggled calf.

"We have to jump the animals out," said Beverley, prying at the tailgate. "We can't leave them here."

Luckily, the slippery bank was fairly close to the truck, so they didn't have far to jump and all made it without injury.

"We've got to get moving, it's soon going to be dark." Beverley sighted down across the rough country toward the river.

How could we put our motley group together? It took some do-

> *We hadn't met anyone since we turned onto that slick Mizpah road. No ranches. No cattle. Only badlands, buttes— and dusk closing in.*

ing to get started.

Beverley decided quickly. She and Anne would lead the bull on the long ketch rope, riding Buck double, while I'd ride Comet without a saddle and chase the others. I led Comet, our most skittish horse, into a ditch to scramble on her bare back.

Finally we set off, each animal objecting in its own way. Buck humped his back and rolled his eyes at Beverley who was simultaneously pulling back on the reins and jerking the stubborn bull along, while Anne held on tight and tried to keep clear of the rope that threatened to scrape her off.

Comet pranced sideways, never happy behind a cow or another horse, determined to run, snorting and nearly unseating me as I hazed along the reluctant cow, calf and bull on the rope. The bull was balky and the cow determined to head off in the opposite direction for home. All of us muddy and wet, the calf covered in sloppy manure from his mom and the bull.

By the time we got our little bunch moving well, darkness was setting in and stars peeked through a break in the clouds. Only a bit of orange lamplight glowed from the distant ranch house, but it was our guide. Thankfully, we made out a set of corrals—just what we needed most.

I doubted we'd make it that night across such rugged, muddy country. Beverley forged ahead, all the while scolding both horses and cattle. Anne and I kept mostly silent.

Finally we all arrived wet and tired at the ranch door, greeted by a pack of barking dogs. The ranch family turned out to be Fesslers, cousins of Anne's friend Sally. They took us in, offered a pole corral for our livestock and fed us supper.

We slept in their bunkhouse that night, and for two or three more nights. A litter of pups quarreled in the dirt under the bunkhouse floor all night, keeping up a growling, barking and yipping. Anne and I were almost asleep when Beverley threw her boots on the floor one after the other.

"Shut up! Shut up down there!" she shouted.

They did, for one startled moment—then yipped all the more, just like the coyotes did when she yelled at them.

Anne and I snuggled deeper into our quilts, turned over and tried to get back to sleep.

Next morning we were able to alert Mom and Dad to our plight on the Fesslers' handcrank telephone that connected several ranchers.

"What can we do?" Anne implored Beverley. "Even after the rain quits, we're going to be stuck here until the road dries out."

We spent a few days in the bunkhouse with the yapping pups

*We set off, each animal objecting in its own way.*

We kept one bull at home to breed heifers, then later trucked him out to our main herd on Powder River.

beneath and packrats running the rafters above. We had nothing to do but read really bad dime novels from a stack on the floor.

We tried to help Mrs. Fessler with her chores, but she seemed unsure what to do with our eager help and suggested we go take a rest and read in the bunkhouse. Mr. Fessler didn't want our help out in the rain with his livestock either. Back to the dime novels. So we spent our days and nights battling the rats above and pups below and giggling over outrageous plot lines.

We were vastly relieved when Mr. Fessler said the road was passable and he'd take us in his Jeep to meet Mom in Miles City.

A few days later when things dried off a bit, the Kemps—from our intended destination several miles down the road—came to pick up our livestock and help get the truck unstuck. Dad brought alfalfa hay to the Fesslers to replace what our animals ate.

But there was a new crisis.

The bull had escaped the pole corral and could not be found. One of our prize registered Herefords, he needed to be fathering our next year's calves. We thought that many ranchers might like a few calves from him. And somehow it seemed the bull travelled from one neighbor's herd to another so the Kemps couldn't quite catch up with him.

Rural telephone party lines were a great way of dispersing news. Every time a phone rang, neighbors would listened in, no matter if the call was for them or not. The ranchers learned from the phone calls that we were short a bull, and began reporting to us where he had last been seen. But still the bull mysteriously disappeared each time someone went to pick him up.

"The girls are going to go out there and they're going to stay until they get that bull," Mom finally announced on the telephone in her firm teacher's voice.

Almost at once a neighbor reported the bull in his herd.

Anne and I set out together, this time driving the truck over dry gumbo. What a difference! Once at Kemps, we rode horseback many miles, fording the Powder River a couple of times at a spot reputed to be free of quicksand.

Soon we had our lovelorn bull back mixing with our own cows on the range at Kemp's. But he continued to hang his head over the fence oogling the neighbor's winsome young heifers, and no doubt,

*We rode horseback many miles, fording the Powder River a couple of times at a spot reputed to be free of quicksand.*

Hunting our bull, we rode cautiously across Powder River's broad, shallow stream bed. Famous as 'too thin to plow, too thick to drink' the river was known for treacherous quicksand.

dreaming of his long and delightful adventure.

Anne and I wondered if any neighbors had new calves next spring that were unexpectedly bigger and stronger.

## Lunch for Lawrence

Jeanie, Francie and I had just finished the disagreeable job of helping Mom clean out the chicken house and rested on the back steps watching the grasshoppers feeding and destroying her flowers.

We had a bad infestation of grasshoppers that year. These inch-and-a-half, big-winged, long-legged insects were eating everything. Those that I caught spit disgusting brown 'tobacco' juice all over my hand.

When we walked through the pasture they jumped knee high all around our legs, clinging to us and spitting brown juice. We brushed or pulled them off.

More tobacco juice—ick. I disliked these revolting creatures intensely.

Periodically I got up and chased them off Mom's zinnias, stomping as many as I could underfoot. I made a pile of those I killed.

"Don't they eat grasshoppers in Mexico?" Francie asked idly.

"They put them in tequila and drink it," I replied in my young wisdom.

"What do you think we can do with them?" Jeanie suddenly got a certain gleam in her eye, and arched an eyebrow.

*Oh my! What do you suppose Jeanie has thought of now?* Francie and I looked at each other, wondering.

Before we had a chance to speculate further, Mom came out on the porch.

"Girls, I need you to take some sandwiches out to Lawrence. He's working late and said if we send him out some sandwiches he'll finish that field before dark."

We groaned. We knew Lawrence was an eager worker, but we did not look forward to walking a mile through the hot fields with sticky

*They clung to our legs and spit disgusting brown 'tobacco' juice when we brushed them off.*

Anne catches grasshoppers that cling and chew on Mom's flowers during a bad grasshopper summer.

Western Ranch Life in a Forgotten Era

grasshoppers jumping all over our legs. We thought cleaning the chicken house was enough misery for one day.

Besides, when we got there, Lawrence was apt to trick us some way. We had yet to get the best of him.

"Sure, we'll take him some sandwiches."

It surprised Francie and me when Jeanie replied so positively.

Mom returned to the kitchen while Francie and I turned to Jeanie. What we saw there stopped us cold. Jeanie had a big smile on her face and that particular gleam in her eye.

"We'll make him a sandwich," she laughed, "out of grasshoppers!"

"We'll have to take off the wings," Francie said. "Wings would be too crunchy."

"And the legs," I added. "They'd stick in his throat."

We all giggled at the thought of Lawrence coughing up grasshopper legs.

"We have to cook them first," Jeanie said, starting to collect little sticks to build a fire on the gravel behind the granary. Jeanie loved to build campfires.

She sent Francie and me to look for an empty can. We even washed it.

When we returned, her little fire was burning and we three knelt around my pile of dead grasshoppers. The ones that were squished too much we didn't use; only those with full bodies.

We started pulling off wings and their long back legs. The front ones we left on—they were very tiny. Their dead eyes bulged up at me and I tried not to look at their faces.

*However could those Mexicans do it?* I almost had to stop when I thought of Lawrence crunching down on those heads and front legs.

Francie and Jeanie made quick work of preparing the grasshoppers, and soon our can was sizzling with the stinky smell of insects cooking. Jeanie stirred them with a stick making sure they were all carefully browned. Finally she said they had cooked enough, and we set them aside to cool.

We hid the can of toasted grasshoppers from Mom, but somehow we managed to get two slices of bread and butter and made our own sandwich. We used extra butter to keep the grasshoppers stuck to the bread. Not good if they fell out before the first bite.

Mom should have suspected something when we eagerly made some good sandwiches in front of her for Lawrence and complained no more about taking lunch to the field.

This was the only time we got the best of Lawrence, a returned serviceman. After WWII, a lot of soldiers hitchhiked around the states. Since one of the two main U.S. highways in Montana ran

> *Lawrence was likely to trick us some way— we had yet to get the best of him.*

Jeanie in her prankster mood gets that special gleam in her eye and we know she's thinking up mischief.

through our ranch, hitchhikers often stopped to see if Dad had work.

Sometimes these footloose ex-soldiers first went into the employment office in Miles City. If they claimed to know farm work, the manager contacted us.

"Elmer, do you need help?"

Dad usually did, so he brought home the helper and we had a new hired man.

Sometimes they lasted only a day, like one young man who said he was a seasoned cowhand but put the saddle backwards on the horse. Since he didn't know how to run machinery either, he lasted only long enough for one good meal. Others stayed until they got enough money to take the bus home and then moved on.

Young men just out of the service—often missing their homes and young siblings—fit right in with our family. They didn't mind our pranks and often dished out as much as they got.

One was Lawrence Kruger, a young man who grew up on an eastern North Dakota farm. Lawrence came to us one day, eager to go to work. He seemed to know how to run machinery and unlike many of the men wanting jobs, claimed to know nothing of range cattle or riding horses.

Lawrence stayed with us through the summer. After supper, when the chores were done, he stayed in the kitchen and talked with Mom. He planned to save his money, then in the fall go back to college on the GI Bill. While she finished her chores, he quizzed her about colleges in Montana and North Dakota. He had started college before getting drafted, but the war experience changed him, and he wasn't sure what he wanted.

Mom had put herself through the University of Montana when she was so broke there wasn't even a penny to splurge on a stamp to send her parents a postcard. She and her brother Chet rented and shared one room in Missoula. Uncle Chet worked at night and went to school in the daytime. She slept on their one cot at night and he slept there between school hours and jobs. Education was a high priority for her, no matter how tough it was to get. She encouraged our young hired men and especially Lawrence, to make use of the GI Bill.

Lawrence knew all about Dakota farming, but wanted nothing to do with cow punching. He was the oldest of many children and had no qualms about grabbing one of us and turning us upside down if we prodded him too much. Sorta took the fun out of it.

Anticipating Lawrence's reaction, it didn't seem so bad walking through the fields with grasshoppers clinging to our legs.

When we got to Lawrence, our grasshopper special was the first

*After WWII, a lot of soldiers hitchhiked by on the highway. They often stopped to see if Dad had work to earn a bus ticket home.*

sandwich we handed him—the rest we left in the pail nearby. As he unwrapped it, we carefully moved out of catching range. He must have been really hungry, because his first bite was a big one.

He chewed thoughtfully for awhile.

Suddenly he stopped chewing and a horrified look crossed his face.

"Gaaaah!" he yelled.

"Pto-oo-ee!" He spit out his mouthful and watched its contents drop to the ground.

He opened the sandwich, and there were all those sticky front legs and bulging eyes staring at him.

> *We girls ran across the field, giggling and darting quick looks behind.*

We girls ran across the field, giggling and darting quick looks behind us—making sure Lawrence was not in fast pursuit.

We could hear him spitting as he dug unpleasant pieces from his mouth.

That night Mom looked surprised when, after Lawrence came in from the field, he suddenly grabbed Francie and me and hung us upside down, one under each arm. Jeanie was much too fast; no one could ever catch her.

I'm sure Mom knew we did something ornery and had it coming.

We didn't tell her how we finally got the best of Lawrence until after he had safely left the ranch and was back in college.

## First Hunt

Francie stared at me.

I stared at Francie.

*Oh no! What do we do now?*

It was dusk. The sun had just dropped over the horizon. Soon it would be dark, so if we were going to get this buck mule deer dressed out while we could still see, it was time to get to work. Francie, twelve years old, had just shot her first deer. I was eight and never had. Neither of us had ever cleaned a deer, though I had been with Dad when he did it.

Francie pulled out of her belt sheath a long curved hunting knife.

"Do you know how to do this?" she asked ominously.

*Me? Oh no!* I shuddered.

As I've frequently told my family—it's not easy being the youngest. The youngest child needs to be ready on the spur of the moment in case one of the older ones makes a sudden slip and asks if you want to go along.

Such joy! Such excitement! And such responsibility actually to be asked.

*Of course I want to go along! But that means I can't whine or complain, must pleasantly follow whatever course the adventure may take, and always be very, very quiet.* And definitely be quick to help out—which lots of times wasn't much fun, since that was probably why I was asked to come along in the first place...

Dad's unique way of encouraging his daughters to do what he wanted was creating excitement out of nothing, probably a holdover from the days of riding the range alone, then meeting up with a fellow rider over a night campfire and swapping yarns.

This late afternoon he noticed several mule deer grazing in his alfalfa field.

Cattle couldn't get in the fenced fields and ruin the hay, but the deer leaped any fence, and ate only the best of our crops. Dad unwillingly fed several herds of deer in his prize alfalfa fields all summer, and though hunting season was a month or more away, it seemed just that he should have some fresh venison every now and then. So went the reasoning of most ranchers at that time.

"There are some nice young bucks in the alfalfa field." Dad's blue eyes twinkled as he thrust the rifle at Francie along with two or three shells in the clip.

Amazingly, she turned to me and asked if I wanted to go along.

Often during hunting season the deer we spotted had been running hard from other hunters, leaving a gamey smell that permeated the entire animal. Venison from resting deer was much better, especially when they ate the best food on the ranch. We decided to sneak up to these quietly munching deer and get one.

We followed the fence line that gave us cover from the brush and cottonwood trees, until we stopped a considerable distance downwind of the deer.

I stayed behind Francie, quietly dogging her footsteps as she periodically shouldered the rifle and sighted—decided we were still too far away—then crept a little closer. Francie shot prairie dogs and magpies beginning as a small child, but this was to be her first deer. When she made the final judgment to shoot, she nailed this young buck right between the eyes. He never even jumped—just went down in a heap.

Hearing the shot, the rest of the herd ran for the hills, and there lay Francie's prize. I looked at her in awe. I'd heard Dad and his friends talk about shots like this they'd made and knew my sister was just as good as those crack-shot men!

We ran across the field to the still buck. Francie reached down

Both whitetail and mule deer graze our alfalfa fields on fall nights. Scarce in our early years, deer multiplied and took a huge toll on the alfalfa seed crop.

*Francie pulled out of her belt sheath a long curved hunting knife.*

with the hunting knife and expertly cut his throat to bleed out. After all, she was learning to skin and cut hides with her mail-order taxidermy course, but she didn't deal with innards of the animals she mounted.

We struggled to turn it on its back.

Staring at each other across the dead buck, Francie asked me, "Do you know how to do this?"

I gave a firm shake of my head.

"But I've seen Dad do it. I think you start up here." I pointed to the top of his brisket, expecting her to start cleaning out the entrails.

Francie looked dubious. Then her face lit up.

"You do it."

She said these dreaded words as she thrust the knife toward me.

*Please, no!* I hesitated briefly. *But if I say no, will I ever get to come hunting with her again?*

Not wanting to chance missing the next adventure, I grabbed the knife. Holding it perpendicular to the deer's chest, I jammed it between the front legs straight into him clear up to the hilt.

An unpleasant odor filled the air.

Keeping the knife deep in the deer, I sawed straight down to the bone all the way to his tail. The slashed-open stomach spewed out half-digested alfalfa. Blood gushed—and the mess from the slashed intestines overwhelmed us as it oozed out over his side and all over me.

Francie and I both gagged a couple of times, and I momentarily wondered about all this oozing stuff, and why it stunk so much more than when Dad did it.

Undaunted, I continued. I slashed around the insides of this deer, trying my hardest to free his insides, which now was slimy, unmanageable goo. It all had to come out. I pulled and pulled, then a few things snapped. Something slapped me in the face, and the insides came tumbling out.

*Yuck.*

I remember telling Francie, "This is the heart—we should save it. Where's the liver? Mom will want that."

The liver seemed to be in several unrecognizable pieces, but as I pawed through the blood, manure from the intestines, and other unknown organs, two purple-black lumps appeared which we considered to be liver.

Must not waste anything, was always my mother's belief.

We stood back and surveyed our venture. Convinced that we had removed the necessary organs, I wiped the knife on an alfalfa plant and tried to remove the worst from my clothes and body.

Francie, who had been gagging over on the sidelines, decided it

*Keeping the knife deep in the deer, I sawed straight down to the bone all the way to his tail.*

was time to go get Dad to take the deer back to the garage for washing.

She was mostly clean. Gagging kept her from helping me. She said I should stay with the deer in case a coyote came along to steal our venison.

Much later I realized she didn't want to smell me the whole trip back to the ranch buildings. On further reflection, this was probably for the best. If our parents saw my bloody manure-covered self, without first knowing this was all deer blood, it might have been a shock.

Dad and Francie arrived on the tractor. Since he saw her first, all nice and clean except where she handled the heart and liver, he naturally supposed she had done a clean job of gutting the deer.

When he saw me, he merely said, "Heh, heh, heh."

Then looking over at the deer and pile of innards he exclaimed, "Oh my!"

Deftly he loaded the deer on the tractor, and once home, hung it in the barn, hooked up the water hose and thoroughly washed out the carcass.

Taking the hunting knife, he touched key points on the carcass and offered quietly, "This is how to gut a deer."

I wondered why he kept trying not to smile while saying, "Heh, heh. Heh, heh."

## Riding the Ditcher

"Slam!"

Francie and I flew off as the ditcher hit another huge cottonwood root. It flipped over on its side and tore out the edge of the ditch. Just in time we escaped out of range.

"Wait! Wait!" we called to Dad on the tractor, running to catch up.

He looked around, grinning when he saw we were okay.

He backed up, jumped off, righted the ditcher and we climbed back on and clung again onto the levers at our stations.

Riding the ditcher was one of the jobs we really disliked—and feared—partly because we felt so helpless in our inability to do the job right. Problem was, at age ten and twelve Francie and I weren't heavy enough to hold the ditcher down. But that was our job.

We had to periodically clean the ditches crisscrossing our irrigated land. During the year, debris, dirt and stones fell in and roots grew through the irrigation ditches—big, tough weeds even where trees could not grow. Without smooth clean ditches the water

*The main irrigation ditch crossed over Jones Creek in a metal flume upstream from our ranch. When the ditch ran too full we opened the steel gate and let excess water flow into Jones Creek. Big cottonwood trees grew all along the main ditch, obstructing water flow with their heavy roots.*

JEANIE

*The main ditch circled around the steep rugged hills above our lower pasture and finally drained into the Yellowstone River. We were last on the ditch.*

Anne and Francie guide water flow in field near strawberry patch.

wouldn't flow properly, defeating the purpose of irrigation.

Cleaning the main ditch that ran the length of our place along the upper level of the fields was worst. We dreaded that job. Huge thirsty cottonwood trees grew all along the main ditch, just as they lined our rivers, and they thrust their large roots into the ditch from every direction. The main ditch had to be cleaned in spring before the water came down, since it never dried again until fall.

Ditching the fields went much easier, through softer dirt and weeds—no trees.

Sometimes Dad cut new ditches to manage irrigation water better in the fields. Then he'd get out his survey transit and set up its tripod on an elevation where he envisioned a new ditch. One of us girls had to run across the field with the survey rod, while he peered through the cross-hairs on the transit, waving his arms to signal us where to move the target as he took readings.

Under irrigation, the angle of slope determines how evenly the water enters a field—and just as important how it drains out to avoid building up new alkali spots. It seemed there was always a new plan, as Dad drained one deathly-white alkali patch after another through the years, and found just the right level of drop to bring that soil back into production.

On this day we were cleaning the main ditch across the highway where cottonwood trees grew all along the way. The main ditch carried our deepest irrigation water from where it entered our land at Art Hill's pasture until it circled around the steep gumbo hills above our lower pasture through railroad swamps and finally drained into the Yellowstone River, since ours was the last place on the ditch.

Francie and I kept flying off as we rode the ditcher from one tree to the next.

Dad praised our efforts and encouraged us to get back on each time.

"You're doing fine, girls," he shouted to cheer us.

We stood on the narrow wooden platform of the ditcher. It was a clumsy, heavy implement about eight feet long by six feet wide with a thick iron cutting blade in front shaped like the 'V' on a plow.

We held onto levers cropping up here and there. At times we needed to adjust them to change the angle of the blade, a difficult problem since they were old, rusty and bent. Most of the time we just held onto them for dear life as our supposedly-level platform heaved, twisted, tumbled and threw us off. Luckily we never fell forward or close to the iron cutting blade.

Dad drove the double front wheels of his heaviest tractor down the dry ditch, the voluminous back tractor tires straddling it, with the ditcher connected behind. He watched ahead to drive precisely on the ditch banks, which tended to cave off on the lower side, at the same time he looked behind to see what the V blade was about to hit and to make sure we were secure. He needed to rev up the power to pull the blade through obstructions, but be ready to stop with each sudden emergency.

He had his hands full. And so did we.

We rode the ditcher to add weight, trying to balance our weight where needed to hold the blade steady in the ground. That was our purpose, as ballast, even though together we weighed less than a hundred and forty pounds — only enough to hold down the ditcher, maybe, when things went well, without obstructions.

We thought Dad could have used rocks or sacks of sand instead of us. But as we debated that, we realized that if they got tossed off, he'd have to jump down from the tractor and load them back on. When we flew off, he only had to to stop and wait for us to scramble back on.

We were more useful than the rocks, if that was some consolation—we jumped off and on again by ourselves. And also it was easier to get the ditcher back in place when empty of ballast.

Besides, theoretically, we shifted our weight as needed to keep the blade positioned in the ditch. However, no way could we hold it down when we hit solidly into a root and the entire implement flew up in the air.

Our method: Hold the ditcher down, balance there as long as possible—but if it's likely to flip, be ready to jump free just before it lunges out of the ground.

We watched out carefully for roots, but since many of them were not visible, we prepped ourselves for violent surprises. Tree roots spreading under the dirt in the ditch bottom grew large with the abundant watering they got.

Dad slowed down when he saw them. But others we didn't see till we gouged them with the heavy blade.

"Lean right!" He yelled suddenly with a glance over his shoulder. "Now left."

We hung on and made it over that one.

But the next root looked too large. Dad stopped, got off and chopped out the root, then climbed back on the tractor.

"Okay girls, let's go."

Away we went again. It was exciting, in a way. I admit I liked that part. But it was always a relief when we finished ditching.

We had nearly reached the lower pasture gate when Dad, seeing

*Humberto, an exchange student from Chile, sets a canvas irrigation dam to direct water from one of the field ditches that skirt our crops.*

*Our method: Hold the ditcher down, balance as long as possible— and get ready to jump clear.*

Western Ranch Life in a Forgotten Era

> We were all four lined up
> on that log,
> one after the other—
> the rattlesnake,
> Dad, me and Francie
> on the end.

Laying out a new ditch, Dad sets up his tripod and survey transit. When we moved to the ranch, much of the irrigated land lay white with alkali from poor drainage. Over the years through much hard work and skill, Dad brought it back into production.

we failed to quickly remount the ditcher, stopped and turned off the tractor.

"Okay girls, we need to take a rest," he said.

We got off and grabbed the water jug.

A big cottonwood log lay just above the ditch, patches of deep-channeled bark gone from one end, and we went to sit there. Hit by lightening, the tree had blown over and lay propped off the ground with its long trunk held horizontal by snapped-off limbs that left branch stubs holding it up off the ground.

We all climbed up and sat there on the cottonwood tree trunk and passed the water jug around. I was sitting in the middle, with Dad on one side, Francie on the other.

As we relaxed, Dad told us a story. It must have been interesting because none of us heard the rattlesnake.

A sudden buzz, and I looked around quickly. I couldn't see the snake. I had to find it before I moved or said anything.

There, about ten inches on the other side of Dad, on this same log, coiled right beside him, was the snake—rattling viciously.

We were all four lined up together on that log, one after the other — the rattlesnake, Dad, me and Francie on the end. In striking mode, it coiled there, its flat, triangular head tucked low into its coils, all the better to spring fast. Its black forked tongue darted in and out, its tail twitched high and the rattles vibrated so fast I could hardly see them.

This big snake was only inches from Dad and all set to strike.

*Why didn't he hear it? Were his ears ringing from the tractor noise?*

Usually, in standing up from sitting on a downed tree, we pushed off with our hands. It was the most natural thing to do. But, if Dad put down his hand, he'd surely get struck.

I thought, *How am I going to get his attention and warn him not to make a sudden move without causing that rattler to strike?*

I slid off the log and stood right in front of Dad until he stopped talking and looked at me. Then I put my finger to my lips.

"Don't move. There's a rattlesnake right beside you."

He darted a quick side glance and in a flash, all at once, flipped backward off the log into a wild rose thicket.

In that moment the snake struck—and missed!

That was a close call. Francie and I kept an eye on it as Dad scrambled out of the briars, coming up with a stout branch in his hand.

He killed that rattler like we did every one we saw, always trying to cut down on the population and the risk that they'd kill us or our livestock.

We took the rattle as our trophy of the day.

*Montana Stirrups, Sage and Shenanigans*

"I wondered what you were doing, Jeanie," Dad said. "How'd we miss seeing him?"

For us not to have seen the snake when we first sat down meant it was too cold that spring morning for it to respond quickly. Maybe it was stretched out, camouflaged by the deep gray bark when we sat down on the log. Or it crawled up there while we were talking.

"Okay, girls, let's get going again," Dad said, the rattlesnake dispatched, and revved up the tractor.

Reluctantly, Francie and I climbed back to our stations on the ditcher.

Soon we hit another tree root, good and solid, and off we flew.

Unhurt, we laughed as we climbed back on the ditcher. Maybe Dad thought we loved that job because we always laughed as we picked ourselves up, brushed off the dirt and got back on. Actually, we laughed from surprised relief that we survived our arial fling once again.

It was like living with rattlers. We expected to get hurt, but we never were.

When ditching season ended, we were joyous that we again escaped that maelstrom. We enjoyed adventure and the unexpected—but riding the ditcher, not much!

*We took the rattle as our trophy of the day.*

## The Goose That Stayed

"Girls! Come quick! Get on your coats and come outside," Dad called as he opened the outside door.

From the quiet excitement in his voice, we knew there was something special out there in the dark he wanted to show us.

JEANIE

We hurried. Nothing better than one of our parent's tales of wildlife or nature. We didn't realize these stories were part of the education they wanted for us.

High above us, we heard musical wavering coming from the sky.

"Look up! Straight above you." Dad said as he bent down beside us and directed our gaze.

He pointed high into the starry sky. For the first time we listened to the musical undulations of geese and saw their 'V' formations as flocks flew south on their migration. They were so high we could barely make out the thin line of tiny specks as they continually changed places in their V flight.

We discovered the wonderful world of Canada geese when we lived for a time near Savage on the Yellowstone River. They were

Western Ranch Life in a Forgotten Era

unique among birds, because of this simple and elegant V flight and their urgent musical cries, growing steadily louder and fainter by distance.

"These geese are headed way down to Texas or even Mexico for the winter," Dad told us.

"They fly all night as high as an airplane and only stop to eat when morning comes." The awe in his voice told us this was something to remember.

How can they fly so high and so far and so long and still have breath to sing their songs?

No other migrating water birds seemed so magical. Sometimes at night they crossed in front of the harvest moon. Other times they swept along against the orange and golden bars of sunset. Oppressive storm clouds sometimes brought them in to feed.

That fall we often went outside to listen to the restless cries of their passage and the excited gabble of landings. We learned of migrations, thousands of miles, and dangers encountered on the way.

Bev was eight and little Anne a toddler at that time, with Francie and me between—we couldn't learn enough about the wonderful wild geese.

Dad, especially, loved Canada geese. The pairs mated for life and he was proud that he had never shot one. He was an avid game hunter and we often ate wild waterfowl, pheasant, or venison for dinner. But he never allowed himself to shoot a goose. If he hunted ducks and a flock of geese landed among the ducks, he shot a duck even though the goose's larger body provides more food, because he didn't want to break up a pair. When one was shot or killed by a predator, the mate was left alone and lonely, to grieve, he thought, for the rest of its life. Not ordinarily sentimental, he felt special affection for wild geese.

Every spring and fall on their migrations, small flocks of geese landed on the beaver pond not far from our house. There, years ago, beavers had dammed up a little creek to form a pond. As they built the dam higher and longer, the pond grew until, by the time we moved there, it was a sizeable lake. Since beavers are efficient engineers, about as much water filtered out through the front of the dam as entered the lake. A quarter-mile below, the creek flowed into the roiling Yellowstone.

Mom watched that pond and alerted us to wildlife there.

"Come girls. Look out the window. Those are beaver swimming out there!" We eagerly sighted along her finger.

"Let's go out quietly and see," she said.

When outside by ourselves, we had orders to stay away from the beaver pond, but sometimes our parents took us to watch the

*Sometimes at night they crossed in front of the harvest moon. Other times they swept along against the orange and golden bars of sunset. Oppressive storm clouds sometimes brought them in to feed.*

beavers at their incredible work. Seldom could we sneak up on them, but when we did it was amazing to watch. We had to keep still, but had lots of questions. One movement from us and they slapped their broad flat tails in warning and instantly disappeared underwater.

Dad came in worried one fall afternoon.

"A flock of geese just landed on the pond and one of them is crippled. It's dragging one wing. Maybe it's been shot."

The flock stayed several days, but the crippled one didn't get better.

We watched them and one morning they took off on their southern migration, but the injured goose—a female, couldn't take off. Another goose circled back, calling to her and trying to get her to fly. He landed beside her, encouraging her to lift off. She tried valiantly but her broken wing kept her on the water. She struggled, but finally the other goose flew away and left her. Mom said it was probably her mate and he eventually overtook their flock.

When a day or two later, the flock didn't return, Dad caught the injured goose. Then he found and removed two or three shotgun pellets that had cracked her wing bone and splinted her wing.

"She'll heal, now," he said. "When her flock returns next spring she'll be able to fly away with them. But we must all watch for her. If any of you kids see a fox or a coyote come around you must open the door and yell as loud as you can. It'll scare them away so they won't kill the goose."

The goose lived alone on that lake all fall. When ice formed around the edges of the water Dad broke it and fed her grain on the snow. Finally, she made short flights around the lake and he removed the splint. The goose became quite tame and came up to us when we fed her.

The open water had to be large enough so she could stay in the middle and a coyote or fox couldn't get her. Occasionally Dad went over to the lake and chopped the thin ice away.

"Oh, Elmer," Mom pleaded, "please be careful. What if the ice breaks when you are standing on it chopping? If you fell in, you could freeze to death so quickly."

When the temperature dropped below zero, a person who fell in could easily freeze walking home.

Once when Francie was about twelve, she stood on the frozen ice of our stock tank, trying to break it for the cows in the corral. The tank heater had quit during the night. She couldn't break the ice with a stick so she got up there and jumped on it, hanging onto a cross board above the tank. Suddenly both the ice and board broke and she went completely under water—it was twenty-five below

*Canada Geese migrated over, but few nested in Montana.*

"A flock of geese just landed on the pond. One is crippled and dragging a wing."

zero. She made it to the house but her clothes were frozen solid.

When finally the ice got too thick to keep a safe place open for the wild goose, Dad brought her into the chicken house. Mom and we girls fed her, and she seemed to enjoy that and the company of chickens.

In spring he took her back to the beaver pond. Flocks of geese began stopping overnight on their way north to their breeding grounds along the Great Slave Lake in Canada.

Our goose watched, but her flock didn't come. We felt sorry for her lonely vigil and worried they had forgotten her.

"Why doesn't her mate come for her? When is he going to come?" we kept asking as we watched out the window.

Geese have acute memories. Every day our parents reassured us that her flock would come back for her. Geese came and went, but our goose always stayed behind, swimming in the water alone, gazing toward the southern sky.

Then one night Dad came in with a big smile on his face.

"Our goose friends have come," he said. "Look out the window."

**Some literally ran across the water with their wings outstretched, rushing toward her.**

He saw her flock land. The geese had made an unusually loud honking as they came in for the landing, flapped their wings and called to her. Some of them literally ran across the water surface with their wings outstretched as they rushed to her. Beside herself with joy, she flapped her wings and ran toward them.

It was almost dark, but Mom dressed us warmly and we went up to the lake. There they were, gabbling happily, swimming around each other in the semi-darkness. Shadows of party-goers on the water who weren't quite yet willing to call it a night. We couldn't see which one was ours, but clearly the flock was joyful.

The geese flew around the countryside for a week or two, taking our goose along. Testing her strength. They found food in grain fields, leftover from fall harvest. Always they returned by late afternoon to the beaver pond.

Our goose might not be strong enough to fly with them and they all might spend the summer with us. That's what we hoped for. Or they might fly away and leave her. If they did, we could put her back with the chickens for company.

*We girls didn't want any of the geese to go and we didn't want her left alone, either.*

Early one morning we heard the normal commotion of the flock leaving—only it sounded louder and more raucous.

"The geese are on their way," Dad announced when he came in. "They left for Canada and they took our goose with them. They headed straight north. I watched them as they flew higher and higher, formed into a V and disappeared."

"But I want her to stay! I want all of them to stay!" Francie cried.

"Me too," I said. "Couldn't her mate have stayed with her? It would be such fun to have them make a nest and raise baby geese here!"

"It's nature's way," Mom told us. "If they're dependent on us they won't learn how to get along on their own. Then someday a coyote might get them. Or they might not go south soon enough in the fall and starve or freeze to death. Geese learn from each other, the entire flock teaches the young. They all need to stay together."

We girls felt sad that our goose was gone. We wondered if she got too tired. If the flock got all the way to Canada with her or if she dropped along the way.

"Don't worry about our goose anymore," Beverley finally told us. "The rest of the flock knew she was strong enough. They knew she could make it all the way to Canada or they'd still be here, waiting for her to get better."

She said our family did a wonderful thing, taking care of that goose all winter, then letting her go free with her flock. Beverley was older and wiser than the rest of us, so we believed her and felt proud we'd let her go.

That summer we happily imagined our goose nesting in Canada with her mate. Then, after the eggs hatched, trailing a string of little ones to the water's edge, dropping into a beaver pond, and swimming serenely off in a line, one after the other—mother, five or six little ones and father bringing up the rear.

That fall we watched each flock of wild geese land on our little lake, wondering if our lame goose was with them. We wondered which of the young geese might be her babies, and if she still remembered us.

Mom helped us put out a little grain when each flight of geese came. We knew our goose was in one of those flocks.

> *"It's nature's way,"
> Mom said. "If they're
> dependent on us
> they won't learn how to
> get along on their own."*

## Move to Brink Ranch

We moved to the ranch ten miles east of Miles City early in the spring of 1939 when patches of green grass were just starting to show under the snow. How wonderful the land looked to Mom and Dad who so recently left the dry, depression-parched land of the Missouri Breaks.

FRANCIE

Never again would they have to wait hopelessly for rain to water the crops, watching them wither and die before harvest. Irrigating was hard work, but rewarding, with dependable moisture when needed for growing fields. With efficient farming practices Dad was able to drain alkali bottoms and eventually more than triple our acreage.

To us girls it was an exciting time, exploring the big two-story ranch house and all the many barns and outbuildings.

This new ranch was to be ours for the next thirty-three years and, as kids, we wasted no time picking out which of the ten rooms were to be our bedrooms. All the six large bedrooms even had walk-in closets! We could get lost in the house and in fact, Anne did just that. She was almost two and we heard her crying, but it was hard to figure out the right combination of stairways and closed doors to find her.

For the first time we had running water—piped from a free flowing spring up in the hills above the buildings—and a real bathroom. In 1939 this was one of a very few ranch houses with indoor plumbing. I remember Mom explaining that we had to use real toilet paper, because catalog pages would plug up the plumbing. I was only six, but quite alarmed by the extravagance of having to buy rolls of paper for this purpose.

We came here after two years in a log house at Savage on our first irrigated place farther downstream on the Yellowstone, a move that came after living several dry years in the Hell Creek area of the Missouri River Breaks. That in turn followed government confiscation of the original Brink ranch on Missouri River bottomland,

*View of our ranch from the rimrock hills near the spring. Large cottonwood trees line the main irrigation ditch and Yellowstone River into the distance at right—with Miles City ten miles beyond. Irrigated fields lie to the center, with grazing land in foreground. Road at bottom of photo shows our driveway to highway at right. This was part of the original graded Yellowstone Trail, which extends off to left, going east, as it wends its way through the hills to a plateau above.*

taken for building Fort Peck Dam in 1931. The homestead there with its rich topsoil was now under deep water.

When we arrived at our new ranch, Sam Undem, a sheepman, was there wintering a band of sheep with the help of his friendly sheepdogs. He looked pretty grubby, but was friendly and nice and lived with us several months. We girls enjoyed his easy-going humor and wanted to call him 'Sam' as did everyone else. But we weren't allowed. Calling him 'Mr. Undem,' was a courtesy our parents required.

Once we had explored the house, barns and sheds, the huge sandstone rocks that topped the nearby buttes enticed our explorations. One to the south we named Battleship Rock, with Bathtub Rock just beyond. With steep sides, the rock looked just like a battleship. Even bigger. Its front resembled the prow of a ship sailing through rough rolling prairie seas, while the back reared up high above the other rocks.

ANNE

Jeanie, who was even more daring than Beverley, found several deep fissures on Battleship Rock that we could climb, boosting each other to the top. We had to climb cautiously, as rattlesnakes sunned themselves or hid in the crevasses of the rocks. We learned never to place our hands on a rock without first making sure no snake waited there to strike.

Bathtub Rock was an oddly-shaped sandstone formation—a bowl-shaped 'bathtub'—cresting the next high butte, not as interesting to us as the endless nooks and crannies of Battleship Rock. However, we spent time carving our names into the nice smooth sandstone and enjoyed the mystery of the perfectly-shaped bathtub at the very top. During wet weather, the bathtub held rain water, though since it never grew stagnant, it must have leached through the sandstone.

When town kids came out to the ranch with their folks, Battleship Rock was a favorite place we took them. For most, one hike exploring Battleship Rock and they quickly became friends. Others, though, with scant outdoor experience, were shaken by the country, rock climbing and the threat of snakes and never wanted to venture there again.

Our visitors might be terrified at the idea of walking carefully to avoid dangers like rattlesnakes and the steep pitch of a gumbo butte. Or they could be reckless, without caution or understanding. One teenage girl horrified us by suddenly running full tilt off the end of Bathtub Rock's steep gumbo butte.

"Stop, don't run down there!" Jeanie called.

But too late. We were sure it would end in disaster—and it did.

*Battleship Rock, our favorite hike, above, was a huge sandstone outcropping that crowned a big butte not far from our house. Below, Beverley climbs the face of nearby Bathtub Rock.*

Western Ranch Life in a Forgotten Era

*This mysterious carved face high on a sandstone pillar looked down on Buford Trail ruts. We speculated it might mark an early grave of a fur trapper, Indian scout or freighter.*

*Our garden grew lush that first summer when we arrived at the irrigated ranch— Francie, Jeanie, Beverley and little Anne, at lower right.*

She wouldn't, or couldn't, stop when we shouted. Soon she fell, tumbling over and over rocks, and broke her ankle in several places. A glistening bone protruded when she came to a stop. Her screaming sent Jeanie running for help, while Francie had to stay and give comfort, all the while listening to her blood-curdling screams.

Running down these high steep buttes was great fun and we ran down them too, but we learned how and where to run without stumbling. Needless to say, she didn't return for another run.

When the new US Highway 10 was built along those buttes in 1952, the highway department crushed some of our great rocks for fill. In 1976 when Interstate 94 cut in front of the two buttes, all the remaining rocks went to build the road base. The only remains of Battleship Rock and Bathtub Rock were parts of the gumbo buttes they topped, cactus and the ever-present rattlesnakes.

Soon after we moved to the ranch, our running water dwindled. Dad, understanding how springs worked and realizing this spring needed cleaning, set out to remedy the situation. He took us along to explore the area.

Surprisingly, there was a deeply rutted road up to the spring and beyond. Our neighbor Grandpa 'Bill' Hill later told us this was the Buford Trail, an old military and freight road that climbed out of the valley and up on the plateau above through this break in the rimrocks.

This point of rocks drew our attention while Dad cleaned the spring. Names, dates and cattle brands were carved in various places on the rocks. The sandstone rimrocks there ended abruptly in a massive pillar and, on the very end of the pillar, a large human face was deeply carved. We scrambled up the rocks to investigate.

The face, about three or four times larger than a normal human face and three inches deep into the sandstone, was weathered; perhaps carved fifty years or more years earlier. The nose was large and broad at the nostrils. Two deep frown lines between the eyes and two deep slashes from cheek to jaw depicted a harshness indicating a man. But what kind of man? A freighter? An early trapper? An Indian warrior?

Many questions remain about who carved this figure and why. Was it carved as a grave marker by an early Mountain Man who buried his partner nearby? By a soldier from a wagon train camped in the shelter of the rocks and cedar trees, digging his knife into the soft sandstone to pass the time? Or could the face have been carved by an Indian? Seems unlikely, since they seemed to carve stick-like figures with shields and without face detail. Or did it warn of hostile tribes lurking in the rimrocks? There is a purposeful look to it, compared with the randomness of other carvings on the same rocks.

We came to believe 'the Face,' high up on the sandstone pillar as if carved from the back of a wagon, was a landmark—a road sign—for freighters and stagecoaches traveling to military forts farther east on the Buford Trail. The Face told them they were on the correct road, and it was time to take a sharp left.

As soon as Dad was ready to go, we went back to the barn and checked out the water free-flowing into the water tank.

Success! There stood our Holstein milk cow Bossy drinking cool fresh spring water gushing from the pipe.

We brought Bossy with us from Savage and she proved to be a treasure.

The fall before the move to the Miles City ranch, our parents felt great despair when, with four small children to feed, the range cow they were milking went dry. Dad wrangled another wild range cow that had calved that spring but she, too, soon went dry. Range cows, beef breeds often mixed in with lean southern longhorns, produced little more than enough milk for a calf, in a dwindling supply for five or six months.

During that time, we lived in the irrigated valley near Savage. Coming there in those dry years was a blessing, but it turned out to be a mixed blessing. Most devastating, hordes of mosquitoes swarmed off the standing water along the swampy back waters of the Yellowstone River and spread disease to our livestock. All our horses died of sleeping sickness *(equine encephalitis)*. Our sheep came down with *vibrionic abortion*; the ewes aborted and we lost the entire lamb crop. The three hundred ewes were bought on credit, so we lost them, too.

Mom talked about the fear of not having milk for her children.

Then she saw a small advertisement in a farm paper listing for sale, up on the dry benchlands, a Holstein dairy cow and her two-year-old cross-bred heifer.

Our folks went to see the owner, an elderly Norwegian trying to scratch a living from a treeless dryland homestead on that windswept plateau. Dryland in the 1930s in eastern Montana meant no grass, hardly a spear to be found.

They admired the large, beautiful Holstein.

"She's a vunderful cow. Look at that big bag! She'll milk nearly all year. And this vun, too," he pointed to the black white-faced heifer.

"She'll be a good milk cow, even she's half Hereford."

Built around the turn of the last century as headquarters for a large cattle ranch, the right half of our ranch house was designed for the owner's family, the left half for workers. Back door faces the barns and corrals at left. Originally, long porches lined both sides of the house.

FRANCIE

*Mom told us about the fear of not having milk for her children.*

Western Ranch Life in a Forgotten Era

*"We can't buy two cows. Just one," Dad said, glancing at Mom who was shaking her head warningly.*

"How much do you want for the Holstein?" Dad asked. "We just want her."

All the money they had was $50 from selling the wheat crop—not much to pay for a good Holstein cow even then, but maybe he'd take it. Mom was hoping.

"It's $50. But they both got to go together. I got no feed. I can't keep that vun either." He regarded the two sadly. Their ribs jutted out from bony backbones.

"We can't buy two cows. Just one," Dad said, glancing at Mom who was shaking her head warningly. "We want the Holstein. We need a good cow that will milk all year. We have four small children. We need the milk."

"No. I can't separate them. I got no feed—you can see that."

The farmer waved a hand forlornly around his desolate land. They followed his gesture, taking in the small wooden shed, flimsy fences, the small unpainted house, and most of all, the dry and desperate land.

"Chust $50. They got to go together."

"We thought he meant $50 each, but couldn't be sure by the way he said it," Mom explained—this was one of the stories she told us at the supper table. "We of course wanted to have both cows. Then we could have all the milk we needed, sell cream in town and sell calves, too, every year. But we couldn't afford to pay for two."

They dickered back and forth with Dad insisting they could just take one cow and the dryland farmer repeating they cost $50 and had to go together.

Hard-up people had a lot of pride and didn't like to expose their desperation. But finally Mom said, "We only have $50. That's all we can spend."

The farmer brightened. "Good! Both for $50."

Everyone laughed with relief, the misunderstanding resolved. The men shook on the deal, then loaded up the two cows.

"We took them home, turned them out on irrigated land. They fattened up, kept us in milk year around and raised good calves, too," Mom told us.

Bossy, our first Holstein milk cow, and her offspring played a big part in Mom and Dad's success in rebuilding their herd of range cattle at the Miles City ranch. The extra milk and cream sold for cash, allowing them to buy more livestock.

Bossy was indeed a great treasure and the start of a new and better life for our family. Hard work, yes. But following our parents' example, we girls didn't mind, and always found adventure and felt we played an important part in our ranching endeavor.

# 2  Horses We Knew

### Rider Over Backwards

Dad surprised me one morning at breakfast.

"Francie, you're old enough to have your own horse to break. How'd you like to earn the money this winter by riding out and checking my coyote baits after school?"

FRANCIE

Speechless with excitement, I could only nod my head. Was my long-term dream—a horse of my own—about to come true?

"Bring in the coyote carcasses and you can have the money when I sell the pelts."

He pushed back his plate. "Then we'll go to a horse sale in the spring and pick out a nice two-year-old off the range."

I looked at Mom. She didn't say a word, so I guessed the plan was cleared with her.

Wow! I could hardly wait.

That winter I rode the draws and hills around the baited dead horse through the brief after-school daylight and on weekends. Poisoned coyotes could travel a half-mile or more, and their tawny coats were hard to spot when they lay in long yellow grasses or clumps of brush. Over the months I brought home, tied on the saddle, four frozen coyote carcasses for Dad to skin and stretch along with the others he trapped, shot or poisoned. Enough for a young horse with dollars left over for a new bridle.

Flexi was a red sorrel, fresh off the open range from an early spring round-up north of Miles City, somewhat impetuous, but sweet-natured when she tamed down.

An eighth grader, I owned my first horse—and soon after, experienced my only serious riding accident.

*Was I pushing her a bit too fast?*

I spent the first month taming and breaking Flexi and then every morning before school rode her out to bring in our three milk cows. Again after school I rode her.

> *Speechless with excitement, I could only nod my head. Was my long-time dream about to come true?*

Western Ranch Life in a Forgotten Era

Dad's technique in training a green bronc was to 'sack 'em out' by flapping a saddle blanket over the back, working it around and under the belly till the horse accepted it calmly. After a few days of this he eased on the saddle, cinched it up tight and jumped on.

So I broke Flexi that way. Only problem was to get on her first thing in the morning when she was feeling nervous and no one to hold her. So for my benefit he showed me how to tie up a front foot before opening the corral gate if Flexi jumped around too much.

Dad always got a kick out of riding a bucking horse and would laugh uproariously, waving his hat until he 'took the buck out' of him. But I didn't have that confidence.

That one frosty March morning Flexi humped up and snorted as I slid the saddle blanket onto her back. I tied up a front foot to get her saddled and when she relaxed, cinched it up tight. She didn't like it, but no time to spare—I had to bring in the cows and get ready for school.

Pushing open the gate, I jumped on, released her left foot and set both boots in the stirrups before we cleared the open gate.

Flexi felt frisky that morning. Prancing sideways instead of heading out, her front hooves slipped on the ice. A perfect excuse to pitch down her head and come up bucking. I pulled in the reins hoping to stop her before I lost a stirrup.

Flexi reared up on her hind legs, skidded on the ice, went over backward on top of me, then scrambled to her feet just as I grabbed for the reins. I missed and away she galloped, reins flying, her head high and turned to the side to avoid stepping on them.

I struggled to rise, but couldn't.

*Just my breath knocked out of me. It'll come back if I lie still a minute.*

I lay back, waiting. But still couldn't catch my breath. Couldn't shout for help. Couldn't move. And an awful pain roiled around my stomach.

I peeked under the truck that happened to be parked in the middle of the yard between the house and where I lay, hoping to see someone come out looking for the milk cows.

Seemed like I lay there a long time on the icy ground, unable to move. Then peeking under the truck, I saw Dad come out on the steps. I tried to shout, but only a whisper came out. Seeing neither me nor the cows, he went back inside and shut the door.

That morning was crisp and cold, the kind of morning when horses like to buck and the ground's hard as concrete. I was eating breakfast when Jeanie came downstairs and into the kitchen.

ANNE

"Where's Francie? I saw Flexi galloping up the road. She's

> *I jumped on, released her left foot and set both boots in the stirrups before we cleared the open gate.*

saddled, but no one is on her," she said. "She was holding her head to the side with the reins flying loose."

"Oh my," cried Mom and ran out the door.

Just then Dad came running to the house yelling that he found Francie hurt. Mom and I dashed out around the truck to where Francie lay still and white on the ground.

Dad had thrown his coat over Francie as he ran to the house to tell us and then brought the car. He thought she'd been lying there at least half an hour. They carefully placed Francie's still form on the backseat of our gray Plymouth and covered her with blankets. Mom knelt on the floor and held Francie so she wouldn't roll off the seat while Dad drove.

I cowered in the front seat, terrified to see her so white and quiet. I knew my folks didn't want me to cry and tried hard not to. But I was so scared. There was no blood.

*Why isn't Francie saying anything? Is she really alive?*

She didn't even moan.

Dad drove the car so fast—I'd never ridden that fast.

He dropped me off at school, a mile and a half toward town. Before we got there, he told me to get ready to jump. He wouldn't come to a complete stop as they had to get Francie to the hospital fast.

All day at school everyone worried with me. No one had telephones then, so we didn't know what was happening.

Finally, just as school was about over, Mom came from town to take me home. She told me when Flexi slipped and reared over backwards on the ice the saddle horn had caught Francie in her right kidney, splitting it open. She lost a lot of blood internally. Dr. Garberson, the surgeon, a horse lover himself, removed the injured kidney through a front incision and called for blood transfusions.

Luckily, Dad's blood was compatible.

They took a pint and he said, "Take more, I'm strong and healthy."

So they did, not sure of what the effect would be on him. But they were desperate for blood because Francie had lost so much and was weak. This was 1946 and blood banks were in their early stages, not yet arrived in Miles City.

Dad went to every business he could think of downtown, asking for help. Calls went out over the radio for blood donors. People started coming to the hospital to see if their blood matched Francie's. Dr. Garberson said they had never seen so many people as came that day to donate blood.

Years later, I still met people who said they had gone to the hospital in hopes their blood type would match Francie's. All seemed

*Francie trains Flexi, her young horse purchased with money from selling coyote pelts.*

*Calls went out over the radio for blood. People started coming.*

proud to help this little girl in the hospital fighting for her life.

Parke Krumpe, owner of the welding shop, told me, "All of us who donated blood for your sister listened everyday to the radio report on her progress."

When Mom and Dad came home from the hospital that first day, they brought Francie's cowboy boots and set them by the kitchen stove. Our dog Jimmy went over to them, sniffed, and then lay down with his nose on them and whimpered.

We all felt like crying.

Francie stayed in the hospital three weeks and was not allowed to get up in all that time. When she did, she'd been lying there so long she couldn't walk for a few days and feared she'd never walk again. But she did; better yet, she started growing.

Always short for her age, Francie grew four inches the first year after her surgery. We told her the accident did it.

"You needed a kick-start to start growing!"

In those days before antibiotics, doctors kept patients immobile a long time to promote healing and ward off infection, which could kill even when surgery was successful.

She lay in the hospital so long she could hardly stand it and started carving horses into the plaster wall by the bed with a metal fingernail file, low down where it wouldn't show. One day our minister walked in, saw it and realized she needed something to do since she was normally so active. He brought her pieces of soft wood and showed her how to carve animals.

After that she worked on cute little horses, with shavings all over the bed, along with her pencil sketches of horses and cowboys inspired by the Smoky books, which she gave to Dr. Garberson and others on request.

When Francie came home at last, Jimmy whined with joy and barked a happy greeting. To all of us, the world seemed brighter.

*Mom and Dad set Francie's boots by the kitchen stove. Jimmy went over, sniffed, lay down with his nose on them and whimpered.*

## Facing the Fast Freight

My saddlehorse, Buck, suddenly stopped. He shivered and I knew he felt vibrations through the rails that I couldn't feel.

He stiffened and froze up. I looked up and ahead, at the point of the curve, the arm of the semaphore was coming down. We were out beyond the middle of the railroad trestle, high above Deep Creek. Nothing but air under and between the wooden ties, and a fast freight coming.

Crossing the trestle was a bad choice and going back at this point impossible.

I was riding home after helping trail cattle the first day to Uncle Chet and Aunt Margaret's ranch, on the way to summer range. I needed to get back that evening.

That day we trailed ten miles over the roughest part and I either had to ride back over that rough stuff or think of a shortcut home.

The main line of the Northern Pacific Railroad ran right there along the banks of the Yellowstone and past our ranch. The tracks were actually built on a little shelf routed out between the river and the high bluffs that rose steeply above it most of the way.

When railroads were first laid out, if there were rivers it was an advantage for the railroad to follow them because the grade ran fairly flat there. But often they cut through some very rough, washed-out country, as here.

It was too tempting.

*Along the tracks can be a quick and easy ride home if Buck can stand the fast freights thundering by. Besides, it's a new and interesting way to ride.*

Obviously riding terrain I'd never seen before in the right-of-way through those high bluffs wouldn't be a piece of cake, but I couldn't resist.

A vehicle road crossed the tracks near Uncle Chet's, leading me easily into the tightly-fenced railroad right-of-way. There were no other crossing gates I could open if I needed to escape, but I didn't know that. They were stretched so tight I couldn't open them.

My saddle horse was big Buck, our quarter horse cowpony, very reliable, calm and courageous. A little kid could ride him, although kids needed help to get on because he was so tall.

He was a good saddle horse and great for roping, a horse that did everything we asked of him.

I knew Buck could handle the sudden appearance of freight trains and probably cross the rough cut-coulee tributaries, usually dry, that went under the trestles and spilled into the Yellowstone.

It seemed a cinch. I decided to go for it and entered the railroad right-of-way.

The fenced railroad enclosed about twenty feet or so on either side of the tracks to keep livestock out. Called the railroad right-of-way, this looked fairly flat at that end.

Right-of-way fences were much higher than normal range fences. Woven wire reached four feet high, topped by two strands of tight barbed wire between steel posts.

I figured there was a gate every mile or so where we could get out, if it proved too dangerous to take Buck down the right-of-way with all those trains going by.

*I thought Buck could handle the rough cut-coulee tributaries, usually dry, that went under the trestles, and even a freight train thundering overhead. It seemed a cinch.*

*If we get in a pinch I can always let us out one of the wire gates and go on home through the hills.*

However, three or four miles later we couldn't cross the dry creek and cutbanks that ran under the trestle at Deep Creek. The great sandstone boulders in the creek bed would kill us for sure if Buck slid down the cut bank and rolled. The only choice was the long trestle that crossed it, where just beyond, the tracks curved around a hill and disappeared.

A semaphore at the point of the curve held its one arm up, showing no train on this section of track. Good. Maybe Buck would cross the trestle with me, one hoof at a time.

Then I had second thoughts—*I better get out of this trap.*

I rode back several miles, but couldn't force open any gate to escape the railway. I should have ridden all the way back and left the tracks where I came on them. But it would waste time and I'd still have a long ride back to the ranch across all the rough country—an extra six or eight miles. Buck and I were both tired from the long day of chasing cows.

Then I made the worst decision of my life. I decided to take Buck across on the trestle.

I dismounted and coaxed him onto the wooden ties.

Carefully he put each of his hooves on the tie, feeling the location. He moved one hoof to the next, dipping down with it into the empty space between ties until he felt the edge, then slid his hoof up and onto the center. He had to feel because he couldn't see his feet. Sometimes he checked the farther edge of the tie the same way, then moved his hoof back to the middle.

He lifted each hoof with care because we were high above Deep Creek on the trestle. He could see the rocks in the creek bed far below between the ties. There was no railing and the space around us was all open air.

Even though he turned squeamish, he continued to move as I walked around him, checked each foot and urged him gently.

"It's okay, Buck. Everything is fine. Just keep moving your feet. We'll get across."

It was slow going and I didn't dare hurry him.

"You're doing fine, Buck, lift this hoof. Good. That's good, Buck."

I was scared. We were going too slow. It was taking us forever. One slip of a hoof and he'd panic, His thrashing legs would drop between the ties and he'd be caught. Seemed like we had a slim margin for success.

"Keep going, Buck. Good boy!"

> Then I made the worst decision of my life. I decided to take Buck across on the trestle.

*Jeanie and Buck put in a full day trailing cattle and then, both tired, with a long ride ahead, she chooses the railroad right-of-way as the shortest way home.*

The trains were mostly fast freights with sometimes a hundred boxcars pulled by a big black steam engine with an engineer leaning out the cab, trying to see the rails ahead. The freights left Miles City gaining speed and by the time they went by our ranch they were high rollers, moving fast.

I had reason to be scared. Crossing the trestle was stupid and there was no going back.

Then Buck felt vibrations through the rails and froze up.

I saw the arm of the semaphore coming down. We still had lots of track to cover. I was as alarmed by the semaphore as Buck was by the vibrations.

He knew a train was coming and the rush and noise of trains panics horses. Worst of all, engineers blew their whistles at the worst possible times.

To get him to loosen up and move, I picked up each hoof and placed it on the next tie. I went around his body. He leaned against me in fright, maybe drawing courage from my body. His heavy bulk almost pushed me down, but I let him lean on me, comforted him and we kept moving.

His whole body trembled. Round and round him I went, trying not to see the rocks far below or the open space between the ties. I moved each foot one after the other.

My plan was that if the engine came screaming around the curve ahead, I'd have enough time to try to push Buck off the trestle down to the dry creek, then jump myself. Within the few hundred feet, before the engine hit us I'd try to save us both. If I couldn't, I'd save myself. It didn't look like a soft landing down there with all those piled rocks, hulking over each other, but I'd do it. I didn't want to wreck the train, or have it land on top of me.

We hurried on, one hoof at a time.

"Come on, Buck. Good boy! Pick up your foot, Buck."

The grass at the top of the cut bank met the rails just ahead. Not far now.

With only about five feet to go—around the curve came the fast freight, the shiny black engine looming high above us. The cow-catcher grill gleamed silver against the black and it bore down on us to scoop us off the tracks. With its shrieking roar it loomed so close, the engineer couldn't even see us.

I screamed at Buck to jump and yanked the reins. But I didn't need to.

Buck leaped over the last few feet of track, over my rolling body and down the grassy embankment.

I clutched the reigns tightly to keep Buck from stampeding down the right-of-way with the train roaring above us. Even though

*Buck felt vibrations through the rails and froze up. I saw the arm of the semaphore coming down.*

Western Ranch Life in a Forgotten Era

panic-stricken he had the good sense to run side-stepping to keep from stomping me.

*Good old Buck*. I jumped up and fought Buck over to a fence post and snubbed his head tightly, holding his flaring nostrils to calm him.

The freight cars roared by, ten or fifteen feet away.

In the screaming click-clacking of iron wheels on the steel track, the rushing wind of boxcars, I caught sight of the engineer looking back at us. With a big grin, he waved his striped engineer's cap.

He didn't even know how close he came to trapping us on the trestle and killing a good horse.

## Buck—Mr. Personality

"Mom, here comes Dad and he has a new horse in the truck!" I ran to the house to tell her, then back to the barn.

The Winningham brothers and their Uncle Charlie had been with Dad at the horse sale that afternoon and now they leaned against our corral fence expecting a rodeo.

Dad backed up to the loading chute. Off jumped a big well-built three-year-old, bucking and kicking fiercely with both back legs.

"Buck. His name is Buck," Dad told Mom as she came up. "See that dark stripe down his back. He'll be a buckskin when he sheds off."

*Buck*. A new horse was a special event in my eight-year-old life.

"Elmer, you've got yourself a bad one there," Louie called out. "Maybe you can sell 'em back for a bucking horse!"

"We'll see," Dad laughed.

When he brought out the saddle Mom saw he was not going to break this horse in his usual manner—by gradually, over several days or weeks, introducing the horse to the halter, blanket, saddle, bridle and rider in turn.

"No, Elmer, don't you do it!" she begged. "You're not as young as when you used to ride those wild horses out of the breaks."

But the Winninghams stood around grinning and she knew Dad wanted to give them an old-time show.

Buck didn't disappoint. He kept rearing and leaping away, jumping and snorting, as Dad saddled and bridled him.

In one quick motion he swung into the saddle and released his hold on the cheek strap. The surprised horse stood rigid for an instant.

Then he exploded. He sprang up with all feet in the air and came down hard, blatting and squealing. He was a straight bucker,

*"You've got yourself a bad one there,"* Louie called out. *"Maybe you can sell 'em back for a bucking horse!"*

but each time his four hooves hit the ground, stiff-legged, there came a terrific slam.

With each jump, Dad whooped and laughed and flailed his right arm up and down in best rodeo form. His hat flew off and his red hair blew in the air.

Then suddenly, the show ended.

Buck stopped, took a few deep breaths, crow-hopped once or twice, then broke into a nice trot, looking around at his audience.

Was Buck looking at our neighbors, even then, with that amused glint in his eye?

As he gentled, Buck thrived on attention and expected—even insisted—on being petted. If we rubbed behind his ears, he inclined his head and moved blissfully into just the right angle, taking great pleasure in it.

This didn't stop him though from jerking his head up suddenly, to crack a jaw or hit someone's nose. Or stepping on our toes…

A dark golden tan, with black points, Buck never did shed off light enough for a true buckskin, as Dad predicted. But he retained the promise with that dark stripe down his back from black mane to black tail.

Dad bought Buck for an irrigating horse and he turned into a good one. A big strong horse, he willingly carried Dad with shovel and portable canvas dam. He never spooked—even when canvas flew out like a sail in the wind and flapped against his flanks. He waited calmly on soft ditch banks with reins down while Dad set a new dam in the ditch and directed water onto corn rows or alfalfa.

More than that, he became a useful all-around ranch horse, intelligent, sound and dependable—though admittedly somewhat lazy.

Buck liked to tease and was always ready with a new trick.

He learned to nibble the brim of our straw hats. He'd pull them off if we didn't shoo him away fast enough.

One day he plucked Dad's hat off by the crown and held it high in his mouth, nodding his head solemnly.

Dad roared with delight. "Hey. You give that back!"

Buck regarded him innocently.

*Who? Me?* he seemed to say.

Dad had a pet hen that came and flew onto his shoulder when he went to the granary, eating wheat from his hand.

Buck took great interest in that white hen. It was pretty funny to see the three of them out there, the chicken on Dad's shoulder and horse at his side, both maneuvering for Dad's attention and snatching nibbles of grain. The hen scolded and pecked at Buck, never

Buck with his great endurance was a sound horse for any job and Dad's favorite.

## Buck liked to tease and was always ready with a new trick.

## 'Breaking horses' Dad's method

Haltered and tied, a two-year-old mare, half-wild, just off the range and still in her long winter hair, shies and jerks away from the gunny sack Dad holds for her insepction.

Francie spends a few days, with soft rope halter looped over the mare's neck, 'sacking her out' with saddle blanket until she stands quietly, allowing the blanket to be flipped and dragged over her back, tail and under her stomach.

Days later Dad introduces bridle and bit, while the horse objects and pulls to the end of her rope. Dad prevails, forcing the bit between her teeth and slipping the headstall over her ears. He cinches the saddle in place and she is ready for her first ride.

quite reaching him, but teetering and almost losing her balance.

Buck liked to tease Mom, too.

He'd hang out in the yard where he could intercept her going to the chicken house or garden, assured of her pats, caresses and friendly conversation.

One June, Buck discovered Mom's yellow roses. Our lawn was fenced with woven wire and just inside the fence she had painstakingly nourished two yellow rose bushes to a height of about four feet. Just high enough for Buck to reach over the fence and pluck off a bloom.

*Why would a horse want to risk thorns to pull off a blossom?*

Buck figured out how to do it by carefully pulling back his lips and nipping it off with his teeth so the thorns never touched his skin. He looked so funny, holding the yellow rose in his teeth like a gypsy dancer, tossing his head and rolling his soulful eyes coquettishly.

Mr. Personality didn't eat the roses. Just stood there with a bloom in his mouth until someone noticed and called Mom.

"Get away! Buck, you get away from that fence and leave my roses alone," she'd scold, flipping a dishtowel at him or swinging a broom.

"And look, he's stretching the wires, sagging the entire fence down. Buck, you're such a nuisance!"

Buck's eyes opened wide. This meant business. He threw his head in the air and jumped back a few feet, looking at her reproachfully. But he gave it away with that mischievous gleam in his eye and the raised quirk of an eyebrow. Then he'd trot away, triumphantly, swinging the rose in his teeth, sashaying his hips.

He enjoyed having the last laugh.

When we opened the granary door, we often felt someone coming up behind. Soon Buck was breathing down our necks, leaning in with his great bulk, nosing in, pushing us out of the way and reaching past to munch on scattered grain.

When I led him, Buck rubbed his head against me, pushing, bumping along. If I responded by petting he came alive, participating vigorously, rubbing and jerking his head up and down against me, gouging his bridle buckles into my stomach.

He stepped on my toes—then took his sweet time moving off them.

"Buck! Move Buck! . . . Get off! Get off!" I yelled that a lot, pushing on his leg.

As I howled and pushed, he'd swing his big head around and regard me with those dark melancholy eyes. Then finally, finally, move

*Montana Stirrups, Sage and Shenanigans*

his foot. Because of Buck, I spent some of my hard earned 4-H fair money on a new pair of steel-toed boots.

He knew exactly what he was doing, I'm sure—and did it for attention and to get my frantic reaction.

After all, he never stepped on a kitten.

Buck liked to gently sniff and nose kittens he caught out in the yard. But he never hurt them. He stepped carefully while they played at his feet. Grown cats he didn't disturb, probably already having felt the slash of their sharp claws on his nose.

Buck had great endurance and could have alternately trotted and loped all day if he wanted. But he was stingy with the effort. Some days only when he feared being left behind by the horse and rider ahead would he break into a gallop.

His rider seemed always to be kicking those well-padded ribs.

"Come on Buck—let's go!"

Yet he could be frisky. Heading back to the barn, that's when he kicked up his heels. He'd run wide-open all the way home if we let him, anticipating his oats at the end of the ride.

One winter Buck enjoyed a special feast.

Dad made sure our livestock wintered well. The horses ate with the cows when hay was fed every morning. They had plenty of protection from bad weather, and he gave them extra grain to get through a cold snap.

Then he noticed a strange thing. While our other horses maintained their usual weight, Buck seemed to be getting fatter.

"Don't anyone give Buck extra oats." Dad told us. "He's getting too fat. Probably he needs a good workout."

Dad couldn't figure it out. He watched Buck more closely.

*Was Buck stealing oats from the other horses?* He started feeding Buck his ration in a separate place.

The horses spent many of those cold, windy winter days standing in the lea of the granary, which was a large rectangular building with four grain storage rooms.

"Buck's still getting fat." Dad said to Mom one day when she opened the granary door for a bucket of wheat for the chickens.

"And it's odd—he doesn't seem terribly hungry," she said, holding out a handful of wheat. "See he doesn't really want to eat out of my hand."

Buck nosed at it in a finicky way, delicately nibbling a few grains.

Then Mom noticed that when the other horses went out for hay, Buck spent his time standing by a certain corner of the granary.

There, in the corner, they found a small break in the wood siding. When Buck pulled on the siding with his teeth, it wiggled and spilled oats out on the ground.

*Buck enjoyed having the last laugh. He'd trot away, swinging the rose in his teeth, sashaying his hips.*

Dad put an end to that. And Buck's belly returned to its normal size.

Poor Buck. For a few weeks, he stood there at the corner, trying to wiggle the board loose, mourning the loss of his secret cache. The fun was over.

Buck's life ended in a terrible accident.

One summer afternoon our family and Margaret and John Grauman and their two little girls, friends from town, picnicked at the new highway rest stop.

That year, when the new U.S. Highway 10 cut its route across our pasture, it isolated a corner of irrigated land below the main ditch—well watered, with the grass a deep green under tall cottonwood trees. This small, shady area along the road was no longer useable for either pasture or field. So Mom and Dad donated it to the highway department. The highway and right-of-way had already carved out a huge strip of land through our pasture and fields. In return the department agreed to make it into a picnic site, add tables and keep it mowed.

For a picnic spot it was wonderful. In shade beneath the cottonwoods the air always stayed fresh and cool.

Dad and Buck came from their irrigating and joined us at the picnic area.

While the mothers fixed lunch I led the little girls on Buck around the cottonwoods. They had such fun riding and feeding him tasty alfalfa.

We tied Buck to a small tree off to one side as we ate.

The sun dropped lower in the sky and the low clouds in the west turned orange and gold. We finished eating and sat at the table visiting.

No one noticed that Buck had worked his reins loose and started for home. Dragging the reins, he headed up the highway toward our driveway.

He almost made it. Probably didn't even see the car coming.

For the woman speeding down the long hill toward the picnic site and town, the sun might have been in her eyes. She swept past the *Slow—Cattle Crossing* sign three or four hundred feet above our driveway without slowing down.

*Scree-ee-eech! . . . Whoomp. Thud!*

At the sound of the crash we looked up horrified to see Buck thrown into the barrow pit, blood gushing, struggling to rise. A blue car, its front smashed in, was crumpled sideways in the middle of the highway. The saddle lay demolished in the ditch.

Uncle Will jumped in the pickup and sped home for a rifle.

> *Dragging the reins, Buck headed up the highway toward our driveway. He almost made it.*

*Anne works with canvas dam and irrigating shovel that Buck willingly carries, along with rider, on soft ditch banks and wet fields.*

We rushed to free the occupants. The windshield was totally ravaged, glass shattered in tiny bits held together by the gummy inner layer. Like a piece of floppy canvas, it laid inside across the dashboard.

Dad was about to pull open the driver's door when a woman leapt from the car and ran over to the ditch looking up and down the highway.

"Where is the man I hit?" she screamed. "The man who was on the horse? I've hit someone! Did I kill him?"

"Are you all right?" Dad asked.

"Yes, yes, I'm fine. But where is the man who came up over the hood and hit my windshield, and then bounced off into the ditch?"

"No one was riding the horse," Mom said gently. "It must have been the saddle that flew off. Are you sure you're okay?"

The poor woman was distraught, sure she had killed both horse and rider.

Buck lay with head stretched out, lunging upward, then falling back.

Uncle Will arrived with the .250 deer rifle to put him out of his misery. He took careful aim and fired.

As they mourned poor Buck, Mom and Dad were most concerned for the woman. But she insisted she was not injured.

Later that woman sued us, claiming personal injury, even though she had not bothered to see a doctor at the time. She found a questionable doctor who testified that she had permanent injuries. Researching this woman, our lawyer found she had filed unsubstantiated lawsuits before, some so frivolous as to be thrown out of court.

She had hit Buck at the entrance to our driveway between two cattle crossing signs, which gave the right-of-way to livestock crossing there. If she won her case it would mean cattle crossing signs had no legal value in the state of Montana.

Concerned local ranchers gave us their support. Unfenced public roads cut through many of their ranches. These signs—and others that read *Open Range: Livestock at Large*—were their only protection. Open range was then a sacred right in Montana.

Ranchers from across the county filled the courtroom seats at the trial. Buck made headlines for a week in the *Miles City Daily Star*. The trial testimony, including details of Buck's sterling character and endearing personality quirks, was reported in front-page stories.

Part of the defendant's case was her dramatic statement that Buck was a huge killing machine. That he attacked her car, rearing up to stomp her through the windshield. That she wouldn't have hit

Beverley with Buck—he became an all-around mainstay horse for anyone who needed a solid ride without trouble catching a horse.

> Concerned local ranchers gave us their support. Unfenced public roads cut through many of their ranches.

Western Ranch Life in a Forgotten Era

him if he hadn't charged her. She insisted he meant to kill her.

In truth, as the car slammed into him, Buck probably was thrown up over the hood into the windshield. We hated to think of that impact and his last moments.

I came back from teaching in Billings to testify on his easy-going nature. With a cattle country jury, we eventually won. Fortunately, even then Mom and Dad held liability insurance, so our costs were negligible.

Thus Buck triumphed, in a way. He ended his days enjoying the caresses of two little girls feeding him alfalfa spears while he nuzzled their hands. Then in the courtroom, neighbors and family testified to his excellent traits and temperament. Ranchers quietly cheered as his case prevailed.

If there was any victory in it, surely Buck had the last laugh.

We imagined him trotting away from the courthouse that day, swinging his big hips back and forth majestically, swishing his tail and peering back over his shoulder with an amused glint in his dark eyes, making sure everyone watched his exit.

*Dusty shied and made his infamous 'get-rid-of-the-rider' move, ducking and swooping sideways, then coming up bucking.*

## Hung Up in the Stirrup

"If only he'd held onto the reins," Mom said sadly. "Then he would've had some control of the horse."

ANNE

As kids we heard gruesome stories about riders dragged to death when a foot got hung-up in a stirrup. It was our cardinal rule when getting bucked off a horse: hold tight to the reins.

I shouldn't have been riding Dusty that day, but we only had two horses in the pasture close to the corrals—the others turned out on the range. Francie was a much better rider than I, and as the youngest I obviously couldn't ride the other horse—a wild three-year-old that Dad wanted us to work out. This was one of the few times I gleefully pointed out my younger status to Francie.

So we caught up the horses and she took the young horse while I saddled Dusty.

Our job was to corral a bunch of yearlings, the wiliest of cattle. Like teenagers, they are bristling with energy, but not much sense to match. Francie rode back at the rear hazing the yearlings toward the corral, trying to cut them off from their escape routes. One yearling evaded her and dashed up a side hill. I quickly swung in pursuit.

Coming in for the noon meal, Dad stopped to watch the ruckus from the wrap-around porch on our big, old ranch house.

In the midst of chasing that yearling up the hill, Dusty saw a suspicious movement in the grass out of the corner of his eye. He shied and made his infamous 'get-rid-of-the-rider' move, ducking and swooping sideways, then coming up bucking. Even though I felt him tense up beneath me and quickly squinched my knees together, I felt myself flying through the air as he continued to buck.

Only my shoulder and head and right foot hit the ground. The other foot stayed with the saddle.

*Oh, no! The most dreaded predicament in horseback riding!*

My left foot was hung-up in the stirrup. My heel had slipped entirely through the too-big adult-sized stirrup and became tightly wedged. Since I fell off Dusty's left side, my back was to his head, making it impossible to control him. No one was close to dash to my aid. This mess I had to get out of on my own.

Tightly I clutched the reins, but Dusty jerked one loose as he jumped and whirled, dragging me on the ground. Frantically I twisted my left foot, but this only wedged it farther into the stirrup.

My head banged the ground, rattling with noise. A rock sent lightning flashes through my brain. Mouthfuls of dust and gravel choked me. I tried to reach Dusty's head so I could pull him around but he leaped away.

I pulled my rein as close to his head as possible so he had to circle me and stop bucking. But if he stepped on the rein it could break and I'd lose all control. I figured then I'd be kicked unconscious, giving Dusty a free head to light out at a dead run—with me flopping along behind.

Meanwhile Francie saw what was happening and fought her green-broke pony, kicking and slapping him with the reins, trying to get up the hill to circle Dusty and calm him down. But the more she kicked and urged, the more he fought the bridle, threw his head, pranced and danced, and refused to go in my direction.

Dad, watching from the porch, sized up the situation and dashed into the house. He reached behind the door where the rifles with shells in clips leaned and grabbed his .250 Savage open-sight deer rifle. He levered in a shell and aimed, but held his fire.

Dusty kept whirling and jumping around me, my shoulder bouncing on the ground as I clutched the rein and tried to keep my head up.

Dad was known for his accuracy in what friends called impossible shots on fast moving game. Living in the Missouri River Breaks during the early days, he and others there depended on wild game which, between coyotes and hungry homesteaders, became scarce and hard to hunt. No matter how far off and no matter the running speed, he led his shot in front of the deer and brought it down.

One of his best-known feats was dropping in its tracks—but not injuring—a good-looking wild horse from a bunch of wild

*Dad dashed into the house and grabbed his .250 deer rifle. He levered in a shell and aimed.*

horses. A nerve runs along the top of a horse's neck near the mane, and Dad could nick this nerve with a shot from his running mount, momentarily stunning and knocking him down. A missed shot could kill the horse.

If he shot now he'd have to place the bullet to avoid me as the horse whirled and I tried to lunge upright on my free leg. And if he missed the nerve and killed him, Dusty might kick me to death on his way down.

Then, suddenly, my foot was released!

My stocking foot pulled free from the boot and I fell down—right into a cactus patch.

Without the drag on his stirrup, Dusty stopped. I picked cactus off my sock, pulled on my boot, scolded Dusty and rode off to retrieve the yearling, unaware of the drama on the front porch.

Would Dad have pulled the trigger? I don't know, but he was ready for the right opportunity—and a dead eye with open sights.

He didn't shoot. I lived and so did Dusty—for the pleasure of dumping me another day.

I don't want to seem unappreciative of a fine horse, but through the years I gained an intimate knowledge of the hazards of hitting the ground hard and tangling with jumping cactus, prickly pear, yucca needles, rocks and rattlesnakes—all due to our best cow horse, Dusty. And it wasn't just me—my sisters had the same problem.

Dusty, a tall long-legged, dark sorrel, could turn on a dime even while running hard. His long, rangy gait covered miles of ground at a trot. He had stamina and, more important, he had cow sense.

He loved the challenge of cutting cattle. Flick a rein against his neck and lean in the saddle toward the critter to separate from the herd, and Dusty stuck out his long neck and maneuvered that animal off to the edge. If a cow was particularly devious, he'd reach out and give her a nip. Riding Dusty made quick work of the hardest cow cutting jobs. When there were cattle to be worked, we all wanted to ride him.

However, there was a problem with Dusty....

Once in a while a local rancher, after seeing one of us cutting cattle on Dusty, came by to borrow him in the midst of working his own cattle. Dad always mentioned Dusty's problem. But the rancher assumed this horse, ridden by teen and pre-teen girls, was, of course, a horse he could ride. A tough old rancher always thought he could ride anything a girl could. If he borrowed Dusty, he soon learned otherwise, and we noticed that no one

*Anne enjoys a quiet moment on Dusty, while he decides at what point he'll throw her off.*

ever asked for Dusty twice.

When riding Dusty, I never let my boots slip too far into the stirrups so my feet might be hung up when I got thrown off. I was often bucked off Dusty and usually managed to death-grip the reins. If I dropped them, he might wait for me or he might head home and if I was alone I walked—sometimes for miles. Another rider could easily catch Dusty, as a horse trailing bridle reins can't travel very fast—just fast enough so his angry rider on foot can't catch him.

Dusty loved to dump his rider, but he awaited his chance. While he rarely bucked, he had this plan. I know full well he did it with calculated forethought. He'd trot along nicely after we left the ranch yard, then at some later moment suddenly go into his spooky mode.

*Was that a grouse lurking behind the sagebrush, just waiting to fly into his face? Was a rattlesnake coiled beside that clump of cactus? That tall bunch of crested wheat grass whipping in the wind? What a great excuse to shy!*

No horse had better perfected how to slyly unseat a rider.

*Trot calmly along till the rider relaxes. Then abruptly stop, trembling, stiff-legged, and snort at some supposed danger. While the rider is distracted looking for the problem, duck your body, slide sideways, and come up bucking. Leave that rider hanging in midair with no horse beneath.*

*Look at her sprawled on the ground, as if to say, What a surprise! What are you doing down there?*

I'd jump up and grab Dusty's bridle, look him square in the eye and scold him severely. He always looked so hurt when I accused him of spooking deliberately—but I swear there was a smirk on his face.

I noticed when I stayed particularly alert, Dusty continued to shy at the least little thing until I let down my guard, and then—*whoosh!*

Sideways he swooped and left me alone up in the air. After he dumped me, he behaved just fine.

Sensitive to human moods, Dusty liked to get the rider mad at him. He knew that throwing one of us into a big patch of prickly pear—or a hillside of jumping cactus—was a good bet to increase our wrath. A smirk lit up his face and slightly curled his upper lip.

Jumping cactus grows with many spiny balls, barely attached to the main plant and covered with barbed hooks. At the slightest touch they flip off and grab onto clothing or flesh. We tried to avoid riding through those patches since hooves easily flick off those little cactus balls and flip them through the air onto a horse's tender underbelly—a good excuse for a Dusty-type bucking job.

And there we went, right into the patch of jumping cactus.

Or beside a rattlesnake....

One bright fall day, late in the afternoon, I decided to go hunt-

**A tough old rancher always thought he could ride anything a girl could. If he borrowed Dusty, he soon learned otherwise, and we noticed that no one ever asked for Dusty twice.**

Western Ranch Life in a Forgotten Era

*A few inches from my nose lay the flat dusty-gray triangle of a rattler's face.*

Under Beverley's forceful demeanor, Dusty behaves himself and doesn't try his usual monkeyshines.

ing. I disliked riding Dusty when hunting because it was such a problem staying on him while holding a deer rifle. I sure didn't want to get bucked off and ruin the rifle. But Dusty was the only horse near the corrals that day.

Dad encouraged us to use his gun scabbard, but by the time I'd see a deer, get off the horse and pull the gun from the scabbard, the deer was gone. So I carried the rifle in my hand as I rode. Of course there was never a bullet in the chamber. Nor were we allowed to shoot from a horse—Dad said it would damage their ears.

I was riding through the badlands with the sun warming both Dusty and me. Always beautiful, the badlands are spectacular late in the day when the sun casts long shadows on cutbanks and eroded gumbo shapes. A steep bank reveals a thin layer of coal, layered with red scoria from long-ago burned coal, then yellow and rust-colored soil, all separated from each other by a gray seam of gumbo clay. Nothing grows on these buttes, and when it rains, torrents of water mixed with gumbo cascade down, pouring into washouts below. Between the washouts and buttes grow nutritious high protein grasses.

Here I found several large-antlered bucks.

Dusty and I saw the deer at the same instant. He immediately shied and ducked sideways, bucking. I dug in my heels and hung on. Darned if I was going to get thrown off with my hand full of rifle. We were upwind from the deer and they hadn't spotted us.

"Bzzzzzzz." Suddenly I heard a telltale rattle.

Dusty jumped in the opposite direction. This was too much. I felt the stirrups fly out and knew I'd soon be flat on the ground. I clung tightly to the rifle and landed it first, with me falling nearby. I couldn't see where I was landing, just concentrated on not jarring Dad's favorite rifle too much as it hit the ground.

As I looked around, a few inches from my nose lay the flat dusty-gray triangle of a rattler's face. Its hooded eyes loomed extra large and its slim black-forked tongue flicked rapidly in and out. I quickly moved back, abandoning the rifle.

"Bzzzzzzz," came from behind me.

Not the right direction to move! Carefully I eased away, then discovered I was in a patch of rocky shale, a perfect spot for more snakes to soak up the sun from the warm rocks. But luckily there were only two and they didn't strike.

Landing eyeball to eyeball with a rattlesnake is a nightmare. Maybe those rattlers were as surprised as I was. Safely out of range I bombarded them both with rocks till they were smashed, then tweaked off their rattlers for souvenirs.

Dusty stood quietly without trying to run away. It seemed he wanted those rattlesnakes dead too. All the time I was throwing rocks, he watched with interest, and then waited for me to collect the reins and get back on.

The deer had disappeared and shadows were lengthening on the hills, so I headed for home, deciding to save hunting for another day and a different horse.

Definitely a different horse.

The problem with Dusty? It was his attitude. He enjoyed far too much his talent for alerting his riders to rattlesnakes, cactus and the orneriness of critters.

*Leery of being caught, Gypsy is tempted by pile of oats.*

## Corralling Spooky Horses

Our horses were always hard to catch. In the corral my sisters and I could slip a rope on them, but to get them in the corral was a challenge.

Dad didn't like us to leave horses in the corral because they were such gluttons for hay. So when we needed to ride we seldom had a horse handy, or any way to catch one except on foot.

We schemed to fool them into the corral when they came in for water. Otherwise we had to go out in the pasture to get them.

Taking a bucket of oats, we walked out there, shaking the oats and calling. They came tantalizingly close—but not close enough. We couldn't let them see the rope or bridle and hid it, tied around our waists under a shirt or jacket.

Sometimes they just didn't want to be caught, especially if we'd ridden them hard recently. And some horses are naturally suspicious and jumpy.

We had our little tricks—and they had theirs.

We'd hold up a handful of oats, letting a few kernels trickle enticingly back into the bucket. One pony might nose in, ears laid back, to grab a nibble while keeping his neck out of reach. It's amazing how long a horse's neck can stretch.

We piled a little oats on the ground and stepped back, so without getting too close, they took just a taste and wanted more.

If we could get a horse to put his head in the bucket, we might slip the rope around his neck. Hanging on tightly, we sometimes got a whirl through the air as he jerked back, until he realized we weren't going to let go.

Then we could slip the rope around his neck and tie Dad's favorite nonslip bowline hitch.

*Once a horse is caught, the others run all over us trying to get a nose in the bucket, but Francie is quick.*

Western Ranch Life in a Forgotten Era

For gentler horses, seizing a fistful of mane was usually enough to hold them. With others, we had to grab the mane under the neck, on the far side, so they were encircled by an arm and shoulder. Every horse wanted oats and when he decided he was caught, he gave up and ate.

Then if we weren't overrun by other horses trying to get their share from the bucket before it went empty, we got the rope on and headed for home. Once we had a horse caught, or the greediest one following, the others tagged along.

*Anne brings in the horses, led by a too-eager Comet, heavy with foal.*

But sometimes our gentler horses were out on the range, and we were stuck with horses that loved running free more than they liked oats. They came only so close. And if we chanced to grab a mane, they whirled and escaped. Sometimes, by using all our tricks, and enormous patience, we got them to follow the bucket or tempting little grain piles along the way. Maybe the horses we wanted followed into the corral and we shut the gate on them. Maybe not. Some horses we had to run into the corral on horseback.

Every day our horses trailed in to drink—a chance to catch them.

Our long metal water tank watered three ways—into two corrals and from the long side that opened to the larger pasture and included house and yard.

Catching them drinking would have been a snap had there been a smaller fenced area around the water tank like some of our neighbors used. Then, when they came in to drink we could sneak over and shut the gate.

*Every day our horses come in to water, but not always can we catch one or lure them into the corral by the water tank.*

But this seemed impractical because of our multiple use of the large open area between house and ranch buildings, crisscrossed by roads. And because in winter all our livestock needed plenty of space to drink from this one heated water tank.

So we'd try to lure the horses inside the corral with a bucket of feed strategically placed there. *But when they're feeling snaky, just try to get the gate shut.*

After planning to ride, it was disheartening to watch them run past, all the way to the far end of the pasture.

One day, riding Dusty, I went out to run in the horses. Running horses is a wild adventure that sometimes doesn't end well. In that situation they try desperately to outrun the pursuing horse. Nothing they like to do more.

JEANIE

I rode hard bringing our horses in from the hills, when they dodged through a grove of cottonwood trees. I'm sure that was Dusty's idea. He had the bit in his teeth and he was going to run those horses until they all, including him, got in a good morning run. He was in cahoots with them and avoided the corrals.

I wanted to ride around those trees, but Dusty refused to turn or slow down. You can hardly control a horse when running other horses—they all love it so much and run flat out.

*Sometimes the only way to catch our horses is to run them into the corral.*

Suddenly I saw a big cottonwood branch coming straight at me, about waist high. Just in time, I kicked my feet loose from the stirrups and the branch caught me right in the stomach.

There I was, draped over a branch off the ground as high as a horse. And there went Dusty. He ran on after the horses and they willingly went right into the corral. Without me, they lost their challenge.

With my breath knocked out, I was still hung limply on the branch, gathering my wits.

I saw Dad catch Dusty and close the corral gate so the horses couldn't get out. Dropping down off the branch, I walked lamely up to the barn. Dad was holding the reins, I was holding my stomach and Dusty pranced, all excited because he wanted to keep running.

"You have to get right back on this horse," Dad said when he saw I wasn't badly hurt.

"Why? Why do I have to get back on? The horses are in the corral."

It didn't matter. The range rule was that you always get back on a horse after being thrown off, however it happens. It's not good for the horse to think he's outsmarted you and Dusty was having his good old horse laugh.

It's not good for the rider either. You're supposed to show the horse you're the boss, and believe it yourself, too. He'll be less likely to buck the next time.

At least that's the range theory.

*Hah! Most horses, but not necessarily Dusty!*

It never seemed to make a difference to Dusty. He snapped his rider off any old way he could manage, whenever he felt like it.

A branch was as good a way as any.

*You can hardly control a horse when running other horses— they all love it so much and run flat out.*

Western Ranch Life in a Forgotten Era

# Loading Horses on the Range

Our horses could be stubborn when trying to load them in our truck with its high stock rack. But never Comet.

*FRANCIE*

She made it easy. Too easy, maybe.

One day, to my horror, I accidentally ripped a gash in her side while unloading her on the open range when looking for a lost cow. In wintertime the open range above our place was empty of cows, but some rancher had seen a lone cow way over on Dixon Creek that he thought was ours.

I backed the truck into a depression to unload. Since the truck was high, this was never a good solution and the horse had to jump both loading and unloading. Comet did this willingly, but this time she somehow raked her side on a bolt as she jumped.

Without noticing, I rode out on a snowy trail checking the deer and coyote tracks. When I found cow tracks on a cross trail, I circled back.

Comet seemed to lack her usual fire and I wondered if she was heavily pregnant.

Only then did I see drops of fresh blood on the snow. Not realizing it was Comet's blood, I assumed it came from an animal that recently went down the trail, maybe even the cow we were looking for. I tried to trace it back to its source.

After backtracking some distance, then turning and finding two trails of blood, I realized it was us. Appalled, I jumped off to find a seven-inch gash in Comet's right side, still seeping blood.

Through tears, I loaded her back in the truck and headed for home.

*Forget the lost cow.*

I was terrified that even before we reached home she might die from the wound and the way I had mistreated her.

But Dad examined the cut calmly.

"Get a big needle and some heavy thread."

He headed to the shop for disinfectant.

I held Comet's head, expecting that she'd lunge and jump around, but she stood patiently as Dad swabbed the wound and then sewed up the gash with big whip stitches, using pliers to pull through the needle.

In a few weeks, much to my relief, her cut healed good as new.

*Dad gently unloads the horses against a bank in the sagebrush. Horses are easily injured if they jump too fast or slip coming out of the end gate.*

# Horses We Knew and Loved—
# Most of the Time

A marvelous event occurred when a bachelor neighbor, an old cowboy, gave us three horses—*Dusty* for Jeanie, *Queenie* for Anne and *Comet* for me, Beverley, by then, gone off to college.

FRANCIE

Up to that time it seemed we never had the right horse for the job we set out to do. Our best cowhorse was old Eagle, far past his prime.

Our neighbor, Bill Hickock—no relative of the legendary Wild Bill Hickok, though that's what we called him when out of earshot—liked to hang around the sales yards in Miles City, keeping in touch with the old days. He bought and sold a few horses now and then when an appealing horse came through the ring at a bargain price.

"These three horses are just too good to go for killers," Wild Bill said, as he unloaded them in our corral.

He was right. All three were wonderful horses that we rode and enjoyed for many years.

Two were young. Dusty, a tall buckskin with black mane and tail became our best cow horse, although he had his tricks. Queenie, a small and classy black Morgan-type, was smart and dependable, with a sweet and gentle disposition, just right for Anne. Comet, a tall, lanky Thoroughbred, black, with a white blaze, two white socks and wide intelligent eyes, was about ten years old and pregnant—a gallant steed who loved to run, right over any cow she was supposed to chase. She probably raced on the track in her early days, Wild Bill said.

From that time, we had plenty of horses. There was Spotty, a sturdy brown and white pinto we rode for many years and Buck, a stalwart mainstay Dad bought for irrigating work.

Comet raised one colt after the other—Pedro a tall, splashy pinto, Gypsy, Jinx and Sundance. We broke them as two-year-olds, so always had a young horse or two in training. They inherited Comet's long stride, easy gallop, and were always eager to line out on the trail or, unlike her, work cattle. Jinx raised outstanding colts of her own, stylish, high-spirited, and half-Arab, a delightful mix with that wonderful Thoroughbred blood.

Eagle took this horse invasion in stride.

Already completely white and blind in one eye when Dad picked him up at the sales barn, he was still a good, dependable ranch horse who knew well how to handle cattle and little kids. He had lots of heart and cow smarts.

At first only Dad and Beverley could handle him. Jeanie and I

*Eagle, the first horse we girls could ride, with Beverley, above, is a dependable ranch horse with lots of cow smarts. Below, Queenie became a smart, gentle and dependable horse for Anne.*

rode him too, as little kids, but Eagle's opinion of us wasn't much. He thought us too small and insignificant to bother with, so when we kicked him to speed up, he often just turned around and headed back for the barn.

But he let both of us pile on behind Beverley, riding double or triple. Bareback, that could be a tough ride. We all three slipped and slid first one way, then the other, my short legs sticking out straight as I hung on to Jeanie for dear life.

Eagle was the first horse we girls could ride after we moved to the ranch at Miles City and we learned to love him dearly.

But I yearned for a horse of my own.

That happened a few years later when I broke and trained my own two-year-old mare Flexi who, sad to say, met her untimely end after scarcely a year. That night our horses ran in terror before a crashing thunder and lightning storm and poor Flexi stepped into a prairie dog hole, probably at full gallop. Next morning we found her standing in shock, holding up a broken front leg that bent the wrong way. I was heartbroken, but nothing could be done to help her and Dad did what had to be done.

With his great heart, Eagle, despite being half-blind—with an opaque milky marble in his left eye socket—mostly ignored his disability. Only once did he and I have a near accident because of it. His quick, impulsive action gave me a good scare.

That afternoon Eagle and I were moving some cattle that broke

*Francie, Beverley and Anne on Flexi, Buck and Spotty, ready to work cattle all day.*

through a fence onto the highway, chasing them toward a gate up ahead.

Suddenly one ornery cow dodged back across the highway. Quick as a flash, without turning his blind side to check for oncoming cars, or giving me time to look, Eagle dived after her.

*Screech!*

In that instant we both heard the squeal of brakes, and out of the corner of my eye I saw the red hood of a convertible skidding into view just below my right elbow.

Eagle shuddered and kept going, as if to say, *"Too late to stop! Let's get this critter off the road."*

As Eagle turned the cow back, the red convertible sped off, a hair's breadth away.

Our first horses at the Miles City ranch were work horses, Slim and Jumbo, an oddly matched pair Dad bought when we moved there in early spring of 1939.

Slim was a lanky black with jutting hip bones and a glistening white star on his forehead, mid-size for a draft horse. Jumbo stood much taller, a big broad-backed gray with vast round hips and a mottled coat hinting of blue roan in younger days. Though mismatched in size and strength, they pulled well together. Good dependable horses for feeding hay in winter and field work in summer to supplement the power of our lone tractor.

Only once did I see Dad ride one of those big draft horses.

Sitting high astride Slim, he was laughing and waving his straw hat, flipping a piece of harness back and forth in one hand, like reins, heading home from the field.

At first we had little real need for a saddle horse—Dad and Mom milked dairy cows and fed lambs in winter—until they began to build back a herd of range cows.

Losing all our horses to sleeping sickness at Savage the year before must have been a difficult time for our parents. Both grew up riding horses. Mom's father was a horse rancher in southeastern South Dakota during the 1890s when horses were the primary mode of transportation throughout the country.

Dad helped his older brothers train their dad's string of race horses in Minnesota, riding jockey as a slight kid at county fairs, and breaking wild horses from the badlands when he came to Montana

*Comet with our first colt—a pinto surprise, Pedro, above. At left, Comet proudly shows off her new colt, Gypsy.*

*Pedro, a tall, gentle, classic brown-and-white pinto with blaze face, became a sound cow horse for us.*

Western Ranch Life in a Forgotten Era

## Horse society

Horses are social creatures. They dislike being alone. But they have certain rules to follow. Horse society maintains a strong pecking order.

We often laughed at their antics as our horses came in to drink, all of them thirsty, and plenty of room for three or four to drink at once.

But no. That was not allowed.

Comet always claimed her rightful place first at the water tank, usually with her colt or yearling at side. The others stood back and waited. If one chanced to sneak a thirsty snoot toward the water, she squealed and batted him away with her head, reminding him of his inferior status. He knew better than to try again.

She drank, slobbered and sloshed in the water, taking her sweet time, pausing to admire her colt as he drank, just to assert herself, while the others stood back, stamping impatiently. At last she nickered her colt away and allowed the next set to come and drink.

But they proved no more compassionate and followed her example precisely, drinking deeply, then frivolously spitting and sloshing, while admonishing the lesser horses to wait.

Geldings will pair up as special buddies and hang out together. Uncle Will called them *pards*–as in *pardners*.

"You'll often see a couple of pards off in a corner of the pasture, rubbing necks, nibbling at each other, scratching an itch," he said.

The rule was, pards could drink together, but had to wait their turn in the pecking order. Meanwhile a mare hangs out with last year's colt.

Riding horseback, the protocol of horse society sometimes comes into play too. No matter the riders' preference, certain horses keep nosing into the lead while others hold back letting them go ahead.

as a young teenager.

From the time they were first introduced into the west, horses ran wild through the Missouri River Breaks, a large area of rugged, broken badlands south of the Missouri in eastern Montana. Some ranchers increased their horse herds by roping or corralling the best of them.

Dad regaled us with stories of running wild horses out of the Breaks while working for some of these ranchers. We imagined him in our minds, dashing over boulders, up and down buttes and over cutbanks behind a wild herd, riding like an Indian as if one with his pony without regard for his own safety. Dust blowing, billowing from their hooves, thunder shaking the ground. Dad reveled in half-broke horses.

But this changed with the need to provide for his growing family.

Slim and Jumbo were part of our family's fresh start when we moved to the larger ranch near Miles City with better land and buildings.

Eventually Dad acquired a spooky bay mare who rolled her eyes and lashed out at us with both back hooves when we came near. We weren't allowed in the corral with her. Dad rode that mare, but was too busy to break her well, so she was never good with cattle. He traded her for a sorrel gelding, still too high-spirited for us to ride.

Appearing in our corrals from time to time was a *killer horse*. In those days ranchers were rounding up a lot of the wild horses that had invaded their range and selling them at low prices. These were not wild mustangs. Most descended from ranch horses turned out on the range two or three decades before, at the time tractors and cars began replacing horses.

Ranchers called them killer horses, not because they were killers—though some certainly came close—but because the wild horse roundups spelled the end of the trail for them, most destined for dog food.

Or in Dad's case, coyote bait. He didn't keep them long in the corral, wary that his kids might get attached. They laid their ears back at us, snaking out their toothy muzzles, or wheeling to kick out ferociously in our direction if we chanced to hang too long on the corrals. Mostly they acted terrified, being corralled, and we empathized with that.

We eyed them sadly. Dad and Mom didn't talk much about a killer horse, and it sometimes stayed in the corral only a day or two, then disappeared. Later we'd find a fresh horse carcass up a side hill a mile or so from the sheep corrals where Dad carefully placed his poison baits and coyote traps.

*Montana Stirrups, Sage and Shenanigans*

We saw wild horses, too, glimpsed occasionally on the horizon in the open range above our place, racing along a high ridge, manes and tails flying.

*Can I someday catch and tame a wild horse?* I wondered.

By the time Dusty, Queenie and Comet came into our lives we were trailing our cattle to and from summer range, spring and fall—a four-day trek each way—and needed several good horses.

This was a great place for on-the-job training for young horses. Dusty and Queenie took to it well, as did Comet's colts. Frisky in the morning, they soon settled down to the daily routine. We took an extra horse for two riders, so they got a break every other day—two days on and one day of tagging along.

When not heavy with foal or with a colt at side, Comet sometimes came along, with mixed success. She was tireless on long rides, but never wanted to be a cow horse. Impatient with the pokey, grass-nibbling ways of cattle, she disdained adjusting her long, eager stride to theirs.

A dependable mainstay on the trail, Buck was Mr. Personality, and kept us entertained. He liked to nip at tired calves, but in a friendly way.

No more borrowed horses for trailing cattle.

Eagle slowed down with age and we seldom rode him, saving him for kids and inexperienced visitors. But he rose to the occasion when needed. He'd come alive from a dogged walk, and we had to admire his eager youthful style—right up until his tragic end.

One day a friend of Beverley's jumped on Eagle when he stood saddled beside the corral and kicked him into a gallop. She thought it was fun to ride him at a dead run, but had no sense of what might be dangerous for a horse. Somehow she scared Eagle and he stampeded, running wildly from the corral up into the hills. We could hear her screaming all the way from the barn, even when they were out of sight.

JEANIE

Then her screams stopped and we ran up there.

*Beverly's friend thought it was fun to ride him at a dead run, but had no sense of what might be dangerous for a horse.*

*We use our horses almost daily—sorting and moving cattle, working calves, and trailing to and from summer range.*

Poor Eagle. He stood rigid, in shock, dangling a front leg that had snapped in a prairie dog hole. We would never have run a horse across that flat, peppered as it was with prairie dog holes. Beverley's friend stood there crying, looking scared.

Dad had to shoot Eagle.

When it came time to sell one of our horses—a difficult decision!—Dad trimmed it up, tapering the long hair at mane and fetlocks against his jackknife. Then he 'pulled' the tail—tapering it into a neat switch, again with his jackknife. He always said it was important to pull—not cut—the tails of all our horses once in awhile, so they looked like decent saddle horses instead of shaggy broomtails just off the range, with ground-sweeping tails entangled in cockleburs and twigs.

"Gosh, this horse looks so sharp I hate to sell her!" Dad would say when finished, patting a shoulder affectionately.

He'd stand back and admire the horse, petting and scratching behind her ears while the horse rubbed the top of her head against his chest.

Anne was our best show rider. From the time she was eleven or twelve, she could ride most any horse through the sales ring. And no matter her mount, she made him look good, as she turned him this way and that, kicking him into a canter with ears alert and head held high.

She looked like such a little kid—so cute in her long braids—that the horse seemed to be acting sharp all on its own.

Of course, Anne was expert at handling horses, so showed them off at their best. This may have confused prospective buyers, but not our neighbors, who knew how well she could ride.

*Always they saw me first, from a far ridge, already on the run. The black stallion stood guard on a high knoll, letting the mares race past.*

## Wild Horses

The black stallion stood his ground on the opposite ridge. On high alert, he watched me, head up, mane and tail flying in the wind, front feet stamping dust, while the rest of the wild band of mares and colts ran by him up the trail and disappeared. Then he charged after them and vanished.

FRANCIE

The horse trail led down to water. I knew that. But where it went afterward I could never discover. Somewhere in the six-section pasture the wild horses claimed a secret hideaway. Maybe this would be the day I'd find it.

Spotty perked up. He'd like to follow. So would I. But not yet. *First we need to check cows. Besides we'll never catch up with them on the run. Better to sneak around another way and try to catch them grazing.*

How exciting that spring—when Dad announced he'd leased Art Hill's high pasture for the summer.

"Enough grazing for about half our cows. Six sections of rough country. There's a band of wild horses living in there."

*What? Wild horses! How many? How'd they get there? Can we tame one or two?*

We were full of questions. But Dad brushed them off.

"Just a bunch of broomtails. I don't know—a dozen, maybe more. Inbred. They eat too much grass. I've told you girls—one horse eats as much as three cows."

Wild horses sounded wonderfully exciting and romantic to ranch kids raised on pioneer western tales.

We knew wild horses were tough and they could run. We followed one of the first endurance races from Hell Creek to Miles City—nearly won by the wild white stallion that Tot Robertson, a family acquaintance from the north country, had caught and tamed. Together he and his wild stallion led the race for much of the distance in a mixed field that included highly-bred and expensive horses from across the nation. The white stallion raced into the fairgrounds at Miles City a close second in that famous race.

Western lore abounded with tales of magnificent wild horses tamed by bold riders. *Wildfire,* and the horse books—*Smoky, Scorpion, Sand, Big-Enough*— by my favorite author, Will James, fired my imagination. I yearned for a beautiful wild horse to race like the wind across the open range.

I could hardly wait till we got our cattle into that pasture and shut tight the gate.

Best of all, it was my job that summer to ride the six or eight miles up there every couple of weeks through the open range and several fenced pastures to check on our cows.

Each time I rode there I saw the wild horses—but could never get close. Always they saw me first, from a far ridge, already on the run. The black stallion would stand guard on a high knoll, letting the mares race past him. His lead mare loped along the horizon with the others lined out behind. The stallion waited there on the skyline, a magnificent and haughty sentinel, until the last of his harem dropped out of sight. Then he, too, disappeared in some hidden canyon known only to themselves.

Never was I able to surprise them or find where they vanished. And no matter how I rode the ridgeline trails and hidden draws, I never spotted them a second time that day.

I was fourteen that summer, with a passion for horses. My own Flexi had met her sad fate. With a broken leg, snapped in a prairie

## The circus horses

Sometimes people left horses at our ranch for a few months, always encouraging us to ride them. We girls regarded that as a favor, as we liked the adventure of trying a new horse—but actually, of course, they wanted their horses tamed down or kept gentle—for free.

One time it was a pair of circus horses, slightly injured in a wreck, that Dad agreed to pasture—and persuade his daughters to ride—through one summer and fall.

As Jeanie said, you never knew what those circus horses would do.

"One day up on the range I rode toward a herd of Grieves' cattle. They all turned facing me with their startled white faces, red ears standing straight out like always with those spooky cattle—ready to take flight.

"I raised my hand to send them off and suddenly, while they watched, my little white circus horse stopped and bowed slowly and deeply to them. When he bowed, he tucked his head and went down on his knees as his head disappeared beneath him. Since he did strange things, I prepped for him lying down by sliding my boots out of the stirrups.

"But no, his bow was deep, slow and profound and then he stood respectfully at attention.

"Grieves' wild cattle were so astounded, they just stared instead of stampeding in the other direction. Thirty of them, all with identical expressions of astonishment!"

*(continued)*

dog hole, she had been 'put out of her misery' as Dad put it.

I was back to riding Spotty or old Eagle.

*If I could somehow catch and tame a wild colt—what a magnificent steed I'd have then!*

I liked to think the wild horses there descended from wild mustangs accidentally enclosed when the fence was built.

This was high remote country, far from any ranches. Not yet the Pine Hills, but close—I could see pine trees on the tops of distant hills.

Two long coulees cut through this range with higher plateaus and draws running off either side. It was about three miles long and two miles wide, six square miles of rough country, so there were any number of hidden draws and pockets of grass. But it seemed like I had checked every possibility.

All summer I speculated where the wild horses hung out—and dreamed of catching them grazing in a hidden valley. I entered the big pasture through its various gates hoping to surprise them.

But each time they'd see me first and race off into the distance. I had no problem finding all our cows and calves along their favorite draws and water holes, but could never find the horse hideout.

On that morning Spotty and I checked the spring up a cedar draw and the small reservoir a mile or two downstream. At the same time, I kept scanning the hills and ridges.

In the heat of the day we found all our cattle resting near water. In one bunch, near the spring, I counted thirteen cows and calves, with the new young bull, just learning his trade. Others hung about the reservoir. Cows lay contentedly, rhythmically chewing their cuds, calves dozing alongside. Some stood grazing while a calf sucked thirstily, slurping foam and butting mom for more. All seemed healthy and happy—but some I kicked up just to make sure. Calves were accounted for and mothered-up. And the older bull was doing his job, following a cow.

I checked the water sources, springs all running clean and in good shape.

Finally, our work finished, I focused on the possible hideaways we had discovered on other days.

No horses. Somewhere there had to be one more hidden valley.

Riding below a high plateau on the far side of the pasture, my eye followed the fenceline. It ran along the bench above me, disappeared and came out farther down the canyon. So no need to ride up there.

But wait!

*Where does the fence go for that space when it drops out of sight? Does it frame a small bowl back there where wild horses feel safe?*

---

### Circus horses *(cont'd)*

Usually the circus horses pranced instead of trotting. After several hours of that uncomfortable gait, a rider begged for someone to trade horses.

What a relief when the circus people took back their performers! And it was probably better for the charming white horses. They gave their best efforts to cowponying, but could never really understand how to do it.

Then there was the horse that just ran—as hard as he could go in a straight line. You couldn't stop him. Francie didn't realize that until one fall morning when, since he was in the corral, she took him to chase in the other horses.

He stood quietly while Francie saddled him and mounted. But the minute they were out the gate, he seized the bit in his teeth and took off racing as hard as he could go down through the fields—luckily already harvested. His appearance had given no hint of his tremendous desire to run. Far from a beauty, he was a plain-looking bay, not large, but powerful, with shaggy mane and long unkempt tail. Long black hair swept his fetlocks.

That horse had big round hooves like dinner plates and could he run! Right through a muddy plowed field, flinging great chunks of mud to both sides. No way could she slow him down or jerk him sideways into a turn as he ran. When they almost hit the fence at the far end of the field, close to a mile away, he spun around and charged back as fast as he'd come.

Only then was he ready to go run in the horses.

I realized I'd never checked this out and headed that way, wary of our noisy scattering of rocks.

As we came closer Spotty pricked up his ears. He heard something I didn't, and I knew then we'd found them.

Near the crest of the hill, hastily, fearing he might whinny, I slipped off Spotty's back and tied his rein to a nearby sagebrush. Bushes make a flimsy anchor and a leather rein can easily come loose, but I didn't expect to be gone long.

*If they're there, they'll spook fast and be gone in an instant.*

Silently, crouching low, I crawled toward a clump of sagebrush. Heart pounding, I raised up and peeked through the scraggly brush.

*Yes! There they are, at last!*

Below, in a shallow green basin, six or seven mares with three colts and a couple of yearlings grazed near the black stallion. Off to the side at some distance, stood two dejected young males, apparently defeated in their latest challenge.

I scanned the herd rapidly, half expecting them to sense my presence and flee.

Then, disbelieving, I looked more closely. Disappointment washed over me as I searched from one to the other for the beautiful wild horse of my dreams..

Close up they were a shaggy, scruffy-looking bunch. No denying it. Bushy, matted manes, sweeping black tails. Long hair on their fetlocks betrayed draft-horse blood. Big stout horses with heavy legs, their heads and feet seemed somehow too large for their shaggy bodies. Was this evidence of the inbreeding Dad often mentioned?

No spectacular colors. Bays and blacks, mostly, a couple of red roans, an old white mare. The colts were black like the sire. No flashy blaze faces or classy white legs.

In vain I looked for brands that might denote more recent ownership and less likely inbreeding. No doubt the younger mares were offspring of the black stallion. Their colts would be double-bred from their father, a single grandsire on both sides and maybe farther back. An unsavory prospect in my judgment.

The magnificent stallion of my imagination stood at rest, head down, dozing.

It took awhile for full disillusionment to set in. But I was so close to them the truth could not be ignored. Trying to take it all in, I lay there, my lovely wild horse vision shattered. I tried to pick out

> *Crouching low, I crawled toward a clump of sagebrush. Heart pounding, I raised up and peeked through the scraggly brush.*

*Francie finds the wild horses grazing in a remote basin and aims her Brownie camera through the sagebrush.*

Western Ranch Life in a Forgotten Era

a promising colt, and couldn't. Dad was right. Maybe they were just *broomtails* and *knotheads,* as ranchers called them.

This was the sorriest bunch of horses I'd ever seen.

Hidden behind the sagebrush, I lay there on the top of the slope for perhaps twenty minutes. I studied them, hoping to see some attractive trait I'd missed, taking photos with my small Brownie camera, sorry I'd solved the mystery.

I failed to notice that behind me down the sidehill Spotty jerked loose his rein from the sagebrush and headed for home.

Finally I stood up. The wild bunch seemed so lethargic I half-expected them to glance my way and go back to grazing. But in this they didn't disappoint.

*"Whoosh!"*

They came to attention in a flash. They snorted and squealed and blew through their noses. They threw up their heads in alarm, every eye and ear aimed my way. Then they charged into action. Mares squealed to their colts, highly agitated. The bay mare led off, calling imperiously for others to follow, charging off in a thunder of hooves and flying dirt.

The stallion stood his ground, reared and plunged, though obviously horrified to see me so close. Then he whirled and was gone, snorting, squealing and kicking up dust.

I watched as they raced out of sight, disappointment still a bitter taste in my mouth.

Spotty was nowhere in sight. I expected to find him headed across a flat or down some juniper draw not far off. He'd be dragging his reins, head angled to avoid stepping on them—caught up short from time to time and jerking his head as he tripped over a rein or tangled it in the brush.

No Spotty.

The wild horses ran along a far-off ridge, lined out, manes and tails flying, just as I'd seen them many times. Again I had to admire their single-mindedness of purpose in protecting the herd—the stallion, fiercely defensive in the rear guard, the mares vigilant with their foals, powerful hooves churning up dust.

> *Their beauty resides in their wild spirit, their badlands smarts. These horses knew every draw and butte, every predator watching for signs of weakness, every weather extreme, every intruder, like me.*

Panicking at being exposed, the wild horses snort and squeal, throw up their heads and charge off in a swirl of hooves and flying dirt. In seconds, they disappear over the next ridge.

Then it hit me.

The beauty and excitement of wild horses is in their freedom, their wild spirit, their free-wheeling action and herd togetherness. Not after capture, with spirits broken, hanging their heads in someone's corral, their highest skills rendered useless.

It's in their way of life. Their badlands smarts. The adaptations they make for survival of the herd, a life and death matter. These horses knew every draw and butte, every predator watching for signs of weakness, every weather extreme, every change taking place on their range—a gate left open, a waterhole drying up and potential threats, like me. They endured it all and survived.

I couldn't find Spotty and gave up when it got dark. I had to walk all those ten or so miles home across rough country alone in the dark.

But next morning, when Dad and I came to look for him, there stood Spotty waiting patiently by the gate, still saddled and reins dragging.

The long walk home that night afforded me time to ponder wild horses and their home in these remote badlands and other ranges like them throughout the west. Clearly my view had been too narrow—seeing them in terms of capture and taming, and disappointment that they didn't measure up in looks to a good cow horse.

*And why should they?*

I thought of what Will James, cowboy storyteller and horse lover, said of wild horses, "For they really belong, not to man, but to that country of junipers and sage—and freedom."

Yes. These wild horses belonged right there—secure in hidden valleys and, on high alert, lined out and running free along the distant ridge.

**Wild Stallion Challenge**

One spring day I was riding Dusty alone down a dry creek bed looking for a lost steer. Here and there cows stood or dozed by the cottonwoods along the creek bed. Above us rose a steep cutbank about ten feet high.

Before I left home, Dad drew a map in the dust of the corrals at the ranch—one of those 'can't take it with you maps' he was famous for. I was farther from home than I'd ever been on horseback, in country I'd never seen before.

Suddenly, as my horse and I wended our way along the gravel creek, there came an angry whistling and shrieking, a challenge, from above.

Dusty and I looked up, shocked to see there at the edge of the dry crumbly cliff, a splendid black horse rearing up on his hind legs

## Horses make the landscape more beautiful

Horses added immeasurably to the pleasure and excitement of our lives, growing up. Though we'd never admit such a thing, they stirred our souls.

"Horses make the landscape more beautiful. That's what a wise old Sioux chief said," Mom told us. "And he was right. Just look at them out there."

There they stood, two or three horses together on a nearby hill, tossing their heads, ears alert, switching their tails in a stray breeze, their bright coats shining in the morning sun, so lovely it took your breath away.

It was true. Our landscape was more beautiful with horses. They enriched our lives, soothed our spirits and stirred dreams in a way that cows can never do.

You could go outside and call to them. They wouldn't come—our horses never made it that easy—but they raised their heads, perked up their ears and nickered a reply.

It lifted our hearts.

> *Ferocious, he gnashed his teeth—his mouth wide open and bright red inside. Clearly we were his target.*

and pawing at the air. Ferocious, he gnashed his teeth—his mouth wide open and bright red inside. Clearly we were his target.

He squealed and shrieked, threatening us.

I expected the cutbank to break off, sending both horse and dirt down on us, but the cliff held. The ten-foot sheer drop below him into the creek bed where we were, was the only thing keeping him away.

*That cliff extends as far as I can see in both directions. What a break for us!*

I didn't know why that flashy horse was up there making those threatening moves at us, but wasn't about to find out. Dusty wanted to get out of there, too. We quickly loped off in the opposite direction, out of the creek bed and up a small coulee to the ridge above.

The horse stormed at us while we retreated, then turned and lunged away.

When we came out on the ridge, I saw him catch up with a string of horses galloping away, led by a bay, probably his lead mare.

I realized then this was the band of wild horses Dad told us about. He said each wild band has a lead mare who takes the rest of the herd to safety while the stallion stays in the rear to fight off anything that might be a threat to his mares and colts—or to his leadership.

Lucky for us, my horse wasn't a mare, or the stallion may have stomped me out of the saddle and snatched her from me, bridle, saddle and all, to add to his brood of mares.

This cooled my curiosity about wild horses. Watching that stallion chop up the air above us was more than enough for me.

Looking back from the safety of the ridge, I realized I had just peeked through a window that was slamming shut—but not too quickly for me to encounter the gorgeous black stallion of legend, right above me in front of my eyes. This one was real.

# 3 Trailing Cattle

## First Time Trailing

The first time we trailed our cattle out to summer pasture I was twelve and Beverley fifteen. She was trail boss and I was her very scared cowboy crew. I had no idea how the two of us would ever do this hard job.

Beverley didn't question it. With her usual optimism and determination, she just kept going, through blistering sun, hard rain and facing, on the return trail, a miserably cold sleet storm followed by snow. Overcoming all obstacles, she made the experience fun and, as always, stayed enthusiastic about the new venture to its end.

Our first big challenge came the second morning as we approached the long, high silver-span bridge across the Yellowstone River at Miles City. We'd held our cattle the night before at the stockyards sale barn and started out early.

Our whole family pitched in to help. We funneled our one hundred or so cows onto the highway toward the bridge, fortunately before much early morning traffic interrupted, because we needed to stay right up on the road all the way across. A car coming from the opposite direction into their midst could scatter and turn them back.

"C'mon, cows. Let's go!"

Beverley and I rode up and down each side, keeping the lead cows moving and the main herd crowding behind. Mom, Jeanie and little Anne, armed with stout sticks, pushed from the rear. Dad guarded broken-down fences on both sides leading down from the bridge, careful not to cross in front of the spooked herd or stop their forward momentum.

The lead cows began snorting long before they reached the bridge, seeing its big silver spans looming high above, like a foreboding tunnel they didn't want to enter. They stopped dead, blowing and snorting when they got to the first span, heads up and rolling the whites of their eyes, refusing to take one more step. I never

*The lead cows began snorting long before they reached the bridge, seeing its big silver spans looming high above.*

*Western Ranch Life in a Forgotten Era* 73

thought they'd go across, but knew better than to say it.

*We can't do this. They're not going to step onto that bridge.*

Beverley and I kept riding the sides, shouting, swinging the ends of our ropes and pushing cows that tried to dodge back.

"C'mon cows. Hey there! Let's go!" we yelled.

Mom, Jeanie and Anne shouted, too, discouraging any stragglers that might turn back. "No you don't. Get back up here!"

Then Dad walked ahead, calling soothingly, "C'm Boss, c'm Boss,"

He riffled bits of alfalfa he'd scattered on the bridge earlier.

The cows in front smelled hay and they wanted it. But when they stepped on the bridge, the hollow echoes of their hooves unnerved them. Like drumbeats.

Finally one hungry lead cow took a couple of shaky steps, licked up the hay and moved on to the next wisps. Soon others followed with hesitant steps, snorting and bellowing in low frightened tones, their heads down, ears cocked. None had ever crossed a bridge like this before and it was a long way across.

We urged the main bunch forward, not crowding too fast, keenly aware of the risk from tangled fences on both sides under the bridge.

*All we need is for one ornery cow to dive down and jump through that flimsy fence, and the others will follow and scatter up and down the river banks.*

At last they were all on the bridge. Not even one escaped past Jeanie and Anne—or Mom, by now following in the car.

On the bridge the cattle made their way slowly, tentatively, running a few steps, stopping to sniff the pavement, lifting their heads, breathing hard, leery of the hollow drumming of their hundreds of clattering hooves on the echoing bridge. Silver spans trussed together overhead and seeming to close in from both sides alarmed them, as did the sight of swift-moving water far below.

Our horses were as scared as the cattle, ready to bolt, their ears pointed sharply forward, necks arched, blowing and snorting and dancing sideways. Beverley rode Eagle, our seasoned cow horse, blind in one eye but ready for any emergency, while I rode a skittish bay mare borrowed from a neighbor. I felt as nervous as the bay as she pranced on eggs, high above the water.

Then the cows began to run.

The drumbeat of their own hooves kept them running, and the bridge heaved and swayed with their weight and momentum. Solid ground seemed far away at the end of the tunnel.

With no confidence in the bay, I feared she'd take a sud-

> *One quick, well-directed lunge could send me over the railing into the Yellowstone's sucking currents, whirlpools and undertow.*

Silver span bridges over the Yellowstone spook both cattle and horses—and many times us, as well.

den flying leap off the bridge. Or rear up, or buck. One quick, well-directed lunge would send me over the railing into the Yellowstone's sucking currents, whirlpools and undertow.

Last summer Jeanie and I watched a big green cottonwood tree floating down the river, roots first. It suddenly upended as we watched. The roots tipped down underwater. The trunk and the entire tree with its broad, leafy branches stood up tall—just as if growing in the river, green and beautiful, glistening branches reaching to the heavens. But only for moments. Then just as swiftly, the entire tree disappeared into the depths, as if some giant monster had grabbed it from below and yanked it under. We watched the swirling water disturbance it made just below the surface, moving down river, until the light tan water blended into the deeper brown below the buttes as far as we could see. In the old days river men called submerged logs that stuck in the river bottom 'snags' or 'sawyers,' and warned of their danger to steamboats on the Yellowstone and Missouri. That tree never surfaced in all that distance and we didn't see it again.

*Trailing in rough country through sagebrush flats where we've never been, we rely on landmarks we'd never seen.*

So I watched the water warily. Finally we reached the other side, the broad and turgid Yellowstone behind us.

Then our family turned back and we were truly on our own. Beverley and I alone with our horses and cows traveling deep into the big open north country. Here were rugged buttes, cutbanks, washouts and sagebrush flats, but no familiar landmarks. Sudden drop-offs warned our horses to step cautiously, wary of holes partially hid by sagebrush.

We headed toward the Glenn Cooley ranch in Rock Springs country, the range we leased that first summer, some forty or more miles from home—if we rode in a straight line.

The first day out, Beverley and I had trailed our cattle ten miles into Miles City along busy US Highway 10 right-of-way, very different from the rugged, thinly-populated country north of the Yellowstone where we now found ourselves.

Back there, in steady traffic, drivers often slowed down and regarded us curiously as they drove by, laughing and waving. More than one car screeched to a stop and someone jumped out with camera calling for us to "Look this way."

Beverley responded with her glamour smile for photos, even while impatient with any spectators who interfered with our prog-

Western Ranch Life in a Forgotten Era 75

ress. I found the attention embarrassing and, by contrast, enjoyed even more this wide, open country.

A warm rain began falling after we stopped to eat lunch, and by late afternoon we were drenched. Mom met us in the car, scouted the area and directed us to a long-deserted homestead. Happy to rest and graze, the cattle didn't move again.

The sun came out and flooded the rain-soaked hills with a golden light, highlighting a lonely country that's never more beautiful than after a summer rain. A rainbow arched across the entire eastern sky and second rainbow doubled above it in reverse colors. The fresh smell of rain-washed prairies energized us.

While Mom fixed hot supper on a campfire, Beverley and I explored. We loved investigating the old falling-down homesteads still found here and there on dry range lands, so inhospitable to cultivation. Once plowed and planted in bygone years of hopeful rainfall, they'd gone back to nature.

*Beverley and Francie bring the herd up a sagebrush draw.*

The old house of dark splintered wood had fallen into its cellar. The chicken house a pile of broken boards. An array of broken fences, tangled barbed wire, a rusty bedspring, broken crockery and battered kettles lay scattered about. Two or three stunted trees and a leaning clothesline pole were cattle-rubbed to a bronzed sheen.

But the barn held up a roof of sorts. After supper, in a dry area of the barn, safe from rain, Beverley and I laid out our double bedroll, first checking for snakes and mice. After dark, bats and owls swooped by us and hooted from the rafters. Smells of long-ago livestock lingered, the soft manure aged to dark powder. We didn't mind; we were used to that. The barn felt snug and cozy with gentle rain pattering overhead.

On the fourth day, after another night sleeping out and long days in which Mom and Dad intermittently checked on us, waiting at dirt roads that intersected our trail or driving across a prairie dog flat to find us, we reached our destination.

Dad opened a pasture gate and that night we slept in real beds at the Cooley's. The Cooleys cared for our cows and calves over the summer.

> **The sun came out and flooded the rain-soaked hills with golden light, highlighting a lonely country that's never more beautiful than after a summer rain.**

However, our trip wasn't finished yet. Next morning, Beverley and I got up early to ride our horses the long trek home. Dad thought we could take a more direct route than we'd come, but through rougher country.

He drew one of his infamous maps in the dirt with a stick.

"You'll find Stone Shacks here somewhere, after a rocky ridge cutting through there. You should reach it around noon."

He knocked a pebble out of the way and marked an X.

"Then ride down this dry creek bed a few miles and top out about here. There's a big butte and when you get up there, you should see the road where you brought the cows along Sunday Creek."

"But … how will we know the Stone Shacks? Does someone live there?" Beverley asked.

"Oh no," he chuckled. "Nothing left but foundations. Used to be a store, all built of stone."

Our lunches rolled in jackets tied behind our saddles, we closed the gate and kicked our horses into a fast trot. Again we faced into the unknown, going home a different way than we'd come, crossing rougher badlands and gullies.

The plan sounded hopelessly vague to me.

*How can we ride those many miles across unfamiliar badlands with no trail or road, no knowledge of fences or gates, and no one to ask?*

Beverley led on, laughing, enjoying the adventure. No reason for concern—a lovely spring morning with the rain cleared up, a rangeland washed fresh and green and invigorating. Meadowlarks trilled their melodic song for us.

I decided to let her worry about the route.

Around noon we trotted over a ridge, and there it was—off in the distance—square foundations amid tumbles of rock. Sure enough, as we came near we identified what could only be Stone Shacks, a vital landmark in a broad lonely country of seemingly endless buttes, coulees, ridges and sagebrush flats.

We explored the ruins as we ate lunch, then rode on.

Reassured by the success in reaching our first landmark, I still had no idea where we were, and though Beverley didn't admit it, keeping up her usual cheerful banter, I doubted she did either.

In a ride of twenty or more miles, we had not seen a single person, much less a car or ranch house. We rode up and down hills, over rugged passes, down cutbanks, across recently muddied creek beds and through prairie dog towns. We skirted washout holes and the occasional bleached bones of a horse or cow, always looking for the divide that, when we climbed it, we'd see somewhere in the distance the welcome sight of our outbound trail.

Miraculously, we finally connected with the right trail and the highway that led past the airport and down the airport rimrocks. We crossed the scary Yellowstone bridge on horses too weary to care. They knew they were going home.

In late afternoon we stopped for a rest with Grandma, Mom's mother, in Miles City, tying our horses to the fence behind the Brath Hotel where she lived.

*After supper in a dry area of the barn we laid out our double bedroll, first checking for snakes and mice.*

*Bev saddles up for the long ride home.*

Western Ranch Life in a Forgotten Era

*That's when we ran into trouble. A hard wind hit full face. The cows turned back and scattered, looking for shelter for their calves among the rocks and cedar trees.*

Grandma, happy to see us and not at all amazed by our long journey, treated us to lemon drops and a snack, shared a few giggles and sent us on our way.

The last thing we wanted to do was climb back on our horses. But we still had ten miles to go. While I wondered how we could make it on these tired horses, my determined older sister kicked up her mount, and my bay perked up too.

We reached home before dark.

That summer Dad built a sturdy wooden stock rack for our sugar beet truck, which had drop sides. Bolted in place, with tie rods forward and back, and a rear opening for loading and unloading livestock—then removed for field work—it worked beautifully.

From then on we hauled horses, bulls and cow-calf pairs back and forth as needed.

In October Dad trucked our horses out, and Beverley and I trailed the cattle home.

The first two days went well. On the third day a cold driving rain turned to sleet, then heavy snow. We crossed the highway and pushed the cattle up a rocky gumbo ridge, slick with ice, rain, mud and snow.

That's when we ran into trouble. At the crest, a hard wind hit us full face. The cows turned back and scattered, looking for shelter for their calves among the rocks and scant cedar trees. We dashed back and forth, kicked our horses, shouted at the cows and finally pushed them over the hill.

But on the other side, a broken-down, barbed-wire fence lay tangled across the slope. We soon had cows on both sides of the wire, slipping, sliding on the ice and mud and stopping to hunch up against the wind. Calves ran both ways, bawling, trying to reach their mothers across the fence.

Mom found us, parked the car, and scrambled up the hill to help on foot. She slipped over rocks and fences, helping keep stubborn cows moving, untangling wires from our horses' legs as they slid into them.

Horses are naturally terrified of barb wire. Ours were trembling and skittish, panicked at its first touch. We didn't dare ride across the fence, especially in the slippery, muddy gumbo, so got off and led the horses, trying to prevent wire from springing up between their legs as we crossed and re-crossed.

At last we succeeded in getting the cows into a draw, sheltered by a stand of juniper trees. We tied our horses there, rear ends to the storm and scrambled back up the slope to rescue errant calves, still hung up in crossing that fence.

Once 'mothered-up,' the calves nuzzled contentedly. There in the junipers we waited out the worst of the storm for a few hours, then again headed for home.

We were as happy as the horses when we reached our home corral next day.

Once again Beverly's enthusiasm and willingness to meet every challenge and see it through, along with Mom's steady calm, brought success despite our difficulties.

The amazing thing about that first trail drive was that after trailing our cattle out four days, we had to turn around and ride the many miles back home. This came at the end of a ranching era in which this was common, even expected. When you rode a horse out a long distance, you rode him home again—or someone else did. Trucking was not an option.

Neither Beverley nor I had ever made such a ride before and rarely did again.

## Finding the Way

Growing up, we four sisters often faced jobs too hard for us. Trailing cattle was one of them. Sudden difficulties often lay in our way, and we had to figure out what to do next. Sometimes in split seconds to save our skins. We didn't question what we were asked to do, just often wondered how.

"This is what I need you to do today," Dad would say, and we tried. Sometimes he gave detailed instructions, often not.

Dad had great confidence in our ability. *How could we let him down?*

An old hand at organizing trail drives, he thought nothing of sending his teenage daughters out into rugged, often-unfamiliar country with a hundred head of cows and a vague map, if we were lucky—riding at least one green-broke horse.

Dad, and Mom, too, probably believed challenges like these were good for us—it taught us self-reliance and resourcefulness. Our parents encouraged independent thinking, never scolded, never blamed, which gave us a certain freedom in making decisions. We didn't worry about whose fault it was if something unfortunate happened. Instead, we freely tested out various options, and gained a sense of responsibility along the way.

*Our extra horse hangs out with Francie's when she dismounts for a lunch break. In background, on hillside, is a rural schoolhouse.*

Western Ranch Life in a Forgotten Era

Not always did we choose well. Sometimes our decisions landed us in trouble; other times we could backtrack and start again.

Often we were perplexed by the challenges.

Trailing the first time to Margaret Kemp's range on the Mizpah took Anne and me through ten or fifteen miles of rough badlands with no roads or trails and few gates.

Neither of us had ever ridden there before and neither had Dad, since our first day cut across from Highway 10 on the Yellowstone to Highway 12 in the Pine Hills.

But with great confidence Dad sketched us a map on a scrap of paper.

"Cross a few high ridges, take 'em down a flat sagebrush draw. Fences, gates here and here, I think. You'll see a big bare gumbo butte. Head out on this side of it. There's a windmill about here. And a reservoir down there where the cattle can water."

Anne and I looked at the map and each other. It felt like a long distance without solid landmarks.

"Later on you'll see a couple of posts on top of the hill and a gravel road. That's the gate you strike for. I'll meet you there with a load of hay and we'll bed 'em down for the night."

We took the map and got on our horses. We knew this was rugged and unfathomable country and might prove an adventure we didn't relish. With some foreboding we set off—the job had to be done.

Leaving the ranch, we headed up the old Yellowstone Trail some distance and then cut to the right, trailing through two neighboring pastures—so far okay. As miles went by, the country increasingly became rougher and less familiar.

Then, as happened many times, we had no idea where to go next. We puzzled over the map. The cattle fell to grazing and our horses grabbed chunks of grass and munched noisily.

*Sometimes we can follow a country road part way to our destination*

"What do you think? There's a long ridge up there. Is it this place on the map?"

"I don't think so. The map doesn't show a cross-fence. Where are the fences?"

"Maybe they're just not on this map."

"Yeah, maybe not."

We had to laugh, well aware of the shortcomings of our maps.

"Oh well. Up that draw—or that one?"

"Hmm …Let's try over there."

As it turned out, the fences and gates were wrong on the map, and we never did see the windmill.

In that country the gate you choose makes a big difference and so does down which side of a ridge you push a hundred head of cows. It's an uneasy feeling—choosing one ridge irrevocably over two or three others, each with its own complex system of secondary ridges and coulees—and from that moment growing ever more certain it's the wrong one.

A wrong choice takes the rider many miles out of the way—and handling a herd of cattle compounds every problem. The momentum of the herd in difficult terrain makes it hard to change course later.

We questioned our decision many times, but knew we had to keep going till the job got done. A couple of times we turned back after one of us scouted over the ridge and found an abrupt drop-off ahead, or a canyon too deep to push the herd across.

Since the mostly-fenced open range we travelled through spanned large pastures of many square miles, we could ride all day without seeing a ranch or decent road. We leased summer range through the years in every direction, fifty or sixty miles from home, so the country was frequently unfamiliar.

Often we didn't know where we were, but we knew our destination. We also knew that, even though we might be lost and it turned dark, we'd stay with the herd. Dad would eventually find us. If we didn't they could scatter over those vast pastures and gathering them could take hours.

I remember Mom's concern when Dad once explained vague directions on how to take the herd across eight miles of road-less badlands.

"Dad! You can't turn them loose with the herd with those directions!"

*Looking for gates and landmarks, one of us rides a high point to scout the way ahead.*

*Rough country for trailing a cattle herd—choosing the wrong ridge means wasted hours. Or worse.*

"It's the best I can do." he said. And then with a slight smile, "I haven't been across there, either."

"They've got to have better directions!" Mom insisted.

"They'll figure it out," he said with that mischievous twinkle in his blue eyes. "They will."

Matter settled.

Such was our challenge.

> "C'm boss, C'm boss."
> The lead cows heard it before we did and lifted their heads.

We weren't taught directions, but rather, the lay of the land. We relied on the general sense of where we were heading. Somehow I, for one, missed learning to watch the sun and its east-west course, or to take note of its wintertime southern trajectory.

I remember Uncle Will, Mom's brother, once trying to inject some map logic at the start of a trail drive.

"Is that gate south of here or southeast? How many miles?"

But he didn't get far. Dad was perplexed. He frowned and returned to talk of creeks, ridges and draws. We girls didn't understand Uncle Will's directions, either. Like Dad, we went by the lay of the land. As best we could.

That day, heading through the Pine Hills, we watched for gates, comparing them with our map, always planning how to get out of one big pasture and into the next. Sometimes we spotted gates easily in the distance—ideally, four solid railroad ties set together against the skyline. Other times, no gate in sight, one of us rode the fence line.

"Must be a gate here somewhere—see that cross fence coming up over there?" Anne pointed.

"Want to ride up and check it out? Or should I?"

"I'll go."

The afternoon grew late with still more ridges and hills rising ahead.

The lead cows heard it before we did. They lifted their heads.

*"C'm boss, C'm boss!"*

We crested a rise and there in the distance stood our truck by the set of solid gateposts, angle poles braced against them on the ridge. Dad was spreading hay and calling the cows.

*"C'mon cows. C'mon here!"*

The cows picked up the pace and were soon tossing sweet-smell-

Watering our cattle at midday, Beverley drinks from an unriled reservoir, upstream from the herd.

ing alfalfa into the air as they ate their fill. Tired as they were, they lowed contentedly to each other.

"Oh, this is so good. Mmm… it's sooo good," they seemed to murmur.

## Spring Creek Canyon

When we trailed east toward Terry, after leaving the ranch and getting up into the hills, we came out on the big prairie dog flat in the open range above—and almost immediate complications.

Ahead lay Spring Coulee, the gorgeous, tree-filled and spring-fed ravine that spread out into a deep, wide canyon. Fed by many long draws from above, it contained way down in the bottom a meandering, mostly-dry creek, but rushing waters in flood stage, flowing between high bluffs into the Yellowstone.

Our own little Grand Canyon.

One of our favorite hiking and picnicking destinations about three miles from home, we knew Spring Coulee in its upper reaches well, but this was our first time taking cattle across.

The long ravine offered numerous cool, shady, secretive places with towering cottonwoods, junipers, box elders and chokecherries, along with green brushy alcoves cut by cattle and deer trails.

Birds of all kinds—goldfinches, orioles, magpies, mourning doves, cedar waxwings—twittered through the leaves, calling to each other. Deer slipped through the trees, and a cottontail stole from under a bush and hunched frozen, watching us, nose twitching.

*Crossing the open range, we drop our herd down into the deep and rugged Spring Creek canyon far below.*

Always we found much to delight and explore.

Francie and I debated where to cross as we approached Spring Coulee through the cactus-littered prairie dog town.

"Should we cross at the upper end?" Francie pointed off to the right.

"Then we'd have to go through loads of draws and gullies. And circle way around."

"Yeah, it's farther. But easier going."

The highway circled around that way, staying high, crossing doz-

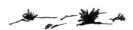

## Open range

The open range just above our ranch was a long stretch of pasture some ten miles long by a couple of miles wide, cut through its entire length by U.S. Highway 10 (later I-94). Locally we called it *open range*, not because it was free or federal land, but because the highway remained unfenced on both sides and cattle ranged freely across it, back and forth. Several big prairie dog towns covered the high plateaus, decimating the grass and leaving sprawling beds of jumping cactus.

Owned by Bill Grieves, a rancher at the eastern end near Shirley, this big pasture was fenced and maintained only around its edges by ranchers like us who bordered his range. Every spring, Grieves turned out his half-wild, dark-red, white-faced Herefords, likely intermixed with edgy Texas strains. They snorted, flung up their tails and fled whenever they saw a rider coming.

Seldom did we get close enough to read a brand, even when hunting a gentler cow of ours, and we checked ear marks on the run.

Helpfully, if one of our cows mixed in, she'd not run far with them in their panicked flight. She'd stop and peer around at us with her ears out, as if to say:

*"Isn't this silly! Why do they run? Have they no self-respect?"*

ens of smaller draws, as did the earlier Yellowstone Trail, the first road built for cars across the northern tier of states.

"Doesn't look like much fun," I said.

"No, it doesn't," she laughed.

"We could go down past the spring."

"Well, it's so thick with trees down there. What if the cattle get scattered in the brush and we can't get them together again?"

"They might like it and want to stay," I giggled.

In fact, we had three choices for crossing Spring Coulee. We could circle right and stay high along the highway. We could cut through the middle to the main spring into the deepest woods and brush, following the hint of an old wagon trail, now overgrown with brush and trees. Or we could take our cattle slightly to the left, down into the lower and deepest canyon, across the bottom and make our way up the high gumbo bluffs on other side.

The spring, our favorite spot, flowed with fresh, cold, great-tasting water from a pipe into an ancient cement water tank leading to a series of small shaded water pools. Around the pools grew mosses and verdant water plants and delicate flowers. When riding the open range we often swung by the spring to rest our horses and enjoy the fresh water. The cool dampness and earthy smell refreshed us after a long, hot ride.

One time we found great globs of grey translucent frog eggs in the main spring and took some home for possible creative uses and…ah, tricks.

The old military Ft. Buford wagon trail, still visible in places, took this middle course past the spring. Here Fort Keogh soldiers watered their horses and splashed cool water on their faces, we imagined, the same as we did on a hot summer day. The now-eroded and overgrown trail, still built up and graded in places, wound on down through the thickest part of the trees midway between the deeper canyon and its upper reaches and eventually snaked up the other side.

Riding or hiking across to visit our friends, the Hill kids, we sometimes followed deer and cow trails that way. But it was not an easy route.

Instead, the lower canyon tempted.

It lay just ahead—deep and wide at its lower end, nearer the Yellowstone River. We surveyed our enticing little Grand Canyon and the rocky gumbo cliffs going up the opposite side.

"It's beautiful down there." I said.

"And I bet there are some good deer trails going up the other side we can follow. We just aren't seeing them from here."

As we debated the question our cattle moved ever closer to the canyon lip. A wooded draw seemed to lead them down into the picturesque gorge.

Crossing there did seem a more direct route. But was it faster? We'd have to take our cattle deep down into the bottom and climb up out of it again along a high, steep gumbo bluff coming out of the canyon.

Still, wasn't that distance faster than going the long way around by the highway?

"Okay, let's do it."

We talked ourselves into it.

Big mistake. Soon there was no turning back.

This beautiful canyon, with rose-colored scoria cliffs, gumbo slopes, lofty cottonwoods, juniper-lined draws, and eagles and hawks soaring overhead, made for a difficult crossing.

The cattle went down easily through the wooded draw, till they hit the final cutbank, then slipped and slid to the bottom. So did Dusty and Queenie with us on board.

The cows weren't yet trail-wise and once on the canyon floor they loitered among the big trees, fragrant junipers and small seeping springs. There were no easy trails and they didn't want to move.

Going up the other side proved too difficult. We could find no good trails at all. After failed attempts to scale the gumbo cliff, we had to move the herd back up the meandering creek bed across gullies and washouts, to seek out the old wagon trail—the mid-canyon route.

Even then, the push up the other side took a long time and we beat the bushes for stray cows.

Hours later we came out on top. We wasted time crossing that canyon, and emerged still only a few miles from home as the crow flies.

Next time, we resolved, we'd choose the boring highway route.

The rest of the day's trek took us across another plateau, down Dixon Creek with its own washout canyon near the mouth of the Yellowstone River, where we crossed on a small highway bridge. We then pushed across more plateau country cut by a series of three rugged washed-out creeks—Deep Creek, Short Creek and Hay Creek—and finally came out on the Buffalo Rapids irrigated valley. There we spent the night at Uncle Chet and Aunt Margaret's ranch, some twelve or fifteen miles from home.

What an adventure! What a beautiful route! But never again did we attempt to cross our cattle herd in the deepest part of that enticing Spring Creek canyon.

We learned that day: there's a good reason why roads skirt a canyon and circle the long way around.

*Deer slipped through the trees and a cottontail stole from under a bush and hunched frozen, watching us, nose twitching.*

# String 'Em Out

FRANCIE

A cattle drive begins in an atmosphere of anticipation.

"Let's string 'em out."

We hurried to open the gate and get them started.

Going south we headed out on the old Yellowstone Trail along our driveway, following the ancient grade up a few miles through Whitmeyer's pasture.

Eager to start traveling, the cattle felt the excitement too. After preliminary sparring, one boss cow established her authority in the fore, unconcerned whether others followed or not. They always did—and we rode up and down the sides to make sure.

This was our favorite system, to string them out up the trail.

Ideally, there'd be three of us—and on some days it did happen. Then two could ride along the sides of the herd, stringing them out, keeping them moving, guarding open gates and broken fences, while the third kept lazier cows and calves coming from the rear. With only two of us—our usual mode—we didn't have that luxury, but alternately fell back to chase up stragglers in the drag.

We had three or four good bossy cows, Roanie, Crumple Horn and Old Yellow—a light-colored Hereford—who just naturally wanted to take the lead. They took turns, each butting the other aside when her turn came. The others fell in behind.

That was fun and we covered lots of miles quickly, moving the herd with less stress than bunching them up. The cows liked it better too, especially in the morning when they were fresh and eager to head out.

Some people seemed always to move their cattle bunched into a tight herd. All the riders behind and no one even trying to narrow the sides.

That works fine for a short distance. Sometimes it's the only way. The cattle just want to bunch up, especially when they're hot and tired and determined to stop and eat. But to us, it made a dusty unpleasant journey for both cows and riders. We felt sorry for calves crowded into the middle of any tight, bellowing herd.

*Even trailing in a blizzard, the lead cow strings them out as she heads toward home and winter feed.*

One rancher who trailed his cattle past our country school each fall used the 'riding point' method. He rode out in front, hoping the cows would string out and follow behind. But they never did. Trouble was, the eight or ten riders he had recruited to help let him down. They didn't seem to understand or care how to keep the sides moving, so they just poked along, all in a row at the back. The chagrinned rancher kept trying to close the gap between his horse and his herd, while the dusty cowboy crew yipped and slapped hats across the broad rear.

Inexperienced friends who didn't know trailing cattle, we surmised.

We didn't even try to ride point. We couldn't waste one person riding ahead like that, while the other did all the work. Besides, we didn't need to. We had our good boss cows taking point.

Amazing how easily we could make eight or ten miles in the morning when the cows were fresh and eager to travel, strung out behind a good lead cow. And what a struggle it was to cover the last four or five miles during long weary afternoons when they only wanted to stop and graze.

*Stringing the cattle out behind a good lead cow keeps them moving.*

In spring because we trailed early, we rarely dealt with calves.

On our fall return trip the calves were plenty big, but some grew tired easily and wanted to quit by early afternoon. They moved a few steps when pushed and then stopped as we rode away. We shouted, swung ropes and slapped our hats trying to encourage them. On a long day even our horses were too tired to keep up an effort.

Then toward evening, sometimes they all got their second wind and picked up the pace.

*Roanie, Crumple Horn and Old Yellow took the lead, each butting the others aside when her turn came.*

"You can't miss school," Mom told us.

So we trailed during holidays when possible, Easter and Thanksgiving. A trail drive took us four days and by coincidence, that's exactly the number of days we got off. Rarely did we get excused from school unless it was absolutely necessary.

*Leaving home for summer range in spring, our cattle eagerly line out on the trail. None wants to lag behind.*

*Western Ranch Life in a Forgotten Era*

> *"If you get caught out in the cold, keep moving—or get off and walk."* Mom told us. *"Don't lie down if you feel sleepy."*

## Trailing through the years

For thirteen years, from spring 1944 though fall 1956, beginning with Beverley in high school and Francie in seventh grade, we trailed our cattle each spring to summer pasture and back in the fall—twenty six four-day treks. Each took us some fifty to sixty miles from home—probably farther as we seldom went straight, but found our route along meandering creek beds, over ridges and across deep canyons. We made ten to fifteen miles a day, depending on terrain.

Over those years we trailed our cattle in every direction from home, spending two or three summers at each place: north to Glen Cooley's in Rock Springs country; east to Orlando Olson's southeast of Terry and in the same vicinity the Brandentaylor range on Ten Mile Creek; south to Margaret Kemp's on the Mizpah and Powder Rivers; and northeast to Harold Cole's Sheep Mountain government pasture, crossing the Yellowstone at Terry. At some, the ranchers branded and cared for our cows and calves. At others, we helped brand or did the riding ourselves.

After the decimation of our livestock during depression years, Mom and Dad gradually rebuilt our beef herd to a point that worked well with irrigated farming. Our herd was about right for the winter feed we raised—but too large for summer grazing at home.

*(continued)*

Even then in the 1940s and 1950s most ranchers trucked their cattle if they went so far. But Dad was old-fashioned about this.

"Trailing is easier on the cattle. Calves and pregnant cows can get hurt loading and slamming around in the truck," he said. Besides, I think he enjoyed the adventure of trailing and probably thought it a good experience for us.

Dad told us stories of early-day trailing big herds of cattle, swimming them north across the Missouri River near Glasgow and loading them onto railroad cars for shipment to St. Paul markets. Sometimes the ranchers rode along in the caboose. As he told them, these adventures usually ended with a funny incident or a joke on himself and his friends. He'd laugh heartily at the punch line.

When trailing spring and fall we could have lovely weather, or any kind of storm. We never knew what the unpredictable Montana weather might bring us. We could go to sleep on the ground by a bed of crocuses and awake in an April snowstorm. Or we could trail through a hot, dry, dusty wind that never seemed to quit.

One afternoon a terrific lightning and thunder storm caught us while bringing a bunch of heifers home from Art Hill's pasture. It struck as we came through the rugged upper Dixon Creek badlands.

Jeanie rode ahead opening gates and scouting for neighboring cattle that wanted to mix in with ours, while Anne and I slowly moved our herd along the ridges and away from the bottoms where the cutbanks and eroded gullies made difficult crossing.

So we were up high, just where lightning hits first—a risky place to be. Lightening hits the tallest objects first. Horses may be the tallest thing around and the person riding is higher still, attracting lightning strikes.

When the thunderstorm rolled in, bolts of lightning began hitting all around. The ridge narrowed and we glanced down either side and hoped our horses wouldn't slide off. We couldn't take the cattle off the ridge; it was the only way home. Then a small butte loomed in front of us and Jeanie rode around to open the gate, uneasy about touching the wires, always dangerous conductors of electricity.

Behind the butte with the drag, Anne and I didn't see her dilemma.

She came racing back in a panic.

"It's horrible! Blue balls of lightening are bouncing across the heifers' ears. Little blue balls that keep running around. The heifers twitch their ears at them like at flies!"

"We need to get off this ridge," Anne said.

But we had to get through the gate first. Jeanie's wild story made us uneasy as the bright flashes of lightning crashed around us, followed by deafening thunder.

*Montana Stirrups, Sage and Shenanigans*

"And more blue balls are running up and down the fence wire. I got the gate open without touching the wire, but I don't know how we'll ever get it closed without getting electrocuted."

Then, just as we reached the fence, the storm suddenly died. The sun peeked out and lightning flashes moved off to the east.

At home, Mom explained the blue lightning balls as 'St. Elmo's Fire.'

A rare occurrence, St. Elmo's Fire once danced around their horses' ears as she and Dad rode along a ridge in the Missouri Breaks in a sudden lightning storm, she told us. Blue-violet balls of a luminous fire bounced and hissed on the tips of the horses' ears and ran up and down their manes. She and Dad hurried to find a way down off the ridge so they wouldn't be the tallest objects. Even their hair felt electrified before they escaped down a deer trail.

Eastern Montana weather can be gorgeous in both spring and fall. I remember many bright crisp mornings and the sharp fresh smell of spring air, with a hint of cedar and pine carried on a soft breeze, while a cluster of returning robins scratched and chirped in the grass. Our cows strung out eagerly, heading for green pastures and open range, somewhere beyond the distant line of hills.

In fall, Indian-summer days brought the sun beaming down warmly. We won't forget those golden days mounted on a spirited and well-loved horse while moving our beautiful white-face Herefords toward home. The cattle trailed eagerly behind a determined boss cow—who tossed her head, happy to be going home. Evenings, the harvest moon rose in a fiery blaze to the east. Huge, red-orange and plump as a tomato, it seemed to bounce on the ridge before liftoff, as if too heavy to rise.

Still, some of our worst storms could hit in April and November, blasting in and pelting us with heavy snow.

"If you get caught out in the cold, keep moving—or get off and walk. Don't lie down if you feel sleepy." Mom often warned. To make the point she told of people who got so cold they just lay down and gave up.

Seemed like our feet were always cold in winter. We wore cowboy boots to prevent our feet slipping through the stirrups, especially risky when slippery with snow and ice. But our feet grew and the boots got tight. Adding a pair of wool socks or two might not be an option. Boot overshoes helped a lot when we had them.

Riding all day in wind, rain, snow, cold or blistering heat while trying to push cows and calves could be miserable.

When Anne grew old enough, she and I did most of the trailing, with occasional help from Beverley and Jeanie, who by then were in college or working. Sometimes we were joined by friends who

> **Trailing** *(cont'd)*
>
> Irrigated cropland made up most of our productive land. Above the ditch, our range broke up into rugged gravel and gumbo buttes. Much of what could be our best grasslands was plucked bare by the city dwellers of two big prairie dog towns. Even with the school section land we bought later, there was never enough good grass for our cattle.
>
> So we leased summer range and trailed our cattle, usually around a hundred head of cows along with some young stock, out to summer range in spring and double that number with half-grown calves coming home in the fall. In addition, some years we took our extra cattle to Uncle Chet's and Art Hill's ranges for the summer.

> *No colder place exists in a blizzard than sitting in a saddle where the full force of the wind hits and chills the whole body.*

wanted a day's adventure or folks who showed up with their horses to help us through their range. Most of the time though, we were just two riders trying to string out a herd of cattle, with an extra horse running loose.

After I turned fifteen or so, it seemed I was the de facto trail boss and Anne, four and a half years younger, the best little cowboy any trail driver ever had. Even faced with a seemingly impossible job, Anne sat her horse sturdily and followed directions quickly, without complaint.

If I yelled, "Stop that cow! Don't let her through the gate!"

She never said, "I can't."

She kicked her horse into a run and did everything she could to cut off the escape.

On Anne's first trail ride, the temperature hovered below freezing one day, with a stinging wind. No colder place exists than sitting high in the saddle where the full blast of the wind hits you square in the face and chills your whole body. Little warmth from a horse seeps through a cold saddle.

We got off and led our horses awhile.

It's true, walking gets the blood circulating again. But you can't walk long when you're trailing an edgy bunch of cattle bent on turning around, putting their backs to the wind and going the wrong direction in a storm.

Anne rode bravely for a couple of hours, but when I turned in the saddle, I could see how cold she was. She sat her horse stiffly, huddling tight against herself, scarf pulled up over her mouth and frosty from her breath. She nestled her hands in their knit mittens down against Queenie's neck under the mane where there was animal warmth.

*Anne and Francie push the drag with Queenie and Pedro. In late afternoon, as more calves and cows hang back to rest and graze, it becomes impossible to string them out. Even the horses are tired.*

"Are you cold?" I asked.

"A little."

"You need to move around more. Trot your horse up along the side then go see what's ahead over the hill."

She did and said she felt better when she came back.

I was cold too, but knew not to say it. Adults didn't complain; just kept going—and told jokes. I tried a small joke and Anne laughed and gave me a big grin.

As the wind died and the sun came out warmly, her happiness shone through.

I could see her delight in just being there and knowing she was a seasoned 'cowboy' on the trail.

## Trailing Troubles

"Woo-woo . . . woo!"

One nice fall afternoon, Anne and I trailed our cattle down a broad right-of-way between the NP railroad and Highway 10.

Across the highway an unfenced field of ripe golden wheat waved enticingly. We concentrated on keeping the cattle moving so they wouldn't notice.

Though we felt a little nervous, they didn't seem to pay any attention to the ripe wheat across the road.

It was too good to last.

Almost past the field safely, we suddenly heard a freight train clanging up behind. As he drew even with us, the engineer blasted his whistle into the very ears of our lead cattle.

"Woo-woo"

*Whoosh!*

In seconds our entire herd stood hip-deep in the golden grain. As the cows guzzled and trampled the tasty wheat, they scattered in several directions, feeding hungrily.

*Traveling down the right-of-way between railroad and highway is easy—until the train engineer blasts his steam whistle in our ears.*

In the distance we saw a man come out on a farmhouse porch, shouting and waving his arms in the air.

"What did he say?" Anne asked.

"I don't know. He's yelling for us to get out of his wheat!"

Those cows had it good and they did not want to move. We thrashed around in the wheat, knocking down even more, shouting and pressing the cows and calves back, one at a time. Our traitorous horses fought us, grabbing mouthfuls of wheat and eager to stay there, too.

For a time it seemed hopeless. But all too aware of that distant farmer shaking his fist and still yelling, in German we thought, we kept pushing the cattle backward and finally got them across the highway and on their way.

We didn't look back.

> In the distance we saw a man come out on the porch of a farmhouse. He shouted and shook his fist in the air.

Headed to summer range, a bridge or two always loomed somewhere up ahead. Crossing bridges was a big challenge.

Long, ominous silver-span bridges crossed the Yellowstone River at Terry and Miles City. The Powder River bridge, shorter, but silver-spanned was just as threatening. The smaller bridges, like those at Deep Creek and Dixon, mostly-dry creek beds by fall, ran brimful of rushing water in springtime.

*Getting cattle onto a bridge is the hard part: there are broken and washed-out fences on either side, traffic interruptions, cars honking, scary silver spans overhead and a hollow drumming of hoofbeats from the first cow to enter.*

**The weather fell way below zero and a terrible blizzard began, the snow and wind blowing in hard.**

We crossed them all at one time or another.

Flood waters washed out and entangled fences on either side, so we had to guard them from errant calves as the cows bunched and milled before setting foot on the bridge.

With its high silver spans, a big bridge resembled a big metal cage, menacing and forbidding. *Why would any cow want to enter that?*

"Clo-omp, Clo-omp!"

At the hollow echoing sound of their own hooves, the cattle kept their heads high, ears pointed, dodging, cringing.

Ol' Crumple Horn, in the lead, would toss her head at any sound, stop dead in her tracks, roll her eyes, and let out a "Whoosh!" snorting and blowing air out her nose. Sometimes when she got about fifteen feet onto the bridge, the cattle behind spooked and ran back and she lost her nerve and turned back too.

One year during Thanksgiving vacation, when Anne and Francie were trailing back from the Brandentaylor place, Dad and I drove the truck out to bring them lunch and help them cross the Powder River bridge. The weather fell way below zero and a terrible blizzard began, the snow and wind blowing in hard. We didn't see the cattle herd, so waited in the truck awhile until we got too cold.

JEANIE

On the east bank stood a bar and dance hall, with no other structure in sight.

"We better go in and warm up," Dad said.

I felt shocked and uncomfortable because he never took us into bars.

As we waited, I kept looking out the window into the swirling snow hoping to see the herd coming while Dad walked down to the bridge and back, searching in the blizzard.

This went on for two hours.

"Well, I'm going to go out in the truck to meet them. They must have had a lot of trouble back there," he said, finally.

I stayed in the bar, even though I didn't like it.

Dad drove back the way the cattle were supposed to come, through rough coulees along Powder River until blocked by washouts, then walked, calling into the blizzard.

"Jeanie, there are no cattle back there," he said when he returned. "They must have gone on ahead."

We drove back across the bridge and sure enough, we caught up with them farther on. Somehow they had slipped past us and crossed the bridge in the blowing snow while we waited in the bar.

Anne and Francie climbed into the truck, ate lunch and warmed up, while Dad and I followed the herd.

Later they told us that with the blizzard at their backs, the cattle kept moving steadily down the road.

"The cows knew they were going home and didn't want to turn into the blizzard," Francie said. "So they just kept on going down the highway and before we knew it, we were halfway across the bridge."

This one time our fears of crossing a difficult bridge failed to materialize.

As we rode out each morning, we checked the near coulees for stragglers and, if it was spring, we might find a hostile mamma with her new wet and wobbly calf. *That's a big problem. Thirty miles from home and a cow goes alone up a draw and calves.* The newborn can't travel on his uncertain legs, but can't be left behind and the herd must keep moving to reach the next camp before dark.

A rancher would hoist a calf on the saddle and hold it on his lap as he rode, but Anne and I never had much luck. Not that we didn't try. We dropped a few calves over one side or the other of a skittish horse while trying to load them on the saddle and hold them there while we mounted. Even if it worked, we couldn't carry a calf there all day.

Fortunately Dad usually showed up at these times and loaded the calf, while the mother bawled plaintively. Still a problem: the baby had to be trucked from one camp to the next, with a frantic mother running through the herd looking for him. We worried and watched so she didn't bolt back the way we'd come, to our last camp where her calf was born.

Traveling down the highway, we frequently tangled with fence issues. All too often we encountered broken-down fences that a whole herd of cattle could get through but our horses couldn't. We'd have cows on one side and us on the other and couldn't find a gate.

Or a young calf would escape through a broken fence into a

A hostile mother guards her newborn, dropped in the sage on a spring night. He needs to be trucked, while his frantic mom all day tries to return to this spot where she left him.

We learn never to trust a fence—cattle break through any weak spot. As we string them out, we ride up and down fences, checking for open gates and broken, sagging wires, while keeping the cows moving. In photo below, cattle follow fence on one side. At bottom, flimsy fences on both sides of a road mean cows will surely break through somewhere, or—there goes an errant calf! And our horses need a gate to head him off.

*Western Ranch Life in a Forgotten Era*

field or the railroad right-of-way and go high-tailing it back the way we'd come. One of us had to cut out his mother and take her back so the calf could mother-up again. This actually needed two riders, but one had to do, while the other kept the herd together and out of trouble.

You can hardly chase a small calf by itself because it goes crazy. One thing we learned, a young calf is quick and unpredictable and with surprising endurance.

Calves can even disappear. One afternoon three of us, Jeanie, Anne and I, drove the herd across a broad sagebrush flat when Buck stopped abruptly beside a sagebrush and refused to budge.

Anne, who was riding him, kicked, slapped the reins and urged him on.

"C'mon Buck, let's go."

A rancher once told us, "Cowboy, if your horse ain't wantin' to go there, neither should you."

Buck was not moving. His erect ears pointed toward the sagebrush, so Anne leaned over and saw, concealed by the brush, a deep straight-sided washout hole—in the bottom a husky calf. He had simply dropped into the hole. One minute he was walking along beside his mother and the next he disappeared from sight. His mother kept on going. Apparently she hadn't missed him yet.

But for an alert cow horse—good old Buck—we might never have found him.

Since he was a big fall calf, kicking and leaping, it took all three of us to get him out. With a rope around his neck, pulled tight against a couple of half-hitches on the saddle horn, and two of us lifting from below, he finally scrambled up the gumbo sides of his trap and out, bawling for his mom.

Sadly, it sometimes happens in washout country, that a deep straight-sided hole like that opens up in the ground from a sudden cloudburst or flooding. A calf falls in and with no one to save him, dies of thirst and starvation.

*The cows didn't want to face into the blizzard. They just kept going down the highway and onto the bridge before they knew it.*

Washouts crossing our trail surprised us the first time we trailed to the Brandentaylor place, southwest of Terry. We followed the highway to the mouth of Powder River, where the road ran close to the Yellowstone, with steep gumbo cutbanks on either side.

ANNE

There we crossed the Powder River bridge and turned right, following along up the Powder, leaving the highway without regrets. We disliked highway trailing—not only did cars and 'tourists' often disrupt our progress, but it failed to provide the sense of adventure, mystery and new country we enjoyed.

After crossing the bridge we hunkered down to discuss directions and Dad drew his map in the dust. Maybe he had 'talked to a few people,' about the route, maybe not.

"Just head up along that fence line beside the river."

He pointed off into the distance.

"There'll be some cross fences. You'll find a reservoir about here on a sagebrush flat where the cattle can water."

He drew an 'X' on his map.

"You'll cross a lot of washouts and cutbanks going into the Powder River, dry creeks—I can't say how many—haven't been there. Then you head up a long draw. I'll circle around and meet you around here later this afternoon."

Rough washouts and cutbanks crossed our trail above Powder River. Jeanie surveys a passable route.

Dad had to drive clear around through Terry and through some hilly flats without roads to cut our trail. He'd walk down the last long divide to meet us.

We started off. Out of these badlands come steep-sided gumbo cutbanks and powder-dry, washed-out creeks that apparently rushed into Powder River in flood stage—spring melt or summer cloudburst—cutting ever deeper into the tough, elephant-grey gumbo.

A wide, shallow, sand-filled, turgid river, broad at its mouth where it ran into the Yellowstone, the Powder was known for its quicksand, crumbling banks, rugged badlands on both sides and rushing waters in flood stage.

It gained fame during World War I in France when homesick cowboy soldiers from this area voiced a catchphrase of their sometimes-savage little river.

"Powder River! Too thin to plow, too thick to drink. Powder River, let 'er buck!"

Soon whole companies of U.S. soldiers took up the cry.

"Powder River! Let 'er buck!"

No question what it meant: "Bring on the enemy! We're ready!"

This was difficult country for moving cattle. We needed to cross each of the washouts, sometimes one cow at a time. Deep and steep-sided, the washouts were cut by a narrow deer or cow trail.

With the first one, we started the cattle across and then both rode back to push up stragglers.

*Whoops! One of us should have ridden ahead.*

This was a one-cow trail. It soon choked with cows and we couldn't ride through. One or two cows at a time descended cau-

*"Powder River! Too thin to plow, too thick to drink. Powder River, let 'er buck!" A battle cry for WWI.*

## A cowgirl dilemma

Trailing brought problems, big and small, some merely uncomfortable—or embarrassing.

Coming home one fall, Beverley and I stayed overnight at the Olson's ranch. They didn't have an indoor bathroom and the morning before we left I felt too embarrassed to look for the outhouse in the daylight. There'd be plenty of opportunity later to go off in the brush for nature's call.

Just my luck, the Olson brothers, Orlando and Lawrence, young ranchers in their late twenties, decided to ride with us. We should have headed for the outhouse right then when we realized this. I believed Beverley would lead the way. She had lots more guts and social skills than I did.

But she swung up on Buck and I had no choice but to climb on Dusty and head out with the others.

Our cattle were held overnight in a small pasture. *Maybe they'll just open the gate and help us get started.*

They did that and then kept riding along, helping us move the herd. After we rode together an hour or so, I was getting increasingly miserable. But they were so protective that every time I started off down a coulee, one of the Olsons would think I was after a cow and come along to help.

Finally, I spotted a haystack off to the side ahead and murmured to Beverley, "I have to go. Let's ride over there."

She agreed, and off we went without looking back. I guess they got the message because they didn't follow.

After that, for the rest of the day, I appreciated Orlando and Lawrence a lot more for coming along. Our cattle traveled easily with their expert handling.

FRANCIE

---

tiously, shoving each other. Another cow lingered in the bottom, and a couple scrambled up the other side.

On our side the cattle bunched impatiently, waiting their turn, butting each other, in a quarrelsome mood.

After crossing, some cows rambled off in the wrong direction. Some blocked the trail, halting to graze right in the way. Others headed up or down the bottom of the dry creek bed.

With the trail full of cows, Francie and I shouted and swung our ropes, without much effect. We were stuck behind the whole herd.

We couldn't believe there was no other way to cross and rode up and down looking for even a narrow place a horse might jump. But the washouts proved impassable both above and below, except for that one narrow crossing, apparently made by deer.

Finally, we cut off our cow string and Francie pushed Dusty through the washout and rode out in front, keeping the lead moving toward the next crossing.

After that one of us rode ahead, scouting gates and difficult crossings, while the other pushed from the back.

We learned to be grateful even for these narrow crossings as the hours passed, and for the deer that made a continuous trail following on through, jumping the gates—local cattle circled their own pastures avoiding the gates. No doubt after each gully-washer the trail broke off into steep cuts for the first deer that crossed.

Getting through the area took forever, with few places to traverse the narrow chutes and the herd often held up, waiting.

We kept looking for that good draw Dad had talked about. We had no idea when to head up onto the flat above. We'd ride up a distance, only to meet rougher hills and more cutbanks.

Dad had tried to count off the dry creeks on his map before our turn to the left, out of the Powder River valley. But he had it second-hand, and we got it third or fourth.

What had seemed clear at the time was no longer plain. Francie and I consulted each other many times about that perplexing map left behind in the dust. If we headed up too soon we'd hit impassable badlands.

But one way or another, we finally found the right draw and the plateau above. Coming out on top we could see where we needed to go and found Dad waiting.

My sisters and I had many enjoyable experiences trailing cattle. We met plenty of challenges, filled with surprises—or humor, or frustration, or panic. We cherished the country we travelled through, the harsh cry of magpies swooping overhead and the sweet song of a meadowlark, the suddenness of a mule deer leaping from a draw to pause on a hillside and regard us askance, the bounding jackrabbit and, yes, the always-respected rattlesnake that kept us alert and on our toes.

While sometimes too cold, too wet, too wind-blasted, too miserable, and the job too hard, we mostly considered trailing a time filled with adventure in which the unexpected often happened. Usually we could find a laugh in the event, however trying.

When we got home from a trail drive, our faces were burned red from four days in the sun and wind. I never wanted my classmates to know I'd been trailing cattle, because in junior high I went to town school and they all seemed so sophisticated. But I was a country ranch girl and my face gave it away. Little did I know many of my classmates would have loved trading places with me.

Later on, in college it was still embarrassing to come back to school after Easter with a red, wind-burned face. True, we looked much like the privileged college kids who'd spent their spring break tanning on a sunny Florida beach.

We kept quiet about our spring break. Fortunately, the red soon faded into an early summer tan.

## Third Horse High Jinks

An inevitable challenge dealt with our extra horse. We took three horses, riding two and leaving the third free to rest and follow along. Trailing meant long hard days for our horses and they needed that relief after two days of work.

We relied on Buck, Dusty and Queenie as our main trailing horses. If we had a young, green-broke horse we brought it along for on-the-job training with two dependable cow horses. Chasing cows and cutting out strays all day long was great schooling for a young horse, who knew nothing about working cattle. After a few hours this took the morning friskiness and buck out of him. Too tired for orneriness or shenanigans, he ended the trip all the wiser.

However, in a tight situation we didn't want a horse more concerned with fighting the bridle or getting rid of its rider, than turning back a wildly fleeing calf. The third horse gave us flexibility. On a day with tough jobs we could switch to Ol' Dependable.

With three horses we each had a fresh saddle horse every other day. If we planned it right, each day we had one fresh horse, one ridden one day, and the other ridden two days and ready for rotation. If we forgot and both rode our favorite the first two days, then we were in trouble; one of them was stuck with three days in a row.

We often wished we could bring four horses. Then we could

> *While sometimes too cold, too wet, too wind-blasted, too miserable and the job too hard, we mostly considered trailing an adventure.*

*Trailing cattle gives Francie an opportunity of easy, on-the-job training for this young horse.*

pick and choose depending on the terrain and tasks of the day, and ride each horse only every other day.

"No, that will just lead to trouble," Dad said shaking his head.

He was right, of course. We found that out when one of our sisters left us after a day or two and we did have two extras.

Horses are social animals. A lone horse naturally prefers to stay close, socializing with the other horses and cattle, tormenting a calf now and then. By contrast, two loose horses are just fine trotting off together over the hill. They relish their freedom to hang back enjoying a patch of tasty grass, or worse, chasing off with other horses.

Still, problems popped up.

Sometimes the loose horse teased. A favorite trick: stand in the gate. One of us rode up and opened a gate and we bunched the cattle ready to go through. Then here came our extra horse trotting up to block the opening, nipping at cows and chasing them out of the way.

Shouting and waving did no good. The horse just stood there, looking apologetic or arrogant, depending on her personality—until one of us rode over and chased her off, scattering the cows.

"Git out of the gate!"

Looking injured, as if to say, *"Aw, come on. Let's have some fun. Silly cows,"* she bobbed her head.

We had to laugh at the extra horse antics, even as we scolded.

That horse often played games with our favorite way to trail: string them out and keep them moving behind a good lead cow. He blocked the string and turned back the lead.

Other times our loose horse flirted or bickered with other horses as we cut through some big pasture where we didn't expect them. We worried that a stallion might charge up and cut out one of our mares for his harem. Luckily, it never happened.

But we tried to keep other horses and cows out of our way.

One afternoon we topped a hill, surprised to see a bunch of cattle lying down beside a reservoir. And from the opposite direction, eight or ten horses trotted single file down a trail, coming in to water.

Our cattle ran to water. We tried to head them off, while at the same time chasing away the local cattle to prevent them mixing. But the other cattle did not want to move.

Seeing us, the other horses nickered excitedly, bucking, kicking and lunging toward our horses. A few nuzzled up to Jinx, our loose three-year-old. Buck and Queenie whinnied, squealed and danced sideways, kicking up their back feet.

"Get out of here!" we shouted.

> A favorite trick of the extra horse: stand in the gate blocking the opening, nipping at cows and chasing them out of the way.

Suddenly they wheeled and raced away—and away went Jinx with them. Buck and Queenie called her, but Jinx only tossed her head and went off with her new friends. She was having way too much fun to leave them. In a moment they all disappeared over the hill.

We had no time to worry about Jinx.

The local cattle, on their feet now, pawed dust and bellowed as our invaders surged in to drink. Soon the two groups were thoroughly mixed, sniffing, butting heads and bellowing at each other.

After drinking, our cattle turned sociable and circulated around getting acquainted.

*How will we ever get them sorted out and on their way again?* we wondered.

We didn't dare let them get too comfortable with the strange cows, so we pushed them on up the trail, cutting out the locals when we could. To our surprise, within a short time they decided they didn't want to come with us after all. They separated themselves out and moved away, regarding our bunch from a safe distance.

We wondered where Jinx went and how to get her back.

With the cattle safely headed out, Anne rode out to look for her. Luckily no stallion was with them and Jinx wasn't in heat. The other horses, already tired of her company, were chasing her away.

She whinnied when Anne and Queenie rode up and scampered back to safety within our herd. Obviously she was happy to join us again.

> *Suddenly they wheeled and raced away—and away went Jinx with them. Buck and Queenie called but she only tossed her head.*

## Sleeping in the Open

ANNE

"*Whoomp!*"

I awoke when something terribly heavy hit my stomach.

Instinctively, I turned over and brought my knees up under me, trying to push whatever was so heavy off my back.

Luckily, as in quick succession more heavy weights fell, pressing me down. If I'd waited a moment longer to flip over, I'd have been crushed flat on my back.

That windy spring night, Francie and I bedded down in the back of the stock truck with a half load of hay bales stacked high for a windbreak.

*Francie and Anne unsaddle the horses and take a lunch break beside the chuckwagon along the trail.*

Western Ranch Life in a Forgotten Era

> ### What horse will I ride today?
>
> With three saddle horses and two riders, one horse ran loose with the cows, resting, every third day—but only if we planned the right rotation.
>
> For instance, starting out, Anne might ride her favorite Queenie the first two days, while Francie rode Dusty the first day and Jinx the second. On the third day Anne rode Dusty and Francie rode Jinx again. Thus, Jinx ran free the first day, Dusty the second, Queenie the third and Jinx again on the fourth.
>
> But we ran into trouble if we both rode our favorites the first two days. Then we had to ride one again on the third—three days in a row. This was too much and they let us know it. These were long, hard days for the horses, trotting up and down along the herd and across the drag all day long, riding the route up ahead and cutting out strays. They needed a break.
>
> So did we; it's no fun to ride a tired horse.
>
> The extra horse, too, gave us flexibility. A green-broke two-year-old might tire in a half-day, or be unable to cope with trailing rambunctious cattle through difficult terrain, so its rider could shift back to a seasoned cow horse.

Dad came to camp late that afternoon with hay for the cows. He fed them half, then stacked the remaining bales in the truck bed, leaving a space for Francie and me to sleep, well-sheltered from the wind.

He eyeballed the sky and added more bales. Piled up to the top of the stock rack, they blocked the wind for us. They'd be breakfast for the cattle in the morning while we ate ours and cared for the horses.

As we crawled in Francie and I giggled that our bedding was so heavy we could hardly move. That was okay, though, since I slept on my back most of the night. Sometime in my childhood the all-knowing-Jeanie announced that some important study showed that when people sleep on their backs, they don't need to sleep so long, and awaken more refreshed. I believed her, and learned to do this. *Humph.*

Sleeping in the back of the truck felt snuggly and cozy. Because of all the dark clouds, there were no stars to watch. We soon tired of watching the clouds roiling overhead and fell asleep. I remember once waking in the night, hearing the wind howling and rocking the truck. Warm and comfortable, packed in between the walls of hay bales, I immediately went back to sleep.

Dad slept in the chuckwagon, still hooked onto the truck. He always awoke in the morning before we did, started a fire in the little sheepwagon stove, then got us up. We'd rush into the wagon to dress while he fed hay to the cattle. Once dressed, Francie and I started breakfast and fed the horses. Then we ate and began moving out.

When something hit me a hard blow on the shoulder that night, I instinctively knew it was an alfalfa bale from the smell and wisps of hay flying through the air. Instantly I was hit another blow, followed by a tumbling heap.

FRANCIE

Fortunately, most hit with glancing blows, but they piled on top of me before I was fully awake. I felt a moment of claustrophobic panic before realizing I could still wriggle and squirm and heave my way out. The wind was howling a deadly blast.

"Anne! Anne, are you okay?" I shouted into the wind, looking frantically to where I last saw her.

She slept next to me in our double bedroll, but close against the heaviest stack of hay, while I was a little farther out. I missed the full force of the tumbling bales, but they had hit her square and buried her.

*"Help! Help me! I can't move,"* came her muffled cry from under the heavy pile.

Frantic, I broke free with a lunge and struggled to reach in toward Anne. But it was hopeless, I couldn't reach her or even move a single bale.

Our alfalfa bales were hard to handle. Our hay baler cut them in rectangles with two strands of bailing wire binding them the long way. Being so tightly compressed of high-quality alfalfa, they were heavy, weighing ninety pounds or more. It took two of us girls to lift one bale.

Realizing I couldn't move anything, I rushed back along the wagon tongue and pounded on the chuckwagon door, shouting for Dad.

As I tried to shift my body, the bales kept settling, pushing down with even more force. They pressed so hard I couldn't hump up my back anymore to keep an air space. I heard Francie call that she was going to get Dad and I wanted to tell her, *No, there isn't time. My breath is getting pushed right out of me.*

But I couldn't talk.

It was terrifying to know she was leaving. *Can she wake up Dad and get back before my air supply closes off entirely?*

As the big pile of bales slowly crushed the breath out of me, I moved my face to the side, my cupped hand allowing for a pocket of air to breathe. But my chest was so compressed I couldn't inhale.

Suddenly I heard the chuckwagon door burst open and Dad leaped out across the tongue into the truck. I could hear Dad and Francie as they pushed and pulled on the bales. As Dad lifted each one, Francie kept calling out to me and listening to make sure I was still breathing.

Francie told me later that my voice seemed to grow fainter against the howling gale, and she fully expected to find me suffocated. No way could I raise my ribs to bring in air. There was so much weight on my ribs I thought they would break.

Then just when I thought I couldn't get another breath, suddenly I was free.

Dad lifted the last bale off my back. Francie brushed a pile of dry alfalfa leaves and twigs off my face.

I could breathe! I lay there for a moment, sucking in that good air. Then laughing a bit, feeling sheepish and helpless, I popped up, with only a few bumps and bruises.

"Well, maybe that wasn't such a good idea after all, sleeping up against the bales," Dad said, with a relieved chuckle, standing there in his long white underwear and cowboy boots.

"I don't mind sleeping out in the open," I said, climbing over the top of the stock rack and out of the truck, not wanting

*Looking for a place to make camp for the night, Dad finds us with the chuckwagon, pulled behind a truckload of hay.*

> I heard Francie shout that she was going to get Dad and I wanted to tell her, No, there isn't time. I can't breathe.

*After a cold night on the bed ground, the cattle are eager to hit the trail.*

## Bedrolls

Our bed was no sub-zero Arctic mummy bag. To put together a two-person bedroll, we laid out a double-long canvas tarp, with a foot or so extra to pull up over our heads at the top. For the next layer, came a thick quilt to sleep on, then sheets, then more quilts and wool blankets as needed for warmth. Lastly, the bottom half of the tarp was pulled up and over the whole, tucked in at the sides and leaving a generous fold at the top. When we went to bed, we pushed the stiff cotton tarp into a peak above our faces, giving us room to breathe and keeping out most of the cold air.

As a boy Dad learned to make his own bedroll from old-time cowboys and we girls used that same method. Sleeping bags didn't exist in those days. Dad called quilts in a bedroll 'sougans' and he also used that term for the entire bedroll and contents.

The early cowboy layered his sougans and blankets and tucked in an extra shirt, sox and pants, shaving equipment and, if he was lucky enough to have one, wrapped a long canvas tarp around the whole thing. His bedroll was stuffed in a big burlap sack or tied with ropes to keep everything together.

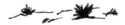

to get back behind that flimsy windbreak. "I really don't mind the wind at all."

We often slept on the ground when the weather was nice on a fresh spring night or an Indian summer evening. We'd spread our bedrolls under a tree, after first clearing the ground of pine cones and sharp rocks, with the scent of pine wafting down and a spring breeze swishing through the pine branches above our heads.  *JEANIE*

It was nice.

It might be a bit hard and uneven, with a few rocks in the wrong places, but there is something grand about sleeping in the open on the prairie. Like jewels strewn across the heavens, the stars never seem so bright, or the silence so secure as when free of unnatural lights.

And there really is starlight.

Even without a moon the brilliance of billions and billions of stars and the Milky Way stretching overhead forever and ever, lighted up the land in a special way, making us feel protected, relaxed and ready for sleep.

Suddenly, far away, a single coyote howled and from another direction came the answer. The next howl broke off abruptly into a yipping, barking musical cacophony and we knew a pack of coyotes was pursuing a jackrabbit.

Since we trailed in early spring or late fall, we never knew what weather might hit—it could change rapidly. A wintry storm might blow up in the night and cover us with a two-inch blanket of snow.  *ANNE*

When we anticipated a cold morning, we learned to bring our clothes, boots and jackets into bed with us at night to keep them warm. Having warm dry clothes to pull on under the covers saved a frantic run for the wagon in freezing boots and a stiff jacket over pajamas on frosty mornings.

If we left the wash basin out, we checked to see if it had a rim of ice around the edge. If it did, we snuggled down a bit longer, dreading to leave that cozy bed, before making the dash to the warm little chuckwagon.

Our chuckwagon was a green canvas-covered cook wagon once used in cattle round-ups and pulled by horses. Shorter than a regular sheepwagon, it had been the domain of roundup cooks. A model of efficiency it included a miniature cook stove with chimney, a pull-out table and cupboard. Under the small lift-up window was the narrow wooden bunk where we slept, with space under the bed for grub boxes and storage. Benches ran along both sides and provided more storage underneath.

Each evening well before dark Dad came to set up camp, hauling our chuckwagon behind the truck to the new campground. He spread hay for the cattle—a full feed in early spring when there was no green grass, and a smaller amount in the fall to settle them down. Maybe he'd help us move the last of the tired cows and calves onto the bed ground, then make us a hot supper while we cared for our horses.

The next afternoon Dad would come back and move camp to the next place. Sometimes our 'camp tender' was Mom. After they set up camp, we ate and cleaned up, it seemed they always had an interesting story to tell of earlier times on the trail.

"It was a cold day like this…." Dad would begin.

Or Mom, "One night, we had a dry camp, no water for the cattle or horses…"

Off to the side, cows chewed their cuds peacefully, belching now and then, while our horses nickered softly and stamped their feet in the darkness.

On several memorable occasions it was Uncle Will, our bachelor uncle, who also had wonderful old-time ranching stories to tell. Uncle Will knew the stars, more than just our old favorites, the Big Dipper and the North Star. Many times we stood under a pine tree with our faces upturned to the sky, brilliant with a million twinkling stars, while he told us about the constellations.

During these times when we had someone extra overnight, that person slept in the wagon while Francie and I rolled our bed out on the ground.

After a full day of trailing we bedded the cattle near camp, up against a fence if there was one. Tired and satisfied, they easily settled down for the night—and so did we. In the morning, we pulled on our boots, coaxed a fire with numb fingers, chopped ice on the water bucket if needed, brushed our teeth and splashed faces in a basin of ice water while keeping an eye on the cattle, all ready to move out. We took turns, one of us fixing breakfast while the other went looking for the horses and brought them in to guzzle oats as we ate.

Later that year in November, on our last night in the Pine Hills, we bedded the cattle behind a fence along the Mizpah Road near where it branched off to the south from Highway 12. The folks had taken the sheepwagon home that afternoon and left us the truck in which to sleep and a hot meal of Swiss steak and fresh cinnamon rolls—*yummy*.

> *Suddenly, far away, a single coyote howled. Another answered. The next howl broke off abruptly into a yipping, barking musical cacophony.*

*Anne taught Queenie to neck rein, cut cows and other ranch skills during long, intense days on the trail.*

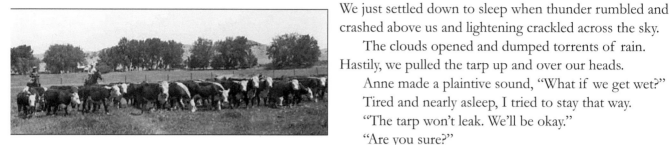

Leaving summer range, our cattle are rounded up and ready for the long trek home.

After they left, we fed the horses, pulled a snug hobble on Queenie so we could catch at least one horse in the morning, and spread our double bedroll in the back of the truck.

We just settled down to sleep when thunder rumbled and crashed above us and lightening crackled across the sky.

The clouds opened and dumped torrents of rain. Hastily, we pulled the tarp up and over our heads.

Anne made a plaintive sound, "What if we get wet?"

Tired and nearly asleep, I tried to stay that way.

"The tarp won't leak. We'll be okay."

"Are you sure?"

"Sure. Pull it up over your head."

Then came more rolling thunder, lightning flashes and driving rain.

"I think it's leaking."

"No, this tarp won't leak. Dad said so. Let's get to sleep." No sooner had I said those words than I felt the dampness seeping through.

Silence.

"I really am getting wet," came a small voice, She was right, of course. We lifted the tarp and peeked out. A flash of lightning revealed where rain pooled in many places across the canvas. Under each pool the tarp and quilt felt soggy all the way through.

"We better run for it!"

We grabbed up our bedroll, sloshed through puddles in pajamas and bare feet, and made it into the truck cab. There we sat an hour or so, huddled miserably, wet, unable to sleep, wondering how we'd last till morning. The rain didn't let up, but kept pouring down.

Then we saw lights flash from the road and Mom drove the car up beside us.

"Come, get in," she called, holding open the door. "I'm taking you home."

Wow, we were rescued! That night we slept gratefully in our own beds.

When Mom took us back next morning, the sun shone warmly, pine trees glistened and red scoria rocks gleamed at their brightest after the good rain.

It felt good to climb in our saddles.

The cattle and horses knew they were almost home after a long summer. Our leaders, Old Roanie and Crumple Horn, lifted their heads, sniffed the clean, fresh air and set off down the trail through the rocks and pine trees.

*Off to the side, the cows chewed their cuds peacefully, belching now and then, while the horses nickered softly and stamped their feet in the darkness.*

## Jerky and Beans

Frog's legs for a gourmet trail supper? It did happen.

For years I longed to be old enough to join my sisters trailing cattle each spring and fall. They returned filled with their adventures. Always tired, faces sun- and wind-burned, excitement sparkled in their eyes.

Finally came the day when my parents announced I was big enough to ride drag—at the back of the herd.

Getting ready was a whirlwind of anticipation. At our ranch, we were always busy, though some jobs were more interesting, and this was one of them.

Preparing the chuckwagon, our bedrolls, warm clothes, and the food that seemed so exotic—canned food from the grocery store—delicacies such as peaches, Queen Anne cherries and pork and beans added to the excitement

Cow camps never lacked for good food. While we enjoyed more variety than early-day cowboys, we liked best the time-honored ranch fare. Cooked on the little sheepwagon stove it always seemed to taste best.

Our favorite supper was steak and potatoes stir fried in the skillet with onions, hot pork and beans from the can, thick fresh homemade bread with home-churned butter, and canned peaches with a filled cookie for dessert. Our grub box, stocked with bread, crackers, cheese, cookies and raisins, also contained canned foods of amazing assortment. Rarely did we have Campbell's Pork and Beans at home, so these were a delicacy for us. And those canned peaches! So smooth they slipped right down your throat!

At home, Mom canned our winter food from her acre-sized garden and purchased bushels of peaches, apricots and pears which she also canned in glass jars. We always ate well, but the food we took on a trail drive was special because most of it was 'store bought.' Mom's glass jars of home-canned food wouldn't survive the rugged cross country jolting of the truck and wagon as we crossed the badlands.

We made and packed tins of cookies. My very favorite were raisin-filled. While the older girls were out helping Dad, Mom mixed and rolled out big batches of sugar cookies. It was my job, even when small, to grind the raisins into a saucepan using our hand-cranked cast iron grinder. Adding just enough water to keep the raisins from scorching, I stirred them to thicken over our wood-coal kitchen range. This mixture I dropped by teaspoonfuls onto cutout circles of cookie dough. Another circle on top, and I crimped it all down tightly with my fingers.

These thick nutritious cookies went with us on every trail. And

*We carried jerky along in our pockets. It tasted great—even when mixed with a few horse hairs and pocket lint.*

> *Beverley's face lit up. "In the south they eat frog's legs. And people say when you fry them, they'll jump in the pan."*

Beverley, sisters and friends brought home fresh-skinned frog legs for supper and pan-fried them in hot oil.

though they were too crumbly to carry in our pockets, Dad always had a tin of them ready when he met us with the truck.

Dad made venison jerky that we carried along in our pockets. Several days before the smoking process he cut the meat in strips and soaked it in his special brine of salt, brown sugar and spices. He always used venison, saying it preserves better with less fat. He smoked it in an old-fashioned refrigerator with legs, by drilling holes in the bottom and building a fire on the ground beneath, using plum twigs, handy from our large plum grove behind the shop. The refrigerator racks allowed heat and smoke to circulate evenly around the meat.

I liked it best when he cut the meat crosswise, then it wasn't so hard to chew off a chunk. Most of my jackets had a few pieces of jerky tucked in and even months later, finding that treasure while riding the pasture, it tasted great—even when mixed with a few horse hairs and pocket lint.

The fall before I married, my fiancé Bert wanted to come along trailing cattle, so Uncle Will came too, as camp tender and cook. Dad drove the truck out daily to move camp. He'd visit awhile and leave, probably happy he didn't have to cook.

FRANCIE

Uncle Will enjoyed having time to cook a good meal and the first night we had it all. Steak and the trimmings—except no beans.

I was perplexed. Anne and I always had pork and beans.

Wanting Bert to have the ultimate cattle drive experience, I accosted Dad next time he came out. When he asked if we needed anything I told him I'd gone through all the food boxes, and no beans in any of them.

"So please bring some cans of beans."

He didn't say much, but the next day after he'd been there and gone I still couldn't find any beans.

That evening Bert and Uncle Will sat hunkered down on the ground eating their steak with relish. I was walking around, looking through the supplies, complaining about no beans.

Finally, Uncle Will had enough of that. He had lived much of his life on the range, in cow camps and sheep outfits, batching and cooking for himself.

He leaned back, set his knife and fork carefully across his plate, and regarded me sternly.

"They ain't no beans," he said. "And they ain't gonna be no beans! I've eaten enough beans for a lifetime and I told your dad not to bring any."

End of discussion.

Beans provided a traditional staple, but occasionally we tried foods of the more exotic variety.

One hot day we trailed cattle up to Orlando Olson's south of Terry and stopped to rest the herd a couple of hours at a reservoir where there happened to be a lot of frogs. That day, all four of us girls rode together—Beverly, Jeanie, Francie and I—joined by Albert Lahn and his friends on their horses.

With all of us laughing and joking, it was more like a holiday than a work day. Amusing ourselves around the water as the cattle rested, we began to notice the frogs. They were huge and seemed to jump in from everywhere.

Suddenly Beverley's face lit up as she studied the large frogs.

"In the south they eat frog's legs. And some people say when you fry frog legs, they jump in the pan." She laughed and rolled her eyes.

Sounded interesting. So we set about catching frogs. Someone killed them and we cut off and skinned the legs and tucked them into our picnic pack.

That night we went home to the ranch for supper. Beverley and Jeanie dipped the frogs legs in egg and seasoned flour and fried them in an iron skillet.

The rest of us paid close attention. Sure enough, when they hit the hot oil, muscle contractions made it seem as if they did indeed 'jump in the pan.'

"Tastes like chicken," was the tongue-in-cheek verdict.

Jeanie ate with relish and talked of eating rattlesnake next. She claimed they'd also move around in the hot skillet.

A small taste was enough for Francie. She said she kept seeing those last desperate jerks of legs trying to leap out of the pan.

Our mother, ever the delicate eater and academic, copped out that night. She wouldn't watch or eat and left the room.

That was the only time we ate frog legs and we never, in all those years, ate rattlesnake. Somehow it lost its appeal.

*Jeanie ate with relish and talked of eating rattlesnake next.*

*Anne carries jerky in her pockets for a long day's ride.*

# 4 Creating Ranch Fun

## Fishing the Yellowstone

"That's Goose Island. Do you suppose geese live over there?" I asked.

"Used to, I bet. In the early days."

Beverley contemplated the island thoughtfully. "The water seems shallow. I wonder if we can wade over there. Wouldn't that be fun?"

"What!"

*She surely doesn't mean wade all the way to Goose Island.* Horrified, I tried to talk her out of it.

"Oh no. We can't do that! Mom says the river is dangerous. It gets too deep."

"You're probably right."

"And the undertow can suck us under!"

"Well, let's just go out a bit and see how deep it gets,"

That June morning, Beverley and I went fishing in the Yellowstone. After we threw our set-lines in the water at the mouth of Jones Creek, we decided to explore upriver while waiting for fish to bite.

We were raised with a fear of the river—brought up on stories of river tragedies and warnings of strong undertow in both the Yellowstone and the Missouri, where Mom and Dad had lived when first married.

None of us girls could swim—the irrigation ditch where we often played in the water was too shallow for that.

But we fished in the Yellowstone at its gloomiest and most foreboding time. Every year the typical June rainy weather, along with snow melt in the Rockies, brought on the *June raise* of the Yellowstone River and gave notice for Dad to take us fishing.

It was a chilly and wet experience, but Dad always said fish bite best then. He usually went with us, because he enjoyed fishing so

**We were raised with a fear of the river— on stories of undertow and river tragedies.**

much.

This time it was just Beverley and me.

The water came only halfway to our knees as we waded upstream along the shore. We struck out a bit deeper and still the Yellowstone seemed amazingly shallow.

Towering green cottonwoods grew thick along the bank. It was a new experience to view them and their dense undergrowth from the water.

We came around a point of trees and saw, far out near the other shore, a long sandbar covered with willow thickets. A stand of young cottonwoods grew at the lower end.

This was Goose Island.

Beverley set off straight for it in deeper water.

"We'll just go out a little way and then come back." She chuckled, suggesting she'd given up the idea of going all the way.

But I knew better. I saw the gleam in her eye and that familiar enthusiasm light up her face. I was scared, but had little choice except to follow, reluctantly, a few feet behind.

"What if we step in a hole?" I ventured. "We can't swim!"

"I'll be careful," she laughed again.

By this time, in waist-deep water, I could feel a strong pull against my legs. I wondered if it was the undertow. If I dropped into a hole it would be the end of me—I knew that and felt resigned to it.

*But what if Beverley drops out of sight—right before my eyes? How will I ever help her? We'll be gone, both of us, swept away by the current.*

"It's getting deeper, Beverley. Let's go back!"

It was terrifying to be so far out in the river with nothing to hold onto.

Suddenly a cottonwood tree, root first, torn from a bank somewhere, came swirling down toward us. Beverley hesitated and stepped back to avoid it. Whew! It barely missed hitting her, and swept on past us downriver.

I eyed the circling waters, little eddies and whirlpools breaking the surface here and there. Big trees floated long distances underwater, out of sight. The water, dark and murky in flood stage this time of year, hid submerged trees, a known fear of long-gone riverboat captains on this river.

Beverley surged on.

"Just a little farther. Can you believe we're halfway across the river already?"

I looked back—it was a long way to the other shore. I knew now she was not going to stop. When Beverley embarked on a mission, she was not deterred.

*What if Beverley drops out of sight right before my eyes?*

Francie re-baits the first leader on a set-line thrown into Yellowstone River.

Western Ranch Life in a Forgotten Era

*Anne and Beverley drive the Bug—our old cut-down, run-around truck, to the river fishing. They debate the size of fish caught.*

## Catfish dinner

Mom's catfish dinners were delicious. She cut the fish into inch-thick steaks, dipped them in beaten egg, rolled in seasoned cornmeal and pan fried till crispy on the outside and flaky inside. Without scales and free of those irritating small fish bones, catfish were easy to eat and tasty.

Sometimes we surprised Mom by bringing in catfish we caught in the alfalfa field.

While irrigating, if we heard a flipping and flopping in the alfalfa we ran to look and, sure enough, there was a large catfish or other fish that had slipped past the screen at the dam and come shooting down the ditch. Often several came at once—something went wrong at the dam outlet, people said.

Dad delighted in fishing of every kind. Sometimes he took us fishing with bamboo poles in a reservoir or the Tongue River Dam, with worms for bait. Other times we went fly fishing in a mountain stream or trolling from a small rowboat. None of that, however, was quite as exciting—and scary—as plumbing the Yellowstone's depths in the June raise.

Her sense of adventure was infectious.

"Look! We're almost there," I said, beginning to feel the thrill of exploration. It seemed safe enough.

The water grew shallow and in seconds we were high and dry, up on the sandy beach itself. Step by step, we had walked through the water all the way to the island, three-quarters of the way across the mighty Yellowstone River.

We did it! It was hard to believe. The lacy willow thickets beckoned, their light-green foliage shimmering in the sunlight.

We began exploring. The island, mostly sandbar, long and narrow, grew willows and shrubs hiding the northern shore. Cottonwood logs and driftwood debris lined the beaches.

A rancid odor hung over a few dead fish washed up onshore. Nearby were the hand-like paw prints, small and precise, of a raccoon going down to the water, washing its food—a fish carcass no doubt.

"Kill deer! Kill deer!"

A troop of long-legged killdeer ran along the beach, leaving thin dainty tracks, calling plaintively.

Ducks swimming in the shallows left their tracks, and yes, among them, the heavier webbed tracks of geese. We saw them then, a beautiful matched pair of stately Canada geese, male and female, emerging from the willows at the far end of the island. Four fuzzy ungainly goslings lined out between them.

A rare sight at that time, geese nesting in Montana! Decades earlier, homesteaders had killed off all the local geese. The big Canadians now migrated through in spring and fall, but nested farther north in Canada and Alaska.

"Mom and Dad will be glad to know geese are living here again," Beverley said.

We crossed the narrow island—alarmed at how deeply the bank on the north side was undercut by swift waters. This main river channel, the deep water always sought by early-day steamboat captains, picked up speed as it rushed through the narrow passage, surging with tree branches and snags.

"Steamboats used this channel." Beverley pointed down into the deep rushing water.

We felt we could have almost reached out and touched those ships as they shot through the narrows, smoking their way up and

*Montana Stirrups, Sage and Shenanigans*

down the Yellowstone, paddlewheels churning and flinging snags everywhere.

"*The Far West* was one of them," she said.

"Wow! *The Far West* went right through this very spot loaded with wounded soldiers from the Custer battle?"

We knew the story. On another June afternoon some seventy years before, the *Far West* carried the wounded, Reno's troops, from the Battle of the Little Big Horn.

Taking two weeks to reach Bismarck and the telegraph, the *Far West* brought to the world its first news of Custer's Seventh Cavalry defeat. It shocked a nation celebrating its 100th anniversary on the fourth of July 1876.

Stepping back from the turbulent channel, Beverley and I explored the length of the island, its vegetation and evidence of wildlife. We could see where, periodically, ice jams below backed up water, covering the sandy island with broken ice chunks. Slammed into each other with terrific force, they welded together, forming icebergs sometimes fourteen feet thick that scraped off all vegetation. The next year Goose Island might be a bare sandbar, all brush and trees gone.

We felt as if we were the only humans to ever walk these lonely beaches. Only rubbish in the debris hinted of life elsewhere—a sand-soaked plank, a torn scrap of fish net.

"We better go back," Beverley said, finally.

She grinned at me. "Wasn't this fun? Aren't you glad we did this?"

"Sure, it was a great adventure." I laughed with her.

*Still it looks a long way back to our shore and I don't much want to step back into that water!*

Beverley led off confidently and I wasn't about to be left behind.

To my relief, we made it back to our southern shore. My lucky day.

Beverly's sense of adventure satisfied, we never went wading so far out into the Yellowstone again.

Beverley was our most avid fisherman, after Dad. She didn't care if it was rainy, muddy and miserable. June, during the rains, can be cold—almost freezing and if there's a lashing wind it's spine chilling. The most disagreeable place is on the banks of the Yellowstone.

"It's time to get out the set-lines," Dad, or maybe Beverley, said.

First we mended the set-lines. These were yellow cords thirty feet long interspersed with six or eight leaders. Each leader was three

*Anne digs worms for fishing with pole in a local reservoir, plum trees from orchard are in background.*

*Fishing from small rowboat with bamboo poles in Clark's reservoir in neighboring Prairie County. From left: Anne, Francie, Jeanie and Beverely.*

*Francie, fly fishing in mountain trout stream.*

**Before long, the pull of the river drew us back out into the rain to check our lines.**

feet of clear fish line with a hook on the end.

On the throwing end of the yellow cord was a rock. On the other end, a rough slab of cottonwood bark was used for winding the line and setting the hooks when carried.

We unrolled each cord full length along the river bank—leaders and hooks stretched out at right angles—and baited the hooks.

But first we had to catch the bait.

We seined the mouth of Kelly Creek in nearly waist-deep water, just before it flowed into the Yellowstone, a fertile spot for swimming schools of minnows.

Using a small two-person net, we swept it up and down the shallow creek and dropped the two-inch silver minnows into pails half full of river water. Naturally, we had to get right into the cold water in the rain, so we were good and wet from the start. We lifted the net many times before we collected enough of the shimmering little fish, dancing and flipping their tails against each other.

*How rewarding!*

After baiting the hooks we hurled the rock end of the set-line as far out as possible into the murky swirling waters of the main river channel, minnows flying and flashing silver through the air. The near end we tied to a small tree or bush at the water's edge.

When this job fell to me, because I was younger and couldn't throw as far, we often had to pull the line back, re-bait, and try again. With six or eight lines this took all morning.

Thus, our fishing became a three-tiered project—first we untangled and repaired the set-lines. Then we fished for bait and lastly, we fished for fish.

After setting the last line we couldn't resist going back to check the first one or two to see if the fish were biting. Maybe we already caught a fish or two. Then usually we went home, returning after a few hours to pull in the lines, take off fish—mostly nice catfish fifteen to twenty inches long—bait the lines and throw them back in again.

Often on these days, Mom finished her chores and took the afternoon off. This was a great time for her and Jeanie to make donuts. Many afternoons I remember coming home from fishing, wet and cold, to the wonderful, tantalizing smell of donuts frying.

We came shivering into our warm house, changed into dry clothes and sat down around the long kitchen table with glasses of milk and a big plate of warm donuts.

Dad recounted the day's fishing adventures vividly and humorously, as only he could. We all chimed in and joined in hearty laughter all around.

I wanted that moment to last forever. But before long, the pull of the river—and Beverley or Dad—drew us back out into the rain to check our lines.

## My How Hot

The imposing Miles Howard Hotel billboard, standing possessively on a big highway curve of our pasture, nagged like a festering sore in our family. As cars swept around the last hill down into the valley, it advertised before them the big fancy new hotel.

Mom and Dad didn't want signs littering their pastures, but money was tough to come by in those days and the annual lease agreement promised to ease their hardship a little. The only problem: years passed and we didn't get paid.

Fortunately we couldn't see the sign from the house.

"What plans do you girls have for this morning?" Dad asked Francie and me at breakfast on one fateful morning.

This was a loaded question. We hurriedly conjured up some work ideas we needed to catch up on, but then were pleasantly surprised when he told us what he wanted.

"The magpies have multiplied this summer and they'll soon start pecking on the cattle. I'd like you girls to take your .22s and knock down as many as you can."

Magpies are notorious for pecking—and eating—new brands on calves. It's dreadful to watch, and even worse to come upon a calf with its brand all blotched out by magpies who have eaten live flesh for dinner.

Francie and I took the guns and headed out to the big cottonwoods behind the barn. We found a few magpies, but before we could get a shot they flew off down the driveway with raucous cries, their bright black and white feathers shimmering in the sun, long sleek tails stretched out behind.

They landed in the cottonwoods about a quarter mile away.

Magpies are intelligent, with keen memories. They wise up to guns in a hurry. In years past, Dad bored an inch circle in the bathroom screen window so he could shoot magpies without them seeing him. They landed on Mom's clothesline and scolded their rasping *'Kaw-Kaw-Kaww!'* to chase away our cats from the food bowl so they could dive in and eat the cat food.

When Dad heard that *'Kaw-Kaw-Kaww'* he grabbed the .22, levered a shell in the barrel, poked it just barely through the bored-out screen.

"Bam!" Another dead magpie.

*Magpies are notorious for pecking—and eating— new brands on calves. Dreadful to watch.*

*Schemers—Francie, Jeanie, Anne—planning another shenanigan.*

**Maybe he imagined trouble and expected it, considering the duplicity of his work that day.**

The others flew away, much wiser. They soon learned to keep watch on the hole in the screen and fly with the slightest movement. After a short time, they abandoned the cats' dish.

So Francie and I tried to hide the rifles behind our backs—tough to do—as we walked down the driveway.

Across the highway, we noticed the sign company's truck parked in our pasture with two men working on the big Miles Howard Hotel signboard.

"Look at them painting, fixing it up! We should just march up there and tell them to pay up or get out." I said, mindful that when Dad went to collect from the hotel owner, he said the sign company should pay us.

We laughed. It felt good to say it, but of course we'd never do such a thing. Our folks insisted on politeness and respect for adults.

The magpies were still in the cottonwood but they started to fly as we circled around.

We raised our .22s and shot. The birds flew away to another cottonwood.

While we waited for them to settle so we could get another shot, we saw a man from the sign company walking toward us from the highway.

He waved a hand and yelled, "You girls almost hit us! Can't you watch where you're shooting?"

We stammered that we didn't—we had shot in the opposite direction.

But he wasn't listening, his face flushed red with anger.

"You girls get out of here with your guns. Your bullets whizzed right past our ears! You're being reckless with those guns!"

With that he turned and stalked off, glowering at us over his shoulder.

Francie and I were shocked. If there was anything we had been taught it was careful gun handling and safety. Never had we been accused of careless behavior.

"Watch your background. Don't shoot till you're sure what's behind your target." Dad drilled those words into us until they became second nature.

*Never* would we have shot anywhere near a person. In fact, the two men were behind us and at right angles when we shot.

*He must have known that. No bullets 'whizzed past' his ears! Maybe he imagined trouble and expected it, considering the duplicity of his work that day.*

We stood speechless as he reached the sign and went back to work. Francie and I were frustrated and furious that he falsely accused us and ordered us off our own land. He was the one with no legitimate business there.

We thought of all the caustic replies we might have given, but knew we wouldn't have, so not knowing what else to do, we retreated. No more shooting magpies that day.

We told Mom what happened and felt relieved that she defended us.

"I know you girls would never be careless with guns. Ridiculous!"

She was angry too, that the sign people continued to repaint and update that sign knowing we'd never been paid rent.

For the next several days Francie and I seethed with the humiliation of it all.

"We should have done it last summer when Jeanie was home," Francie said.

"Yeah," I agreed.

We knew exactly what we wanted to do. *But without Jeanie, would we have the nerve to carry it off?*

"We still could—I guess," Francie said, finally.

"Now? But…but they just got it all nicely painted for another year,"

"So? All the better! They deserve it."

"Just us? You and me? Without Jeanie?" I asked.

Jeanie was our fearless ringleader in any tricks, the more outrageous the better—but she was at college.

"Why not?"

"What if someone catches us?"

"We'll do it early in the morning, before there's much traffic."

"We'll need a ladder and some tools and paint."

"Let's carry them up the night before. How about tonight—right now!"

"Tonight?" I shuddered.

*She's right. We best to do it now before we lose our nerve.*

*She's right—we better do it now before we lose our nerve.*

One hot day last summer a remarkable idea came to us. As Jeanie, Francie and I rode down out of the rimrocks, the big irritating billboard confronted us as usual.

Because of our parents' frustration, we girls disliked that sign intensely.

Simplicity itself, the billboard gave a powerful message—even as it defaced our natural sagebrush bluff where it stood. *'MILES HOWARD HOTEL'* it spelled out in big wooden nailed-on block letters. To the right a brilliant red-orange circle, four feet in diameter, said, *'Room Rates $2.50 to $3.50.'* For the grandest hotel in Miles City.

A perfect location from the advertiser's viewpoint. On a major

*Riding down off the rimrocks Jeanie's eyes narrowed at the controversial sign.*

U.S. highway and ten miles—just the right distance—from Miles City, the sign notified visitors as they came around that last hill into the valley of the business men and women eagerly awaiting their dollars.

"An ideal spot for our new hotel sign," the owner had told Dad proudly when he offered to pay $200 a year to lease the site for his billboard. Inspired by a stay at the Hilton in Washington, DC, he built the glitzy new hotel near the end of the war.

The hotel was a great success. The billboard undoubtedly effective.

But as years went by with no lease payment, the sign caused talk and contention between Dad and Mom. Sometimes when Mom worried about paying a certain bill, or finding money for special needs, she lamented, "If only that sign rental were paid…"

In those days only town businesses sent bills. Other people would hunt up the individual who owed them money and request payment. A person of wealth could easily brush off that request. Refusal or a runaround likely proved embarrassing and Dad had a lot of pride.

Every year or two, Mom insisted that Dad go ask the hotel owner to pay his debt. It was a significant amount of money that would be put to good use.

When Dad found him the first time, he wrote out a voucher.

"Here. How many kids you got? Four? Take your kids out for ice cream sundaes at my restaurant."

Dad brought the voucher home and we girls got excited, but Mom was disgusted.

When Dad talked to the hotel man, he said the sign company was supposed to pay. When he talked to the sign company, he was told the hotel owner would pay. Not surprisingly, often neither could be found. Both employed secretaries who said the boss was out or claimed ignorance.

Always generous, Dad disliked hounding the owner—and maybe even felt embarrassed for him because he so clearly violated the rangeland ethic that a handshake represents a man's honor. Apparently he never intended to pay that lease.

We girls couldn't understand why our folks just didn't take the sign down. After all, surely it was ours by then, since in all those years the promised amount was never paid. Mom was ever hopeful of payment some day.

Riding down off the rimrocks that day, Jeanie's eyes narrowed as she studied the controversial sign.

"Hmm. What do you think we could do with that?" she wondered with that certain gleam in her eye.

We knew instantly what she was thinking. Words could be altered to give an unintended message.

In idle moments, during last winter's drives to and from school, waiting for Jeanie and her friends, Francie and I picked up Steinbeck's *The Grapes of Wrath,* a tattered paperback someone left lying in the car.

A distinctive new title began to take shape as we scraped idly at the letters on the cover with a fingernail file.

*'HE ATE A RAT!'* the new title proclaimed, only a couple of letters reshaped from *The Grapes of Wrath*. Kind of summed up the content of that book, we thought.

Viewing the billboard, I waited expectantly for Jeanie and Francie to come up with the perfect answer.

Then Jeanie began laughing, her eyes merry.

"Look! If we pried off the last letters of Howard Hotel, it would read *HOW HOT*."

"Or even *MY HOW HOT!!* " Francie put in. "With a small piece added to the MI. And end with two exclamation points from the extra letters."

"What about the big red circle?" I asked.

"Paint out the room rate with red paint."

"It'll be a sun! And add heat waves coming out from the circle."

"Yeah, that fits. *MY HOW HOT!!* and there's the huge sun to prove it. That's all! Nothing more on the sign—it's perfect!"

"After all, it is a hot summer. People would think it's hilarious."

"Everyone who came down the highway around that hill would see our sign," I chimed in. "Wouldn't *that* give the tourists a laugh!"

Busy with other things, we shelved the idea and Jeanie went back to college.

Francie and I would never have followed through, except… *Yes, those men certainly deserve it.*

So that morning before dawn, Francie came into my room to wake me. Daylight barely tinged the horizon with pink streaks out my bedroom window. Quietly we dressed and slipped out of the house. Once it starts getting light in eastern Montana, it gets light in a hurry, so we had no trouble locating our cached equipment near the sign.

We raised our ladders and with claw hammers pulled off the extra letters, leaving those that read *MI HOW HOT*. Deft with a saw, Francie turned the *I* into a *Y* and fashioned exclamation marks out of a couple of the discarded letters, nailing them back onto the sign.

*Quietly we dressed and slipped out of the house.*

Anne herding sheep near highway. In distance, center, the Miles Howard Hotel billboard can be seen along a curve coming down out of the hills on old Highway 10.

Western Ranch Life in a Forgotten Era

I painted the circle a warm red-orange, covering over the room rates. Finished with the letters, Francie helped me paint oogle eyes and a big gorgeous smile in the center. We added red heat waves radiating outward from the circle.

Only once did we hear a truck coming down the hills, and we had time to scramble down the ladders and duck behind the sign as it roared by. Traffic was light in those days, and a truck could be heard a long way off, down-shifting the gears.

Finished, we got down and admired our handiwork.

*MY HOW HOT!!* it read, and to the right blazed that large, fiercely smiling sun with sweltering heat waves.

We burst out laughing. It looked good. It *really* looked good—all 48 by 14 feet of it!

As we carried home our tools and ladders, we wondered where Dad was and how to nonchalantly sneak by him. Fortunately, scouting from behind the barn, we saw him climb onto the tractor and drive away.

"I just had to laugh," Uncle Chet said. "And I couldn't help but wonder if you girls were involved in some way."

Several days passed before the men repainted over our art work.

Dad returned from town a few days later with an odd tale. The sign owner sought him out, wrote a check for the entire amount owed for five years and commented gruffly that he didn't expect any more changes to his sign.

Stunned, Dad didn't dare deposit the $1,000 check right away, since he had no idea what the man was talking about.

*What was going on?*

About that time Mom's brother, Uncle Chet, came by. He ranched ten miles east of us and drove down the hill past the billboard every time he went to Miles City.

He had made a quick trip to town a few days before, he said.

"I didn't have time to stop in, but I sure enjoyed the message on your sign."

Dad and Mom gave each other a quizzical look. Before they had time to ask questions, Uncle Chet continued.

"My goodness! That *was* impressive. As I came driving down the hill around the last curve on that hot afternoon, right there in front of me was that big bold billboard. All it said was 'My How Hot!' in great big letters. Just those three words and a big, red happy shining sun!"

He grinned. "Yep. That sign kind of summed up this long hot summer for me."

Then he turned to Francie and me, a twinkle in his eye.

"I just had to laugh," he said. "And of course I couldn't help but wonder if you girls were involved in some way."

Dad and Mom exchanged glances.

"Well, I guess I'll drop that check off at the bank next time I'm in town," Dad said with a smile.

Francie and I looked at each other, relieved.

This was one of our best jokes ever. We didn't get in trouble. And because of us, that hotel owner actually paid up.

## Taxidermy Aromas

"See. It's not very bloody. And its skin isn't messed too much. I could easily fix it."

I agreed until I noticed by the way she looked at me, she expected me to carry it home.

"Oh, no," I said. "I couldn't!"

"But you'll have to," Francie said. "I can't carry it and pedal too. Even when we reach the hill and we have to start walking, I still can't carry it and push the bike."

She thrust the bloody bull snake at me.

What could I do? After all, she was my ride to school. My legs were still too short to ride a bike, so I rode behind Francie.

"It's only a little squashed in the middle. The bloody head I'll cut off and tan the skin."

So, until I grew big enough for a bike of my own, I carried home her roadkill. Even though I washed them often, that lingering odor stayed on my hands.

This was the year Francie bought her correspondence course from the Northwest School of Taxidermy. It cost ten dollars—one lesson in the mail every month.

Francie expected Beverley to be as interested in learning taxidermy as she was, so she talked her into paying half. Beverley agreed to chip in five dollars.

Francie didn't dare give her real name on the order because the course was advertised 'for men and boys only,' insinuating—not unusual in those days—that it was not a suitable hobby for girls. So she spelled her name *Francis*, instead of the feminine way: *Frances*.

"Of course you can do it." Mom told her. "You may not have as much strength as a boy, but you'll figure out a way to get around that."

Although Francie tried hard to get Beverley interested so she'd get her five dollars worth, Beverley would not touch the animals until they were finished, stink free and hanging on our living room wall.

When lesson one came, Beverley summed up her feelings

*"It's only a little squashed in the middle. The bloody head I'll cut off and tan the skin."*

*Francie, seated on corral fence, shaves down a deer hide for mounting.*

*She skinned the deer skull and boiled it outside in a tub with antlers still attached.*

Anne decorates for Christmas the antelope head that Francie mounted for our living room wall.

quickly.

"You do it, Francie. There's no way I'm getting my hands in that bloody, messy stuff."

Every spare minute Francie studied her lessons. I eagerly tagged along wondering which exciting project she would start.

"Let's walk up to the springs," Francie said to me one day, "I need to catch some frogs."

This seemed like a fun thing to do, so I cheerfully climbed the hill with her.

She brought along two coffee cans with lids and we filled them with frogs. Of course I had to carry one home—but at least they were alive, in a can with a lid—and didn't smell.

Stuffed, those frogs became a bullfrog chorus complete with concertmaster, as described in one lesson. Each frog had his own instrument she carved from balsa and the soft woods of cigar boxes. My favorite was an upright stringed bass, the bow made of willow and short strands of horse hair.

She mounted smaller animals and tanned larger hides, one a beautiful bobcat hide a friend of Dad's gave her.

Francie started practicing taxidermy in her large upstairs bedroom, but before long Mom banished her and the smell out to the big old horse barn.

A serious problem there was keeping barn cats out of her work. A small door led into one of the manger rooms where she worked, but every so often one of the cats would pry it open and have a field day with her projects. She mounted a beautiful pheasant rooster, but the cats enjoyed it, too, attacking and teasing the feathers. She kept unruffling them until suddenly it fell apart, the dried skin pierced through by a hundred kitty claws.

Dad helped her build the head and neck molds for deer heads. He was good at shaping animal heads after years of observing them. This was in the days before purchased plastic molds. So, like other taxidermists, Francie built her molds from excelsior, a wood packing material, wrapped and tied with string over wood frames and finally smoothed over with a wash of plaster of Paris.

For a deer head, she initially skinned out the skull and boiled it outside in a tub with antlers still attached. She bought and glued in glass eyes and stiffener for ears. Then when cured, the skin was pulled over the mold and sewed back together with large hand stitches and a curved needle.

I loved watching her work in the barn. It was fun playing with those beautiful soft brown glass eyes for the deer.

Francie continued to have a great time with her taxidermy for two or three years. For our walls she mounted a big mule deer head and an antelope one of us shot. She stuffed deer heads for Dad's friends and once a large Dolly Varden trout.

She even got paid sometimes. What an entrepreneur—I was in awe.

However, we did object to the rank, gamy, dead-animal smell of Francie's hobby, when she was working on hides.

"Even the cows and horses notice it," Jeanie said, "and if Francie is sitting on the corral fence skinning out something when we try to bring them into the corral, they spook and turn back."

That was true, but her animals no longer smelled when she brought the finished works into the house. We all enjoyed the taxidermy work that hung in our house as did neighbors and visitors.

Still, stroking the gorgeous bobcat hide that Francie tanned, I wondered again how such beautiful things could smell so strong and gamy while she worked on them.

And I never liked carrying home her roadkill, especially bloody, mangled snakes.

*People were alarmed when prairie dogs charged at them, despite their obvious joy.*

## Prairie Dog Pets

"Ee-eka! Ee-eka!" Our prairie dogs greeted us joyfully.

We named them Eeka and Meeka because that's how they welcomed us each morning—leaping up in spontaneous joy, arching their backs and flinging their little front legs in the air, as they uttered their cries. They were always happy to see us and we rewarded them with snuggles.

Just like puppies, they loved being petted and often crawled into our hair and snuggled down. Their favorite place to nap was curled up in the curve of someone's neck, with a nose in our hair. They crawled into a shirt sleeve for extra warmth as we carried them around.

Both females, they were the sweetest, cuddliest pets we ever had.

Sometimes I sneaked them into bed with me at night, but Mom usually found out. Then I had to put them back in their washtub bed. They were so trusting. Often they didn't wake up as I tucked them into their own blankets.

We took them outside and they followed us wherever we went, chattering if we walked too fast. Soon they started digging short holes on these ventures, but we were careful to put them back in the tub when not with us.

They continued to be loving pets, following us everywhere. Eventually we let them dig longer holes and they stayed in them at

*Baby prairie dog fits into the palm of a hand.*

Western Ranch Life in a Forgotten Era

*Our prairie dog puppies quickly learn to trust us, taking food from our hands and sitting on their haunches as they daintily nibble their food.*

*Eeeka sits up to nibble delicately on a piece of bread. Full grown, she weighs in at nearly three pounds.*

night. They seemed to have forgotten their old prairie dog home and were completely satisfied to live in our yard between the house and corrals.

When visitors drove up, our pets would greet them with a squeal and run to be picked up. We learned to quickly intercept them, since it alarmed most people to have a couple of prairie dogs charging at them, even in their obvious joy.

When a strange vehicle drove into our yard they gave their warning cry.

"Yip-yip … yip-yip!"

They became good watchdogs as they ran to their holes to yip and warn us of an intruder.

They welcomed neighbors with cries of Eee-ka! and threw themselves in the air. Our pets became quite famous in the community.

Young prairie dogs make great pets, at least while they are young. But when they grew older, Mom and Dad had some unkind things to say about that topic.

As young kids, we girls spent considerable free time visiting the prairie dog town that covered the flat below our rimrock hills, getting acquainted with these cute little animals.

"Let's walk up to the prairie dog town," Francie or Jeanie would say.

As we approached, the first sentinel to spot us gave the alarm—several sharp yips as he sat up on his haunches, stretching his neck to see higher. As we moved closer, he ran to the opening of his hole and dived down. The holes had dirt mounded around from their excavations that served as a higher vantage point to view the surrounding area and the approach of predators. Warning yips alerted nearby prairie dogs, and they called to others until the entire town was yipping or ducked down their holes.

A great communications system.

We sat quietly for several minutes and then prairie dogs began poking their heads above their holes to see if we were still there. If we didn't move, they came completely out and sat on the mound. Standing up on their hind legs, they peered at us until satisfied we weren't going to attack. As we patiently watched, the whole town popped out of the holes and continued their normal activities.

Prairie dogs are social animals—each mated pair digs its own hole or den, built close to other holes with a small area of grass for food between. As the young mature, they move to the edge of town to find space for their own home, so the towns expand from inside out. The choicest places are in the center of town—it's safest—as predators prey on those nearest the outer edges.

Prairie dogs are very fond of each other. We watched them run around the town hugging, kissing and embracing their neighbors. Sometimes when a predator appeared, they ducked down the nearest hole—it didn't matter that it wasn't their own.

They seemed so carefree and what fun they had!

Baby prairie dogs love to dig and little five-inch-deep holes appeared all around the town. This taught them to eventually dig their own homes.

Sometimes we carried oats in our pockets and spread it around their holes. When they came out, their little noses twitched as they smelled the yummy food. They watched us carefully to see we weren't coming closer, then picked up a single kernel in their tiny front paws, sat back on their haunches and delicately nibbled the oat kernel.

"Lewis and Clark named them when they explored the Missouri and Yellowstone Rivers," Mom told us.

She said they first called them *barking squirrels*, but when they finally captured a prairie dog —with great difficulty— and sent it alive to President Jefferson, the label read 'wild dog of the prairie,' and the name stuck.

As we sat at the edge of the town, we could see the Yellowstone River below our fields. We imagined Captain Clark and his crew floating downriver on their return trip to meet Lewis at the junction with the Missouri.

It was the prairie dogs' love of our oats and the thought that Lewis and Clark caught one that sent Francie to her favorite battered book, *Fun Things for Boys to Do*. If the explorers caught a prairie dog, of course, we could do it too. The second-hand book had a chapter on building box traps to capture animals alive.

*A favorite place for Meeka to snuggle is in the curve of someone's neck. Here she plays with Francie and our dog Jimmy.*

Her book, published during the depression, suggested the trap be made from small willow branches. Along with the beautiful huge cottonwood shade trees, our ranch grew dense thickets of willows on the ditch bank—unlimited material for the projects in Francie's book.

I watched with fascination as she cut, fitted little willow bars and nailed them in place. When finished, the swinging door pushed in easily, but not out, trapping the animal.

"Time to try it out," Francie said. "Let's go!"

She carried the trap expectantly up to the prairie dog town with me bringing the oats. By this time the prairie dog babies we hoped to catch were nearly grown. We baited the trap with oats, set the door on its trigger and walked back home to wait a few hours.

*Western Ranch Life in a Forgotten Era*

*What a homecoming! It was almost enough to stop our trapping.*

*Anne set Rex on the trail of an irate escaped prairie dog.*

"We've got one!" I cried when we returned.

There inside our trap was a prairie dog. Only thing, he was a full-grown adult. But we took him home anyway—trap and all—hoping to make a pet of him. All the way he chattered angrily and bared his teeth at us.

Mom was skeptical. She said an old dog could never be tamed and he'd run away first chance he got.

"But you can put him inside my old wash tub," she said, "fix it up like his home."

We put sawdust in the bottom, added a bowl of water, green grass and a little box of oats. What more could a prairie dog want?

Next day we built what we thought was a better cage of chicken wire, open at the bottom. This was his new home where he could dig a hole in the ground and when he came up he'd still be in his cage—or so we planned. Mom predicted he'd dig out.

By this time he preferred oat kernels to the grass we picked and we felt sure he'd stay around for this special treat. We pulled on Dad's heavy leather gloves to keep from getting bit and he soon took oats from our hands, sitting back and eyeing us solemnly with his round black eyes as he nibbled.

We watched as he started digging a hole. With his back feet he kicked the dirt out until he got deep enough to turn around. Then he pushed the dirt up with his nose, using it like a shovel, keeping his eyes tightly closed. A mound appeared outside the hole amazingly fast. We left him to his digging and returned an hour or so later.

Mom was right. He was gone.

We called Rex, Dad's hunting dog, a water spaniel who sniffed out and retrieved ducks we shot in the lower pasture swamp. Rex had a soft mouth and never bruised a bird—perfect for hunting runaway prairie dogs.

We took him to sniff the hole outside the cage where the prairie dog escaped.

"Find, Rex! Find."

With his nose to the ground and his stubby tail wagging, Rex took off across the pasture. It wasn't long before he found our escapee, heading straight back to his town a quarter mile away. Without gloves we couldn't pick him up and we didn't want him to get lost, so we sadly followed him back to his town.

As he saw the first prairie dogs, he ran faster. They stretched up tall, threw their front paws in the air over their heads and shrieked with excitement. Then they ran up and rubbed their noses across his.

What a homecoming! It was almost enough to stop our trapping.

We caught several more adult prairie dogs that summer, but even though we tightened our cage, they eventually got out and ran back home. Mom was probably right—they were too old to tame. So we waited through winter for a new crop of pups.

When spring finally came, every time we went riding for the cattle we detoured past the prairie dog town to see if any babies were running around. At last the time came. Francie and I hurried back up there with the willow trap.

"We'll set it here," Francie said. "By this hole where those little ones went down."

Then we baited the trap, trailing a scattering of oats. I wanted to stay and watch from a distance, but Francie said we might scare them away.

What a surprise when we returned! Not only did we have a little one in the trap—we had two!

At home, we put on heavy gloves to pick them up, mindful of the ferocity of the adults. But they were gentle and trusting from the beginning.

We played with them all summer and that winter our prairie dogs stayed snug and warm in deep holes they dug between our house and corrals. They ate all the oats they wanted and developed extra fat for insulation. On warm sunny days they came out of their holes and played. Then came a long cold snap. The ground froze deep down. Still our pets were warm in their holes, which went much deeper than the frost line.

But one day no water ran from our house faucets.

Dad diagnosed the problem.

"Those prairie dogs!" he said.

Since water still ran at the corral tanks, the pipeline was frozen between barn and house—right where the prairie dogs built their underground tunnels. Their open holes along our water line caused it to freeze, resulting in a tough job for Dad digging down six feet in frozen ground to thaw it out.

*One day no water ran from our faucets. It was the first time Dad said the prairie dogs had to go.*

This was the first time Dad said the prairie dogs had to go.

When spring arrived, Eeka and Meeka stayed out of their holes most of the day, foraging through the grass for the tastiest new blades and generally exploring. They never strayed far and initially continued to be friendly.

Then they started getting mean.

When visitors arrived, instead of giving their welcome cries, they gritted their teeth and ran at them, chattering low and threateningly. Not good.

Dad and Mom warned that we couldn't let them bite visitors. So we ran to catch them when company arrived, shouting for them to

## Too many prairie dogs

For ranchers prairie dog towns devastated their pasture lands. Hardly a spear of grass grew on flats taken over by one of the towns. Only beaten earth and raised mounds of dirt packed hard around each hole, probably ten feet apart, solidly across an entire flat.

Entertaining as they were, the truth is that prairie dogs took over the most fertile grasslands for themselves and turned them into barren wastelands. Only cactus grew. No cows grazed there and the hundreds of holes made them dangerous for horses. Many a horse snapped his leg in a prairie dog hole and had to be shot. Two of our own much loved horses, Flexi and old Eagle, met this fate.

When we moved to the ranch near Miles City all the level plateaus that should have been our best grasslands were covered with prairie dog towns, as were those of our neighbors.

There was no way to get rid of them but shooting or trapping—neither of which made much of a dent. Shooting was not easy once the town was alerted, with only an open-sight .22 against those small distant targets. Besides, you seldom had the satisfaction of knowing you hit a prairie dog—it fell down the hole out of sight. This only kept down territorial expansion. There was no way to eradicate a town so the grass could come back, until later when the government used poisoned oats.

So each spring after the young pups came out, Dad set eight to ten traps daily—all he had time for—

*(continued)*

stay in their cars until we captured our pets.

Then they turned on us, biting fiercely. It was comical watching a hired man trying to dodge and outrun these little animals when coming from the barn. Most wore heavy boots that the prairie dogs couldn't sink their teeth into, but once in awhile they jumped high enough to reach a leg above the boot—ouch!

Mom's clothesline bordered the garden on the east side of the house and, with every load of clothes she carried out, the prairie dogs were waiting to attack. After a few ankle bites she started wearing knee-high irrigating boots to hang out clothes. We girls were getting bit, too.

Dad and Mom laid down the law—something must be done.

"Wild animals can be good pets when they're young," Mom said, "but they go back to their wild nature when grown and cause problems. You know we can't have them acting like this."

They left the decision to us girls. We didn't want Dad to kill our pets, so the best solution seemed to be to take them back to their town.

"Maybe not," said Mom. "Their own town is too close to the highway and target shooters sometimes stop there to shoot prairie dogs. Eeka and Meeka would run toward the hunters and be the first ones shot."

One day we sadly bundled them up in a box and set out for a remote new home—a small, isolated prairie dog town on a little flat high up on the bluffs above the railroad, beyond some of our roughest coulees. There was no way to get there without climbing steep cliffs and hiking in, which prairie dog hunters usually weren't willing to do.

A perfect place to leave our pets.

We consoled each other by saying they'd be safe and now could be with others and raise families of their own.

Eeka and Meeka scolded us the entire trip. They wanted to escape the box, but we didn't dare let them out till they arrived at their new home.

As we approached, the prairie dogs there yip-yipped, announcing our arrival, then ducked into their holes.

Our pets grew silent. We wondered what they were thinking as they realized they were about to meet other prairie dogs.

We set the box on the ground and waited for the others to come back out of their holes. Then we opened the box and our pets emerged. They looked around and slowly approached their new neighbors. Afraid they might follow us, we ran to get a patch of sagebrush between us and the town. As we climbed a hill, we looked back and saw Eeka and Meeka touching noses with other prairie dogs. They were being accepted into a new life.

Later that summer Eugene, a former school classmate who worked on the railroad section crew, came by our house and told a strange tale about a prairie dog hunt. Eugene could get quite excited when he told about something odd or unnatural.

He said he got off his rail car, climbed up a steep cliff with his .22, crawled through a fence and came out on a flat where he'd seen a prairie dog town from the railroad. The prairie dogs yipped and down they went into their holes before he could get one in his sights.

"Suddenly, here came two prairie dogs racing straight at me. I got ready to shoot, but they were acting mighty weird. Chattering and mad." He waved his hands excitedly as he spoke.

He said their cheeks puffed out as they charged toward him.

Eugene was not one to calmly level his rifle sights and get them before they got him, as braver hunters might. He turned and fled, barely making it over the fence ahead of them.

"They chased me," he said. "And they almost caught me, too. I was scared! Maybe those prairie dogs had rabies, ya think?"

"Yeah, rabies... Maybe it's not safe to go up there," Jeanie suggested.

"Don't worry, I won't! Never again!" he vowed.

When he left we giggled with delight.

Our two sweet little pets had chased Eugene right out of the prairie dog town, saving themselves and their friends!

> **Prairie dogs** (cont'd)
>
> and urged hunting friends to come shoot prairie dogs on Sunday afternoons.
>
> "We have to get the young ones in spring to keep the population down," Dad insisted.
>
> We girls didn't much like the idea of trapping babies. Worse, this task fell to Jeanie and Francie on our daily hikes after the cows. We set traps, checked them daily, removed dead prairie dogs and reset them.
>
> If they weren't quite dead we had to finish them off with a cow bone or rock. That was the hard part, with a few tears as we did it. Yet we understood it had to be done. A sad task.

*Our pets dig deep along our water line and by spring claim the yard between house and barn, attacking anyone who ventures close.*

White half-grown kitten surveys view from corral fence. One summer, painted with red food coloring, this cat wore a shocking-pink fur coat.

Our contented cats and kittens lounge around the corrals basking in the sunshine, ignoring chickens.

## Tight Fit

Beverley's feet flailed the blackness, kicking dust and dirty straw—choking the small amount of air left for us to breathe. Although she struggled with all her might, she stayed solidly wedged in the only opening.

ANNE

*Will we ever get out? What if we can't breathe in all this dust?*

Jeanie, Francie and I were trapped below in the dark and dusty chicken house attic.

I could hardly get a breath of air. I panicked, but didn't dare show it.

This one place I always hated for a mother cat to hide her kittens—in the attic of the chicken house. Dad built the chicken house out of concrete with a ceiling of heavy mesh wire running between the wooden beams. Over this he added thick straw for insulation. Inside, Mom's hundreds of chickens kept warm even on the coldest winter days.

The roof, wooden shingled on an easy slope, came down to about four feet from the ground on one side. Perfect for crawling up and sliding down, if we took care watching for slivers. The attic was safe from predators and even we girls seldom climbed in there.

At the very top, a square chimney-like air vent rose above the roof peak. It was scary dropping through that opening.

"Mitzi's gone into the top of the chicken house!" I eagerly reported at breakfast that morning.

"Oh, no! Not the chicken house attic!" Francie cried.

Being older than I, she better understood the risks of this exploration and yet, like me, she couldn't wait for Mitzi to bring out her kittens.

Beverley settled it with her customary enthusiasm and decisiveness.

"If we want to see the new baby kittens then we've got to go in there," she said.

Cats birthed their kittens in early spring or summer and as soon as their tummies started to bulge we checked them daily. We followed to see if they were making a bed since this often signaled birth was imminent. For the first few days after birth a mother cat usually doesn't leave her newborn kittens. Then she appears, sleek and trim. At first they are so small with their eyes tightly closed, mewling in their tiny voices.

"A cat with baby kittens needs more food," Mom said as she handed us meat scraps and milk. "And yet she needs to stay with her kittens to feed them and keep them warm."

Sometimes, even though we brought food, anxious mother cats

didn't want us around their kittens. When we saw the mother crossing the barnyard carefully holding a kitten in her mouth by the nape of its neck, moving to a new hiding place, we knew we were handling her kittens too much.

Mitzi was a proud mother—we were sure she'd be glad of our visit.

Searching the top of the chicken house for a new batch of kittens was quite an expedition. Besides food scraps for the mother, we needed a flashlight for that awful blackness.

It was scary dropping through that opening.

I had to wait to go into the top of the chicken house until I was tall enough to get out—boosted by one sister below and pulled up by another.

*What if we get stuck down there and can't get out?* I wondered each time we went inside.

Then came the day my worst fears were realized.

"You go first," we urged Beverley.

Beverley took charge—she was courageous and knew how to do everything well.

Going down the ventilator opening was easy.

Beverley had grown over the winter but gravity helped her slide through. She took the flashlight and went first into the dark.

Mitzi lay in a corner under the eave slope, purring loudly as we crawled to her with meat scraps and a pan of milk. Five mouse-like little kittens nestled up against her, some napping, others nursing, tightly attached to their nipple. We cuddled the kittens gently up to our necks, soft and trusting, while Mitzi ate.

"Oh, look at this one…so cute." I picked out my favorite, white with black spots.

"He'll be a lot cuter when he gets his eyes open," said Jeanie.

"Good girl Mitzi," Beverley crooned to her.

Finally it was time to leave.

Our plan was for Beverley to go first, then once out, pull the rest of us up. She was to tug me up while Jeanie and Francie lifted me. It was such a relief to know I'd be the second one out. All the time inside I worried, *what if we can't get out?*

Then the unthinkable happened.

As Beverley grabbed for the top rim of the opening, heaving herself straight up, her hips jammed into the opening. Her feet flailed the air a few inches off the floor, kicking up dusty straw and filling the air with choking debris where we waited.

Worst of all, she had our only flashlight. It slipped out of her hand over the rim and rolled down the roof to the ground.

"I'm stuck!" she called. "I can't get out."

Our cats love to be held, and reward us with constant purring. Even baby Anne, above, grasps a friendly cat from her high chair, with help from Beverley, along with Francie and Jeanie. Below, Jeanie, Anne and Francie with pets.

Western Ranch Life in a Forgotten Era

*Sitting on the porch steps, Francie and Anne invariably are joined by pets.*

*I despaired of ever getting out. What if we smother in here and no-one can find us?*

A terrible panicky feeling came over me. I didn't dare cry out or my sisters would never let me come back.

It was absolutely pitch black in the attic, with Beverley blocking the only light source.

I froze in horror as Francie and Jeanie rushed to push up her feet. Beverley was kicking wildly, trying to find something to push her feet against to force her way out—but there was nothing. Every time Francie and Jeanie grabbed a foot she managed to kick them and they let go. Dust and straw bits filled the air. It was hard to breathe.

Letting go of the rim with her arms, she tried to push her way back into the attic, but as she settled down into the tight opening, she became even more tightly wedged.

"Help! I'm really stuck now. Help me!" Bev called.

*What if we smother in here and no-one can find us?*

I despaired of ever getting out. Even if we yelled and screamed Mom and Dad would never hear us. Maybe we'd choke to death on all the dust.

Finally, with Jeanie and Francie resolutely holding onto her thrashing legs and pushing from below, Beverley shifted her hips into the diagonal of the square opening and managed to squirm out.

The bright light of day shone in. I could breathe.

"Your turn." Jeanie and Francie turned to me.

Much relieved, I ran to the opening. They lifted me and Beverley pulled me out.

Francie and Jeanie popped out in a hurry.

*Ah, fresh air and freedom! We all felt the joy of it.*

Beverley never again climbed into the chicken house attic. I guess she outgrew her fascination with newborn kittens.

## Feedlot Rodeo

Cautiously I swung my leg up one rail of the feeder, then the other leg.

ANNE

*Had the steer noticed?*

With a mighty jump I leapt for his back.

Success! I dug my heels into his sides just behind his front legs as he jerked back from the feeder—whirling himself into short bucks—kicking his back feet high in the air. We spun around the corral, bucking and twisting.

What a ride! What fun! How I loved those Herefords!

Suddenly I felt myself leaning too far off to one side.

*Oh no! Here we go!*

Other steers, scared away from eating by my bucking steer, stared at me from across the corral as I watched them from the ground. Slowly they returned to the feeder, tempted by the 'cake' I had just spread.

In the fall our calves were weaned from their mothers and finished for market in feeder corrals. Our three main corrals, built side by side, were divided by long wooden feeders. Just room at the feeders for all the calves to line up with their heads poking through to eat hay.

Every afternoon I scattered cake on top of the hay—three-inch-long pellets of ground grains mixed with molasses and other nutrients, designed to put marble on these calves. To them, cake truly was dessert after eating hay all day. They came running, acting like kids at a pie-eating contest, crowding up to the feeders and shoving each other to be first.

In their greediness they didn't notice as I climbed over top of the feeder and made a flying leap onto one of their backs.

Then what a ride!

As kids, we came to believe such things as wild rides and bucking horses were an adventure—even if the ending was a bit painful.

We enjoyed the stories told by Dad, Mom and their friends. Mostly the stories involved jokes on themselves. This was in the days before television when people spent an evening visiting or took a break when they came on ranching business.

Some of the stories had a rough ending like the time Dad's nose connected with a wild horse on the range.

"That green bronc twisted and bucked till I flew off and landed on a tree stump—nose first! Broke it and pushed it to the left. A year later the same thing happened again—only this time I reached up, grabbed my nose and jerked it back straight again."

He and his listeners whooped with laughter. Forever after, Dad's straight nose had a decided right list.

I watched my older sisters take turns riding the milk cow's calf in its little pen. Even the pig was fair game. The fall wasn't far down and the landings fairly soft. Then they went on to the challenge of riding larger animals. I followed in their footsteps.

But now Jeanie and Francie were gone to college and Beverley was through school and a journalist on a newspaper in Council Bluffs, Iowa. Home alone that winter, I missed them and was left to think up my own fun.

Yet I felt a twinge of guilt each time I rode those calves.

*Calves, weaned in the fall, go into the feedlot. Nearly full-size yearlings by January, they give Anne a wild ride—when Dad's not around.*

*Oh no! Here we go! I felt myself leaning too far off one side.*

> *He exploded in a short, spinning twist, ducked his head, bawled and flung me onto a pyramid of frozen manure.*

A spooked calf, frantically trying to dislodge me, scared other calves away from the feeder and ran off maybe a pound of fat, which is why I didn't take a ride if Dad was around.

Having spooked the calves away from one feeder, I moved on to the next corral and spread cake there. Again I waited till all the calves pushed their heads into the feeder, then leapt onto an unsuspecting back and enjoyed a bone-jarring spin. I continued on to the next row of calves.

On a good afternoon—when Dad wasn't around—I could get in four exciting rides.

The red white-faced Hereford calves were sporting about it all. They gave a few quick jumps, galloped half-way across the corral, then maybe a few more quick bucks until off I went.

Then I discovered even more fun in the antics of the rambunctious black half-Angus steers.

Dad started breeding our young, not fully-grown heifers to Angus bulls, which in those days were somewhat smaller. Big Hereford bulls yield big calves, which can mean trouble when a heifer goes into labor. That's why Dad, who liked the latest ideas in ranching, came up with a simple solution: use Angus bulls on the younger cows.

So we had both Hereford and half-Angus calves.

Angus calves? Riding them was a real test.

They were tough—really tough and aggressive. Newborn calves were small, but wiry and strong, and they grew fast in the feeder corrals.

Rudely, they pushed and shoved their heads at the nice gentle Herefords, trying to get more than their share of cake. Instinctively I favored the mild Hereford steers over the tough Angus—with their spitefulness and bad manners.

Still, it was fun trying to ride them. When I needed a bit more action I chose Angus calves.

The Angus proved challenging—more aggressive, and their much shorter hair harder to grip. Nor did they like my feet digging into their belly behind their front legs, giving them a bit more zest as I clung tightly.

They spun, twisted and bucked ferociously, trying to toss me and tap dance on my head. The worst was after they dumped me. Then they turned and shook their heads, sometimes kicking at me and stamping their feet. Those Angus were just spiteful and, once they bucked me off, looked so pleased with themselves.

As the winter progressed the temperature grew colder and the calves grew feistier and bigger until by January they were nearly full-sized yearlings—even harder to stay on. But my calf riding ability was improving too. Now I rode Angus calves as first choice. Someday, I vowed, I'd be able to ride them across the corral.

But my Angus riding came to an abrupt stop after one fateful late winter afternoon.

The temperature was a frigid twenty degrees below zero that day, a point where warm manure freezes rapidly.

What goes in one end must eventually come out the other. The metabolism of a feeder calf is such that what comes out can be a great quantity. In the fall, before freezing weather, if I was dumped onto such a mixture, I had a messy but soft, squishy landing.

"Plop-plop-plop," I heard.

At twenty below zero each plop quickly ices into a rock-like solid on top of a frozen pile of manure. The corrals were domed with these brown pyramid peaks, hard as small boulders.

After spreading cake in the feeder, with the calves' heads down eating, I eyed my potential victim—a big black Angus steer.

I eased over the last feeder and leaped onto his back. As I quickly grabbed his short hair on top of his shoulders I dug my heels into his flanks and hung on tightly.

He exploded in a short, spinning twist, ducked his head, bawled and flung me onto a pyramid of frozen manure.

*Whack!* My thigh hit the icy peak dead center.

Excruciating pain!

Gingerly I pushed myself up on my hands and found I could not put pressure on my left leg.

*Anne entertains herself by jumping on steers at feeding time.*

This was a real problem. I had to walk so Dad wouldn't notice—or he'd ask and I'd have to confess I was riding the fat off his prize feeder calves.

I dragged myself over to the wooden corral fence where I could stand on my good leg while gripping the rail with both hands. It seemed like forever that I stood there in the bitter cold holding onto that fence, waiting for the pain to subside.

Then I hobbled to the house where I off-handedly remarked to Mom that I'd hurt my leg. Without comment, she suggested I sit at the table until supper. I spent the evening there hunched over school work.

By morning, though still very sore, I was able to walk slowly with a barely noticeable limp. My leg finally healed but left a permanent large lump on the side of my thigh where I landed on the pyramid of polar-ice poop. Later I learned there was a greenstick fracture and though I grew an inch or two taller my thigh stopped growing, requiring a lift in that shoe.

All those years Dad never mentioned me riding the fat off his calves, and neither did I. But one day shortly after his 90th birthday I walked into his room.

He lay on his bed grinning like he'd heard a good joke.

"Heh, heh…heh, heh," he chuckled.

"What's so funny?" I asked.

"You sure used to get bucked off those calves, Anne!"

## Cutting the Christmas Tree

*FRANCIE*

Almost as soon as she stepped off the train, home for Christmas that first year after long months away at college in Missoula, Beverley announced her plan.

"Francie, this year we're going to ride up in the Pine Hills and cut a pine tree."

"Uh, you mean ride in the car, right?" I asked.

She laughed her delighted laugh that we had missed so much all fall, waving a hand airily.

"Oh no. We'll ride the horses up there. I'll take Buck, he'll be up to it. And you can ride—who…?"

"Pedro, he's strong and needs the riding I guess."

"Great. We've wanted a pine tree for Christmas. Now's our chance!"

As always, her enthusiasm was infectious and her ideas usually fun.

"That's a long ride," Mom said. "And don't forget its December."

"Yes, but it's warm. See?" Wearing only a sweater, Beverley flung out her arms expansively, laughing.

She was right—the weather had warmed up for the last days before Christmas. Forty-five or fifty degrees at least and no snow, although Jeanie, Anne and I still hoped for a white Christmas.

Next morning we saddled dependable Buck and Pedro, my tall pinto, tied on an ax, saw and our lunch. We each carried a long rope coiled on the saddle—a 'ketch rope' as we called it. As it turned out, we needed both ropes.

We rode the ten miles, through several large pastures and much unfamiliar land—across ridges, rocky buttes and bad-

> *Next morning we saddled Buck and Pedro, tied on an ax, saw and our lunch, with long 'ketch' ropes coiled on our saddles.*

Beverley, off the train her first time home from college, is eager to ride into the Pine Hills in search of the perfect Christmas tree.

134      Montana Stirrups, Sage and Shenanigans

lands—always heading for the higher elevation of the Pine Hills.

Meanwhile at home, Mom, Jeanie and Anne put the finishing touches on house decorations, gifts and preparations for our big Christmas day dinner.

Mom was right. It was a long ride.

But Beverley stayed enthusiastic. Although the University of Montana nestled in majestic mountains, it seemed she missed our closer, more intimate, buttes and badlands—and our horses. She kept laughing, pointing out interesting formations and rocky outcroppings.

Once in awhile, she reached down and stroked Buck's neck, murmuring to him.

"Good old Buck."

This was silent country in winter. No cattle grazed these remote pastures and even the deer had moved down to more accessible feed in the valley. Once in a while a white jackrabbit sprang up and bounded over the hill, highly visible in his winter coat—camouflaged for snow cover when there was no snow. A golden eagle soared high above.

The sun warmed us as we rode.

Near noon we topped a ridge and pulled up.

"Look, there they are! Pine trees," Beverley said.

"Let's go!"

Excited to see pines trees on a ridge in the distance, I kicked Pedro into his easy lope and Buck responded under Beverley's commanding heel.

The pines grew widely spaced. Some crested the very top of rocky hills, like sentinels, as if reaching for the sun—unlike the cedars that clustered along north-facing side hills and draws closer home, sheltered from the hottest sun.

"How about that tree over there?" she pointed.

"Yeah. Looks like a nice shape."

But when we got closer we saw it was way too big and too scraggly, the branches far apart. Beverley laughed, stood in her stirrups and stretched her right arm up high.

"Look. It's three times as tall as us—even on our tall horses!"

"Well, how about over there?" I waved toward a side hill across the gulch. "Seems like a lot of good trees."

We rode up there and over the next hill—and the next— leading our horses up and down the steepest chasms.

These were big, rangy ponderosas. Mostly far too big, with heavy, scaly trunks and widely-spaced branches, leaving large empty holes.

*Setting off for the Pine Hills, Francie on Pedro, above, with Beverley on Buck, below. Together we drag our big pine Christmas tree ten miles, tied between the horses to prevent it rolling.*

"We can fill the gaps with decorations, can't we?" Beverley said.

"Maybe. Strings of popcorn and cranberries?"

"Sure that works. And long tinsel streamers help a lot."

Long needles fingered out some five to eight inches. *Would decorations even hold on those long needles—or would they slide off?*

"Anyway, it's too big!"

Still with those large, heavy trunks, the smaller trees seemed even more sparse.

Beverley squinted toward the top of a lordly pine.

"See how green this one is up there and with thicker branches. What if we climb up and cut ten feet off the top?"

"How would that look? A big beautiful pine tree missing its top?"

We'd done that before with the juniper trees that grew in our hills, but there were so many and they grew bushier. Somehow it seemed worse to top a majestic pine.

"It'll close over, won't it?" Beverley said.

"Maybe. Let's look a bit longer."

Choosing from among the huge, stately, but rough and rangy native Ponderosas proved even more difficult than from our cedars. Every year when we searched for the perfect Christmas tree just above our ranch, we encountered these same problems. We'd hike the hills for hours, dragging ax and saw, searching for just the right tree.

The native junipers—we called them *cedars*—hid their defects till we came close. Perfect specimens seemed always to grow in the distance, on a far slope, across a steep ravine. When we climbed over there it was a double tree with big holes right in front, or too bushy without a strong trunk or leader for our tree-top silver metal star, or just too large. Trees naturally appear much smaller at a distance. And though we'd finally cut a small tree and our ceilings towered ten feet high, we often had to cut off another foot or so from the base when we got it home. If we cut a tree high, above important branches that swept up from the bottom, this exposed fatal flaws.

In late December a bitter cold often penetrated every step and we tried to keep moving, turning our faces from the wind. Climbing up and down hills helped keep us warm. Halfway home, dragging the tree, we sometimes stopped under the rimrocks where Jeanie built a campfire to warm up.

Usually we'd set up and decorate our fresh-cut native cedar the day before Christmas and enjoyed that wonderful pungent juniper smell for a couple of weeks. On the twelfth day of Christmas, January 6, we carried our tree outside and burned it at dusk, a tradition from Mom's family.

*"You're right about the tree tops," I finally agreed. "We need to cut our tree off the top."*

# How We Celebrated Christmas

We girls cherished our two-week Christmas vacations. This was a time of fun and recreation, with everyone in a happy, joyful mood, hurrying to finish last minute plans.

Preparations began weeks earlier.

At school, we planned, practiced and put on our Christmas program the last evening of the last day, as everyone in our community packed the school room on rows of temporary plank benches.

Our Sunday School Christmas program took another evening, with a similar format. We sang all the verses of the Christmas carols by heart. And always as we filed out the door, each child received a brown paper sack holding a handful of hard candies, nuts and an apple or orange. As we reached the age for Methodist Youth Fellowship, we caroled around town and planned our holiday party.

At home, we fashioned gifts for one another and shopped with care, hiding secrets in our rooms and forbidding entrance. We baked cookies and made candy—fudge, divinity, caramels and popcorn balls. Mom kept busy on the sewing machine late into the night, sewing our Christmas dresses, doll clothes and other gifts.

We attacked all the jobs needing to be done in the day or two before Christmas with a sense of pleasure and anticipation, cooking and decorating the house, already spotless, with evergreen boughs and pine cones, silver balls and candles.

Warm smells of juniper, pine and spices permeated our home.

*Mom and Dad enjoy the first electric lights on a juniper Christmas tree cut from our hills. Before 1951 when we wired for electricity, red candles lit our trees.*

On Christmas Eve day we set up and decorated the tree, fresh-cut from our hills. Our special evening meal featured lutefisk from Dad's Swedish heritage or oyster stew from Mom's.

Afterward, we lit the candles on the tree—a couple dozen slender three-inch red candles set in little tin holders pinched onto a branch or twig. We delighted in that fairyland for a half-hour or so—while

Mom and Dad watched the flames carefully because they were a clear fire hazard. Evergreen sap flares up suddenly, exploding into an unbelievably hot and fast-burning fire.

We eyed the gifts under the tree, but saved them to open in the morning as that was our tradition. Some of our holiday customs may have been Scotch-Irish or early American, since Mom's ancestors were around many generations, from before the Revolutionary War.

At last the great day arrived and we opened our gifts. Mostly they were practical or homemade—scarves, socks, pajamas—but also games, crafts, a necklace, maybe a doll or another toy. A homemade gift that Anne treasured and added onto for years was her doll house that Mom and Francie made from two wooden orange crates, one atop the other, fashioned into four rooms decorated with wallpaper and scraps of cloth, windows clipped from catalogs and cardboard furniture.

And always under the tree we each found a book.

We were not taught to believe in Santa. Mom discouraged this as essentially untruthful—and confusing for children when they learned the facts. Better to teach the Christmas story, undiluted, she said. Perhaps too, in those difficult times, she believed it more valuable for children to recognize their parents' love in gifts they gave—rather than to think Santa magically brought them.

We feasted bountifully at our Christmas dinner. Aunt Margaret and Uncle Chet joined us in celebration. The noon meal featured a festive table filled with delectable dishes, a tablecloth over our usual oilcloth covering. Mom baked ham, or sometimes beef or venison roast, wild duck, goose or turkey—if Dad went to a shoot he invariably won a turkey—along with her good sage and celery bread stuffing and flavorful gravy.

Once Francie brought home a live white goose won in a raffle or shoot for Christmas dinner. But as we lifted the burlap sack from the car, the goose took flight with a loud squawking and flew, low and fast, down toward the Yellowstone River, a mile and a half away. We searched for hours, but never saw it again.

We helped mash potatoes and set out cranberry sauce, fresh homemade bread or buns and stuffed celery sticks with pimento cheese and sliced olives. Canned vegetables, fruits, pickles, beet relish, jams and jellies came from our well-stocked basement shelves. For dessert Mom made her traditional English suet pudding, rich and delicious, served with caramel or hard sauce. Plenty of leftovers for an evening meal, along with mincemeat or pumpkin pie.

Always on Christmas morning we gave our livestock special care. We girls helped with the outdoor winter chores in the cold crisp air, our overshoes crunching in the snow, racing and romping with the dogs.

Our animals all got extra food and attention in celebration of Christmas—cattle and horses in the pasture, milk cows in the barn, calves and sometimes lambs in the corrals, hens in the chicken house, pigs in their pen and of course, our friendly dogs and cats.

Two weeks' vacation allowed plenty of time for fun and leisure, enjoying our gifts and creative projects, visiting with friends, sledding, skating, hiking, making snow angels and, if the snow was right, snowmen and snow forts. As we grew older we attended and planned parties. Evenings we played cards, board games and Chinese checkers and read around the kitchen table.

Afternoons, during one bitterly cold vacation when we had to stay home, Dad read aloud to us the entire book of Robin Hood. Another year Beverley came down with the measles and spent those two weeks, as her rash healed, putting together a coyote skeleton for a biology project. She boiled and scraped the flesh off the bones and, with Dad's help on a light welded structure of rods, she somehow got all the bones glued back together again—every segment of the tail and all the tiny foot bones. Her coyote served as an object of student interest and study in the Custer County High School biology room for many years.

As vacation ended, we always anticipated going back to school.

Under a big pine tree, Buck and Pedro stamped their feet and snorted as Beverley and I ate our lunch.

"Mmmm—it smells so good up here in the pines." Beverley breathed deeply. "It's nice here."

A gentle breeze swished through the long pine needles. A couple of hawks swept low over a rocky outcropping on an opposite butte. When finished eating, I climbed the tree. Pines just naturally invite climbing, their branches solid and spaced just far enough apart so we could easily reach up to the next and pull ourselves higher.

I found a sturdy spot to sit against the trunk and called to Bev.

"Come on up. It's nice here."

Together we sat on a large branch surveying the landscape and the Christmas tree prospects.

"You're right about the tree tops," I finally agreed. "We need to cut our tree off the top. Nothing else is going to work."

"Yeah. Looks like a lot of good trees over there… We haven't been on that side."

We rode on, focusing now on tree tops. Finally we selected a forty-foot tree, climbed up and cut off its top, taking turns sawing and chopping, allowing it to crash to the ground. I hesitated to leave it standing there mangled, but in fact, it looked just fine without its top.

We tied the tree behind Buck at the end of a long ketch rope snubbed to the saddle horn. Buck didn't like it much, side-stepping and prancing, trying not to trip over the heavy, solid trunk as it rolled and dragged the rope over Beverley's thigh and down across his hip.

"Behave, Buck!" she jerked the reins and he straightened out.

But the tree kept rolling over and over, breaking off small dry branches.

"Maybe we can string it between our horses on two ropes," I said. "Then keep the good side up."

We stopped to realign the tree. Butt end first, it pulled evenly on two ropes—when the terrain allowed us to ride Buck and Pedro parallel at the right distance apart.

No more rolling and we set off again for home as fast as our horses could manage. We dragged that tree over many miles of rough country across rocks, brush and gravel hillsides.

We still had a long way to go.

The sun dropped lower in the sky and the air grew chill. It gets dark early in December—the shortest days. We shivered, zipped up our warm jackets, tied on scarves and pulled on furry caps and mittens.

"Look at that gorgeous sunset!" said Beverley, watching the sun

*Buck and Pedro stamped their feet and snorted as Beverley and I ate our lunch.*

Western Ranch Life in a Forgotten Era

## Holidays

We are a family that celebrated holidays. Anyone with a birthday got a small pile of presents by her plate at breakfast, gifts from each of us. Mom made a two-layer birthday cake with frothy white frosting, lemon filling and candles, and inside she inserted surprises for good fortune—a ring, a dime and silver thimble—for first married, richest and last married.

April Fool's Day caught us by surprise—Dad came in from the barn with an exclamation that made us run to look out the window for a baby deer or coyote. And Mom fixed something strange to eat, once a pancake for each of us with a cotton milk filter baked inside.

On Easter we hunted colored boiled eggs in the lawn and donned our new Easter dresses for church. The Fourth of July brought picnics and rodeos. No trick or treat candy for Halloween; instead we hiked over to the neighbors looking for mischief-making possibilities. Once we horrified a bachelor neighbor by painting flashy slogans on his shiny new tractor—it was only the new washable poster paints, but he didn't know that. For Thanksgiving we ate turkey if Dad went to a turkey shoot—a good shot, he usually brought one home. Grandma, Aunt Margaret and Uncle Chet celebrated special meals with us.

drop over the horizon in a blaze of color.

Then the cold settled in. Stars blinked on, one after the other, filling the sky. Coyotes howled and yipped from one ridge to the other on opposite sides of our trail.

Finally we rode down the last hill under a cold white moon. We dragged the tree up on the porch and turned it over in the light.

It was completely bare on one side, all branches and needles gone, the wood polished to a russet sheen.

"Oh my," said Mom. "That's a…big tree!"

It looked a lot bigger up on the porch than out on the hillside.

"How come it's so bare and flat on that side?" Jeanie challenged. "Are we supposed to hang it on the wall?"

Dad laughed heartily. "Actually that's a good thing. If the backside reached out as far as the front branches we'd have to set it in the middle of the living room."

Next day we sawed two feet off the trunk and we set up our spectacular towering Christmas tree, the largest we ever had. Our silver star grazed the ten-foot high ceiling.

We decorated with our small collection of delicate brightly-colored glass balls, saved and cherished from years past. Holes in the tree filled magically with strings of alternating popcorn and cranberries, pine cones and popcorn balls and streamers of silver tinsel. Finally, out on the tips of the branches we pinched on the tin candleholders with their tiny clamps and fitted in three-inch red candles.

Pine fragrance filled the house.

Late that afternoon—Christmas Eve—snow drifted down in lovely big peaceful, lazy flakes, laying a soft blanket three or four inches deep over the landscape.

A white Christmas!

That evening we lit the red candles under Mom's watchful eye.

The beautiful pine stood shimmering in its flickering candlelight—all ready for Christmas Eve.

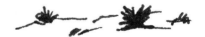

# 5 Riding Summer Ranges

## Comet Runaway

"This isn't working," I told Anne after several attempts trying to cut out two of our cows with calves that had crawled through a neighbor's fence and mixed with their cattle.

"I know," she threw me a sympathetic look.

"I need to ride Buck. You'll have to ride Comet."

"Me? Ride Comet? I don't think I can." She threw me a look of horror.

"Sure. You can do it. Just keep her back at the rear and hold the bunch. That's all you have to do."

"I'll try," she quavered.

Missing a couple of cows at our Brandentaylor summer range, Anne and I set out that morning from camp to ride the upper reaches of Ten Mile Creek. This creek meandered through ten miles of rough country—sagebrush draws cut by steep gumbo gulches and washout holes. Normally dry, except for two or three muddy waterholes, it made fencing difficult because of the rushing water during a cloudburst that cut steep banks.

I saddled Comet. Anne, who was a small age twelve that year, rode Buck, our steady and dependable cow horse, though a bit pokey at times—sometimes we called him lazy.

I enjoyed riding Comet, my big, black Thoroughbred with blaze face and intelligent eyes, especially if we had a long ride ahead, which we did that day. With her great long stride and fast easy lope, she was a joy to ride—most of the time.

However, she came into our lives with all her racehorse baggage. No patience with cows—she'd trot, or even walk, right past them, pretending they weren't there—sabotaging every cattle cutting task.

Anne and I rode the neighboring pasture for a few hours, following narrow cattle trails through cut washes of feeder creeks, checking each bunch of neighbor cattle. Finally we saw in the

*"Me? Ride Comet? I don't think I can." Anne threw me a look of horror.*

Western Ranch Life in a Forgotten Era

distance two familiar looking cows. As we got closer and they raised their heads, we identified them as ours by distinctive ear notches, a quarter cut on the left ear and straight crop on the right, as contrasted with the neighbor's double crop. They also carried our Quarter Circle W-Lazy-A brand on the right hip.

Comet, tall, lanky, determined, and full of nervous racehorse energy, disdained chasing cows.

First we tried to separate our cows and turn them toward the gate, but they'd already made friends and didn't want to come. Anne was too small to do much cutting on Buck—he needed a swift kick to sharpen up—and I wasn't having any luck at all with Comet. I took her that morning because we had lots of range to cover, but what I hadn't figured on was cutting cows and calves out of a difficult neighboring herd.

Nothing for it but to take them all to the corner gate and sort them there.

We crowded the cattle into the corner and I eased alongside them to the gate, slid down and opened it, with Comet prancing back and forth at the end of her rein.

Back in the saddle, I called to Anne, "Bring them up and I'll cut them out at the gate."

"Okay."

Anne moved them up nice and quiet, but Comet was dancing and throwing her head. She refused to notice the cattle right under her nose. Instead of easing ours out the gate, she stamped her feet and snorted at them. Our cows weren't helping. They kept turning back checking on their calves, letting the neighbors push ahead.

I'd bring them up again, get the gate open and go back to chase them through—and Comet would spook them all back past Anne and Buck again.

To complicate our cutting area by the gate, a small, rugged butte crested with rocks and sagebrush rose up right in the middle of it. The cattle ran first around one side, then the other.

Anne could hardly hold them from both sides of the butte, and when she did, they tried to run past me out the open gate. The trick was cutting our cows and calves through the opening without the neighbor cattle escaping.

The problem was Comet and her refusal to do her job. At the same time, Buck resisted getting kicked into high gear. We needed help, but this was remote country, far from any human habitation.

After the cattle had run back several times, I looked over at Anne and told her we needed to change horses.

She didn't want to. But Anne was the best little cowgirl in the whole world and anything I asked of her, she always attempted it, doing her best. A good scout even when the job was too hard for

> As they raised their heads, we identified them by our distinctive ear notches, a quarter cut on the left and straight crop on the right.

her—which it often was.

She slid down off Buck and handed me the reins. We didn't take time to switch saddles—a mistake, of course, as her short legs didn't reach Comet's stirrups.

By the time we changed horses the cattle ran back to the sagebrush flat, so again we started over.

Comet danced and fidgeted while Anne held tight to the reins. She knew that to keep a horse from bolting, she could pull one rein shorter, forcing the horse to run in a circle. It helped to have both feet in the stirrups, but Anne's didn't reach. She did her best with feet between the leathers as we brought the cattle back into the corner.

"You're doing fine. Now all you have to do is hold them here," I assured her.

Buck and I wove in and out among the cattle, separating out our two cows and calves.

*Perfect!* They stood by the gate waiting for me, with the others turned back.

I hurried to open the gate, seeing Anne was in trouble. Comet jumped and threw her head as Anne pulled her around in circles.

"If you can't hold her, just turn her up the butte," I shouted. "That'll slow her down."

This was Anne's best trick when she couldn't hold a horse.

Suddenly, she turned and up the butte they charged. Comet leaped toward the top at full speed, scattering big rocks, sagebrush and gumbo. Anne's feet flew in the air. She gripped the saddle horn and hunched her shoulders and hung on.

*Comet's getting battered by the rocks—she can't keep that up! She'll stop before she gets to the top. Or will she?*

But I watched unbelieving. Comet didn't even pause. She reached the top and plunged right over, coming down at a dead run. Rocks, chunks of gumbo and stirrups flew through the air.

Anne leaned forward, emitting little yelps, hanging on for dear life and squeezing her boots tightly against the mare's flanks.

At every jump I expected to see her flung off head-first on a rock, falling under Comet's flying hooves or, worst of all, tumbling down the boulders in a horse and rider entanglement.

Comet didn't buck at the bottom, as would many horses, but she raced through the middle of the cattle, mixing and scattering them in all directions.

"I can't stop her!" Anne called out.

Comet came to a halt by the gate—right where she wanted to be. Ready to race for home.

What an astonishing sight I'd just seen! The two of them lung-

## Racing an antelope

Covering ground fast and smoothly as a true Thoroughbred was Comet's primary talent. A high-spirited, honest, no-nonsense horse, she never played tricks or tried to outwit her rider. She ran away at times, but never bucked, not even once.

What she liked best was just to run wide open. She disliked being pulled up, lathering to a great white sudsy sweat that covered her sleek black coat from head to hocks, as she danced and fought the bit. Most of all she loved chasing horses, which we didn't do often, but when we did—it was great fun.

Once she even raced an antelope.

Coming around a rocky butte, Comet and Francie met the antelope face to face on the trail.

For an instant all stopped, shocked to be so close to each other.

The antelope snorted and leaped away down the hill. Comet followed, racing flat out, her head reaching out ahead as if on the racetrack.

It happened so suddenly Francie lost her stirrups and fully expected to go flying. Never had she raced so fast before, much less down the side of a rocky ridge. She tried to stop Comet, but it took a quarter-mile to get her pulled down to an indignant trot.

The antelope disappeared over a knoll.

But just before Comet slowed down, Francie noticed that while they weren't gaining much on that fleet antelope, they weren't falling behind one bit either. And on his turf.

If she had another quarter mile, would she have caught up? More to the point, if Francie flew off in a cactus patch or pile of rocks would Comet have stopped? Or gone on to finish the race?

ing fiercely straight up to the top of the butte and leaping back down again in a mad clattering of dirt, rocks and stirrups.

Anne sat there, still astride the big horse, now standing calmly by the gate.

The crisis over, I laughed with relief.

"Great job, Anne! What a ride!"

She threw me a wan look. Then she cracked a tight smile and finally relaxed into a grin.

We finished the job with Anne on foot, leading Comet firmly by the reins. Soon our two ornery cows and calves dodged out the gate and up the hill to join our cattle.

As we turned back to camp, I couldn't help enjoying Comet's long-striding gait, though she had behaved outrageously. Anne happily rode good old Buck, perked up and giving her his best now that he headed toward the corral and his feeding of oats.

*After her wild ride on Comet, Anne is relieved to be riding dependable Buck.*

## Snake Bite

"Anne, get in the barn and get your stirrups shortened," Dad called to me.

It was then that I had a serious run-in with a snake.

Twice. Same snake.

ANNE

We were branding calves at the Brandentayler ranch we had leased for the summer. This place, double the grazing we needed for our cattle, allowed Dad to pasture cattle from the State Industrial School for Boys at Miles City. This meant we girls rode for their cattle, too, when we looked after ours. When branding time came, several Ag men and teenage boys arrived from the school to help brand.

Dad spent the day before repairing the large but inadequate corrals at this place, but we still had trouble holding the cattle. Francie and I had the job of keeping the branding fire going, the syringe filled for vaccinating, and otherwise helping keep things running smoothly while Dad branded and worked the calves. The boys brought calves up the chute and stretched them out on the branding table.

Two men from the state school wanted to do the riding and handle the cattle on horseback—but they let the calves run past them. The portly head man moved too slowly, his stomach overhanging the saddle horn. As we learned early in life, those who said they could ride horses often couldn't.

The cattle were continually trying to break out of the corral and once in awhile one or two succeeded, taking unbranded calves with them. After a short time of the 'cowboys' trying to chase these errant cattle back into the corral, Dad decided the riders needed replacing. The men came in to work the chutes and Dad told Francie to get on one horse and me on another.

One of the men had used the saddle I always rode and lengthened the stirrups four or five inches. I had to unlace the leathers and reposition them, a ten-minute job.

He unsaddled in the old barn and conveniently tossed my saddle over the edge of a manger—a much handier place to shorten the stirrups than on the ground. Only it was dark in that barn and as my eyes adjusted to the dim light I worked mostly by feel.

"Hurry with your stirrups and get out there to bring in those calves," Dad called.

I worked as rapidly as possible, knowing Francie was by herself trying to hold the cattle. The two of us worked together like a team, Francie always said. Even though I was young and small, I could anticipate her needs and give her lots of support.

But my! It was dark in that barn.

As I fumbled in the blackness, I dropped a stirrup and reached down to pick it up. I felt a sharp pain on the knuckles of my right hand, and figured they'd been scratched against the rough old manger boards. As I glanced at my hand, sure enough, there were two closely spaced pricks about a half an inch apart. I ran my other hand over them, trying to feel if there were splinters from the old manger boards but didn't find any.

My knuckles weren't bleeding, so it couldn't be too bad even though they sure stung. Two little pricks—I tried not to think about them. Francie was out there by herself and needed my help.

*Why did those scratches hurt so bad? Must have been really dirty old wood.* I wondered if I might get lockjaw, a secret worry for us girls when scratched by a rusty nail. But there was no tetanus vaccine then, at least not in our budget. When we finished branding, I'd soak my hand in Mom's home remedy of warm salt water.

Outside the cows bawled incessantly for their missing calves, creating a noisy racket. I hurried as fast as I could. I

> *I worked as fast as possible, knowing Francie was trying to hold the cattle by herself.*

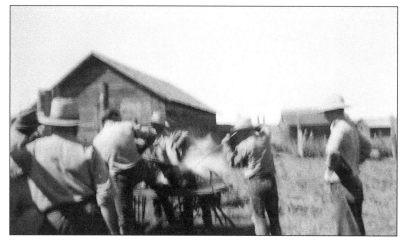

*Men and boys from the State School hold a blatting calf on the calf table.*

> *I kept myself from running or screaming—the increased flow of pumping blood would spread the venom.*

almost had the first stirrup in place, when darn! I dropped it again.

Again I scratched my knuckles reaching down to grab it. This time when I looked at the damage, I saw two more pricks on the next knuckle, exactly spaced as the first set of pricks. Odd.

Just as quickly, the thought struck me what had caused these evenly spaced pricks.

Snakebite!

I jumped back and frantically searched the barn floor in front of the manger. It was covered with dirt and dry manure mixed with old straw in a fifty-year accumulation. My eye caught a quick movement near the end of the manger as the tail of a dull colored snake slithered through a hole in the barn floor.

*Were there tell-tale diamond patterns of a rattler or was it a bull snake? Were there rattles on the end of that tail?*

I strained to see.

Too late, the snake disappeared in the dark.

It had made no sound. No rattling or hissing. *Or did I miss hearing the rattle? With the cows and calves bawling so loudly, maybe I just didn't hear it?*

Bull snakes also have fangs and they look much like a rattler. *Had I been struck by a harmless bullsnake—or was it a rattlesnake?*

In this country, people respected rattlesnakes and feared their bite—a few even died. Stacking hay, we often picked up rattlesnakes with the buck rake. Once, as the hired man stood on top of the stack, pitching the hay into the proper shape, a rattler came flying through the air toward him.

"Watch out, rattlesnake!" Dad yelled.

The hired man whooped and jumped backwards off the stack.

We always watched for rattlers. Which was why I stood there stunned, torn by deciding what to do. If I went out and told Dad, would he be angry because I let myself get bit, or because I was taking too long in the barn? Or if I ignored it and told him later would it be too late for help?

I inspected my knuckles.

Yup. There were definitely two sets of fang marks. I certainly had been bitten. Otherwise my hand looked fine. No bleeding, no redness, no swelling.

I left the partially finished saddle and walked slowly to the branding chute to find Dad. Mom cautioned us not to move rapidly if we ever got bit, since the increased flow of blood pumping through the body would spread the venom.

I kept myself from running or screaming. How often had we been told not to make a fuss, and to think before we acted?

Dad was busy with the calves, and as I hesitantly approached he stopped and appraised me.

"What happened?" he asked.

I thrust my hand at him.

"I got bit by a snake. Twice. I don't know what kind. It disappeared down a hole in the dark."

Dad quietly took my hand and looked it over carefully.

"It should be swelling by now if it was a rattler. Go back and search the barn," he said. "If you can't find it, watch the fang marks. If they start to swell come back and let me see it. Otherwise, get out there and help Francie."

I finished the saddle—taking care not to drop the stirrup leathers again—while taking horrified looks at my punctured hand, but it never swelled or turned red. So I saddled my horse and joined Francie, who gave me the sympathy I craved.

She and I kept checking my fang marks, but since there was no change, the branding continued.

"Come over here," Dad shouted at me a time or two as I rode near. He seemed pleased when he checked my hand.

The scars from those fang marks were still readily visible many years later. I was so relieved, whenever I noticed them, because it must have been 'just an old bull snake.' We learned not to sweat the small stuff, don't complain, but if there's a real problem, take care of it.

*Dad pulls the iron from the branding fire and holds it in the right spot for mere seconds to fix a neat, permanent brand.*

## Chuckwagon Camp

Through the dimness, beady black eyes of little mice examined us without fear. As we stood there in the schoolhouse, several inched closer, twitching the long hairs on their little noses. They'd probably never seen a human and were curious.

ANNE

"Ugh. How'd you like them crawling over us all night?" Francie whispered in a disgusted tone.

Dad pulled the chuckwagon out to our new summer range at the vacant Brandentaylor place and parked it near an ancient one-room school with its fenced yard.

The abandoned ranch house, big barn and many outbuildings stood nearby, in a terrible state of disrepair and grown over with weeds, since no one had lived there for quite awhile. Respectful of their property, we never ventured near and used only one corral for our horses.

The schoolyard suited our purposes fine. Once a playground, it provided long grass for our horses overnight. A hand pump well gave

## A visitor

When we arrived back at the chuckwagon after a day's ride Anne and Francie were always tired and hungry. Often we grabbed up one of the cans of Queen Anne cherries that Dad bought for us, a special treat only for our stay away from home checking cattle.

Sitting in the sun all day, the chuckwagon was hot, so we sat out on the wagon tongue with an open can of cherries between us.

True to rangeland tradition to keep a clean camp, we had dug a small square garbage pit a couple of feet from the wagon tongue. An occasional shovelful of dirt thrown on top kept odors and flies away.

It was a natural that we aimed our cherry pits at the garbage hole and turned it into a spitting contest to see

*(continued)*

### A visitor (cont'd)

who could hit the hole most often.

Late one afternoon when all the cherries were gone, we had just arrived back from a hard day's ride when, riding up through the sagebrush, came a stranger. A shy would-be swain, he was probably goaded into it by friends. After all, single girls were scarce in that country.

"Have you eaten?" we asked, as rangeland courtesy demanded—hoping we didn't have to feed him.

"No," was the dread response.

He came right in and sat at our empty pull-out table, looking expectant.

Anne's eyes grew big. She knew we had nothing to feed him, short of starting a fire and cooking a full meal. But we were so tired.

Inspiration! Fresh radishes just brought from home stood crisping in a bowl of water. They looked inviting: bold red contrasting with bright white where we had trimmed the ends.

*What a treat! Or were they?*

Anne thrust them at him.

He uttered not a word as he munched on the radishes. Apparently he was stunned speechless.

Neither of us had much to say either. We watched him eat them all.

After finishing he rose, not meeting our eyes, then tipped his hat and rode off into the sagebrush. The last we saw of him.

us fresh water and there were boys and girls outhouses—we used the closer 'Boys' with glee. A straggly row of Russian olive trees gave some shade, but the smell of those trees in bloom—a pungent, cloying odor of yellow flowers—imparted a distinctly foreign aura to the open range.

Apparently the original ranchers donated land for the school, eliminating a long drive twice a day to another school site. Likely the teacher lived and boarded with the family during the week and thus provided a little income for the ranch wife.

Dad said we could sleep in the schoolhouse instead of the chuckwagon if we wanted to clean the dust and cobwebs out of one corner. But we nixed that idea when we saw the mice and smelled the schoolroom mustiness.

The bed across the back end of our wagon was narrow, just right for a single man, but not wide enough for both of us, the width of our shoulders side by side. We usually slept on one side—the same side like spoons—and tried not to move too much.

Then we devised a better plan, with Francie's feet beside my head and my feet beside hers. We each had more room to move that way. As long as we kept our feet still, it worked—sort of.

And, narrow as it was, the chuckwagon bed was infinitely better than the schoolroom, so we left it to the mice.

Sometimes, though, I wished I had my library books to read. I had to leave them home in the middle of the most interesting stories.

Francie said we couldn't bring library books out to the chuckwagon camp for fear they'd get damaged. We couldn't always be totally clean there as we didn't take off our boots to come into the wagon or immediately wash our hands.

We were much in awe of the big Carnegie public library in Miles City where we borrowed books every week in summer when we went to town for piano lessons.

What if we returned books with barnyard smells or smears of rain, mud or manure? Would the sharp-eyed librarians refuse to lend us any more books?

*After all, look what happened to Australia!*

Francie hadn't wanted me to take *Australia* home from school that day, but finally gave in when she saw how much I wanted it. The book was one of an armload our teacher periodically brought out from the library—a second-grade level book that fascinated me with its photos and easy stories of strange creatures.

I promised I'd carry it all the way home. A thin book, it was still awkward for me to handle because of its big format.

All went well, until the next morning when we went to school. We varied our routes, exploring along the way, and this day we chose to cut through a corner of our neighbor's pasture, so we needed to cross the main irrigation ditch.

Usually without much water, that morning the ditch ran full.

We stood there at the edge contemplating the best way to cross. It was only about two feet deep, but fairly wide and muddy. If we only had Jeanie along, she'd jump the ditch easily, with the book in hand. But she had left earlier and taken the cut-across, in a hurry to get to school.

Francie knew she could jump across and I might try and get only a little wet. But how could we get *Australia* across? If she jumped with *Australia*, she might lose her balance and fall in, dragging the book with her. At seven, four years younger, I was sure to drop *Australia* in the mud or moving water.

Could I cross first and Francie throw the book to me? No. I'd be sure to miss the catch and the book would fall into the muddy water, maybe getting swept away.

The best answer seemed to be for Francie to jump across first and I'd throw the book to her. I had no doubt whatsoever that she would catch *Australia*.

Francie seemed confident in my ability to throw the book to her. I wasn't so sure, but it seemed the only way. If I didn't do it, would I ever get to bring home another library book? Little did I know *Australia* would be with us a long time—like the rest of our lives…

The big moment came. Francie successfully cleared the ditch and stood waiting on the other side.

"Come on! You can throw it!" she urged.

I hesitated. Francie took her best first-baseman stance and held out her hands.

How could I miss? Giving a mighty heave, I hurled the book.

*Ooops! Should I have thrown it spine first?*

The book launched perfectly above the ditch. Then, oh horrors! It spread open and the fluttering pages caught the wind, stopping its flight. Down it went—far from Francie's outstretched arms.

Disaster.

*Australia's* pea green cover hit the murky water with a smack and sailed into the current.

Quickly, before it sank, Francie pushed her shoe into the mud, reached out and retrieved my book.

What a mess. We wiped it off as best we could and soberly continued on our way to school. Our stern teacher that year sent a note home to Mom. Her daughters had ruined the book. We must buy *Australia*.

That book sat for years on a book shelf in our home, pages

*The chuckwagon, our little home away from home, offers a narrow bed under the window across the back, benches along both sides, a pull-out table under the bed and small cook stove beneath the stovepipe.*

## Dolling up the chuckwagon

Dad acquired our little chuckwagon in a run-down condition. Originally it was a cook wagon for early-day cattle roundups, built in the sheepwagon style with canvas cover.

Dad rebuilt it for us to use trailing cattle and overnight stays on our summer range, putting truck tires underneath. He stretched new canvas over the top. Helping him, Anne got excited about this tiny little house.

"Can I take the chuckwagon as my 4-H Home Living project?" she first asked Mom and then her 4-H leader. It must have given the leader pause to hear this odd request. Usually a member first took her bedroom to redecorate.

Getting her okay, Anne set to work. First she sewed curtains of red and white daisies for the spring rods on the little front door and single back window. With Mom's left-over calcimine, she painted the wooden benches that ran along both sides. Behind the small cook stove she helped Dad build two rows of shelves for canned food and painted them, too.

Across the back end was a narrow wooden plank bunk bed, built for one adult. It was where Anne and Francie slept—one at each end. Smelly feet, but it gave each of us more shoulder space and room to turn.

The table was a wide plank that slid out from under the middle of the bed.

"How am I ever going to get this clean?" Anne asked Mom. The wood plank table, dished in from years of wear, was crusted over and

*(continued)*

rippled and water marked, pea green cover stained with brown mud splashes we could never remove.

Always I could hear Francie's ringing accusation in my ears. "If we had to buy a library book, why couldn't it have been a *good* one? Why did it have to be *Australia?*"

We certainly inherited Mom's love of reading and learning. At our chuckwagon I came to understand her thirst for reading. She told us about the winter she spent at one of their sheep camps out on the Big Dry near Jordan. With Dad gone most of the winter at line camps, she was left tending the hospital band—the sick and injured sheep they kept in a small corral.

Her only reading material was Jehovah's Witness tracts left by zealous missionaries. She became an expert on their doctrine and after that pointed out fallacies in their Biblical interpretations when they came proselytizing to our ranch door.

Mom also taught us a deep respect for the books themselves. We never turned down the corners or marked up the pages—or brought library books to our chuckwagon camp.

Once we went back to the old schoolhouse and found a few old papers to read and a dusty, musty novel with the back partially nibbled away by mice. Francie read the novel. But she said the first half was dull with its long old-fashioned descriptions and, just when it got interesting, too much was missing from the mouse nibblings to discover how it ended.

So we sure did miss reading that summer when we were tired and it was too early to go to bed. When we'd read everything available, we sat out on the wagon tongue and watched the mice scurrying through the grass watching us.

We never seemed to have any in the chuckwagon. When Dad rebuilt it, he managed to get it tight enough to keep out mice. We were careful to keep the door closed so they wouldn't run up the wooden wagon tongue and come inside. Our food stayed in tins or glass jars.

I don't remember ever having a mouse inside the wagon.

*Except once.*

That night, exhausted from a long day, we were glad to be snug in our own crowded little bed across the back end of the wagon. We both quickly fell asleep—Francie's feet at my head and my feet at her's, trying not to kick each other in the face.

Morning light woke us. I was lying on the outside nearest the door, with Francie behind me against the back canvas.

"Let's get up," she said, all too soon. "You first—or should I crawl over?"

I groaned, hoping she'd snuggle back into her pillow.

She was wedged behind me, so if I didn't get out first, she'd klonk me in the head with her feet as she climbed across. Sleeping one at each end had its disadvantages.

I sat up, careful to whirl my legs over the side. Pausing there a minute, I glanced back longingly at my cozy bed.

*But, oh dear!*

There in the very middle of the pillow indentation lay a flat round pink object. I could see it had been a living being.

"What's that?" I squealed at Francie.

I examined the object, careful not to touch it.

"It's a nude baby mouse!" I cried.

A pink hairless baby mouse was curled up tight—about the size of a fifty-cent piece and squashed just as flat.

I recoiled. All night I'd slept on my back with a naked baby mouse under my head. I had killed it and squashed it flat with my head.

Francie twisted around for a closer look as I sat there horrified. She broke into gales of laughter, setting the chuckwagon and the dead little mouse to shaking.

Avoiding her feet and my flattened baby mouse, I gingerly escaped the bed.

While I was rubbing my head and grimacing, Francie blanched and pulled my hand away.

"Look what you're doing!"

"Yikes!"

My hand was smeared with slime and ick from the squashed mouse. I ran out to wash under the schoolhouse pump, rubbing a bar of soap through my hair.

Carrying the baby mouse gingerly in a Kleenex to the garbage hole, Francie was trying not to laugh.

But suddenly a strange expression came over her face. She ran back into the wagon and started throwing back the covers on her end of the bed.

"What are you doing?" I asked.

"Where," she cried, "are the rest of the babies?"

## Riding the Wide Open

"Oh no!" I groaned as the car started chugging—then stopped dead.

Dead. Our 1942 Ford came to an abrupt stop on a curve halfway up the dread Ten Mile Creek hill south of Terry, as Anne and I headed up to the Brandentaylor place.

---

**Chuckwagon** *(cont'd)*

well-greased with particles of years-old food. Camp cooks were none too clean. Dutch Cleanser and plenty of elbow grease got off the worst of the embedded stains.

To enter we walked up the long wooden wagon tongue, used for a hitch when moved, but slippery when wet. The door opened with top and bottom half-doors.

The chuckwagon became Anne and Francie's home-away-from-home for several summers—a forerunner of modern campers. After working cattle all day, it was a treat to spend the evening under lantern light in our snug little home on wheels.

*There in the pillow indentation lay a flat round pink object. I could see it was once alive.*

Western Ranch Life in a Forgotten Era

Hoping to make it up this long, steep, curving hill before the car vapor-locked, I had sped across the last valley and we fairly flew across the Ten Mile Creek bridge.

Now we were stalled. The motor didn't even turn over.

A hot, hot afternoon, heat mirages shimmered off the badlands in the far distance over the Yellowstone River.

I shoved the car into first gear, hoping to prevent its backward slide and set the emergency brake. But we couldn't depend on the brakes. Often as not, they didn't hold.

"Here, I've got the water jug," Anne thrust the gallon jug into my hands.

She propped up the hood and waited expectantly for me to pour water over the water line and fuel pump the way Dad had done last time this happened.

Where did she get the idea I knew what to do? I had no confidence in myself or that method. I knew cold water could crack a hot engine.

Mostly I worried that we were stuck in the middle of this gravel road in near one-hundred-degree heat, couldn't see around the curves up above in case someone came barreling down on us, and at any moment the car could slide backwards toward a drop-off on either side.

But this had happened before as the motor toiled uphill in hot weather, and Dad poured on water.

So I carefully poured a little stream of water over the line. If this didn't work we'd have to sit in the blazing sun and wait for the engine to cool.

"At least we don't have the truck—with cows and horses in the back," Anne said, ever looking for the positive.

Last trip, driving the truck, bringing out a heifer and our young horse Gypsy for some on-the-job training, I'd felt the livestock lurch and sway on every curve. We couldn't take a run at Ten Mile Creek hill for fear of knocking them off balance. Part way up the hill, the truck chugged and lost power. I tried desperately to shift down to a lower gear, with one foot on the gas and the other pressing the clutch. The gears ground and wouldn't shift. I was frantic, knowing if it came to a total stop I'd need a foot on the brake, another on the clutch and one for the gas.

*How can I handle that?*

I had glanced over at Anne. She was gripping the door handle and it made me laugh.

"Don't jump—yet!"

*What if we start rolling backward and the truck tips off the edge?* If it slipped out of gear and the brakes didn't hold we could shoot

*Dad tightens bolts on stock rack, getting truck ready for hauling bulls to summer pasture.*

straight backwards down the long winding road with its abrupt curves and steep drop-offs.

Anne had grinned in relief as the gear caught and we gained power.

We made it up the long hill okay that time.

Miraculously, this time the cool water did the trick and we made it again without having to back up.

Our job that summer was riding herd on our cattle at the Brandentaylor place, a large and remote pasture of rocky buttes and prairie cut by dry creeks.

Every two weeks, Anne and I drove out there and stayed a few days, making sure none of our cows or calves were missing and all our bulls were in the pasture and working.

Francie and Dusty get a count on grazing cattle, checking to make sure all calves 'mother-up' with their moms.

*Any cows lame? Caught in barbed wire? Rattlesnake-bit with a swollen jaw? All mothered up with their calves? Any stray cows that didn't belong?*

Grass grew lush and green in the lower parts of draws, but somewhat sparsely elsewhere. We rode each coulee searching out our cattle. Then, making sure all were accounted for, with plenty of grass to eat and water to drink, we rode on.

There were no telephones to call Mom or Dad if we had trouble. We were truly on our own.

Fences could be a problem in that rough country, where wires stretched across cutbanks and washouts. Flashfloods in a sudden summer cloudburst often left those fragile fences sagging in a tangle of debris that cows could crawl through.

Dad's friends the Olson brothers, Orlando and Orrin, who cared for our cattle the summer before, owned an adjoining pasture where they kept their yearling heifers. Because they bred them as two-year-olds rather than yearlings, to get more growth on their cows, they didn't want any bulls in there.

Once Anne and I found a lost bull of ours in that pasture. We chased him out quick and repaired the broken fence. But apparently we weren't fast enough. Next spring a couple of bright white-faced calves born to young heifers told Orlando our bull was there.

Riding another rugged pasture of theirs a couple of weeks later looking for a missing cow, Anne and I rescued a big half-grown calf that had dropped into a four-foot-deep washout hole.

As we rode through a sagebrush flat that day, I happened to catch a flash of movement from the corner of my eye. Then…nothing. Only the expanse of giant sagebrush tall as our horses and long

**There were no telephones to call for help if we had trouble. We were truly on our own.**

## Hunting the brand inspector

After each session of riding that summer range at Brandentaylor's, with some relief that we hadn't failed our mission, we set out for home.

But when we hauled back a sick cow or calf we couldn't just drive off down the highway.

First we had to hunt up the county brand inspector in Terry. To cross the Prairie County line with livestock in the truck we needed a brand release, as Montana long ago enacted tough range laws to thwart cattle rustling.

Highway patrolmen could stop us and check our permit anywhere along the road. This gave us pause. But they never did. Maybe they knew of us. Or maybe they weren't much concerned that teenage girls with a Custer County license plate would rustle their neighbors' cattle.

When we found the brand inspector, he climbed up on the stock rack and checked the brand on each animal. Anne and Francie watched uneasily because if he couldn't read one, we'd have to unload at the sales yard or railroad corrals so he could clip the hair and read it from the scar on the hide.

To complicate matters in locating him, he spent many hours away from his office. It seemed he often spent that time 'gathering information' in Young's Bar and Cafe or the Yellowstone Bar next door.

Young's was the local rancher hangout, with the bar open to the street and a cafe attached behind. In those days, only a certain type of woman went into a bar. We girls were embarrassed walking through to reach the cafe where we

*(continued)*

grass cut through by a meandering mostly-dry creek bed with steep banks.

"What was that? Did you see something?" I asked Anne.

"Where?"

Just then another flash and a tall sagebrush whipped wildly for a second.

"Over there. Something…"

Anne stood up high in her stirrups, searching the brush.

"We better take a look."

We approached cautiously—and at first saw nothing but big, healthy sage all around.

Then—a heart-rending sight.

Imprisoned in a deep washout hole stood a gaunt Hereford calf. The ground was bare of grass for six inches or so all around the hole.

He regarded us with big sad eyes from his steep-sided gumbo hole. Then as we watched, he leapt up clawing with front hooves and flicked out his tongue to pull a thin spear of grass toward his mouth before tumbling back into the hole. By jumping up he had pulled out the grass to its roots as far back as he could reach.

"Can we lift him out?" Anne studied the forlorn calf, watching us hopefully.

"He's half-grown—he's too big. He'll kick."

But if we left the calf to ride back and get our car, then drove the many miles around to Olsons, it might be too late. Besides, we weren't sure we could find that spot again in the broad sagebrush flat riddled with hidden washouts.

"He's so hungry. And he must be terribly thirsty," Anne said.

"How many days do you suppose he's been in that hole without water?"

I was remembering a hot summer day when I was desperate for water. My horse had taken off for home and I walked seven or eight miles. I had followed a dry creek to a small stagnant pool of water. There, with throat so parched I didn't care, I sipped dark red water from a cow track.

The calf stared at us, desperation in his eyes.

I jumped into the hole, but no way could I lift this big calf. Anne slid in, too. Together we fit the hole with no room to spare and heaved, but could not lift him.

He kicked us—hard—but we got a loop of our ketch rope around his chest behind the front legs.

Luckily Anne was riding strong, dependable Buck and we snubbed the other end on the saddle horn. With Anne leading and Buck pulling, I lifted and pushed from behind.

Eager to help, the calf jumped at just the right time, as we exerted our greatest effort, pushing and pulling and Buck lunging ahead.

Surprise! Suddenly he scrambled up the wall to freedom.

*"Blaa—aat!"* He looked around, stunned.

His mother came running, her bag tight with milk. In seconds, he was guzzling and burbling, bucking against her lustily. One happy calf.

The mom lowed gently and licked his flank.

Another day, I lost Anne after we split up to check cattle in opposite draws off a long ridge. We agreed to meet at the bottom, but when I reached there she hadn't come. I started back to see if she'd run into trouble and came upon Buck tied to a tree.

Topping the next rise, I saw a strange sight.

There lay Anne in the tall grass, waving her hat on a stick in the air. Back and forth, back and forth, she waved, while watching something just over the hill.

"What are you doing?" I whispered, crawling up alongside her.
"Look!"

Anne had discovered two antelope grazing upwind from her, and was trying to lure them closer.

"Guess what? It's true what they say—antelope really are curious. That buck came almost right up to me."

She continued waving her hat, but they saw me and shied away.

Antelope were just beginning to repopulate the more remote plains of eastern Montana, after being hunted almost to extinction.

True creatures of the open plains, antelope rely on sharp eyes and speed to evade predators, rather than hiding like whitetail deer. Good defenses in nature, but it made them vulnerable to the rifles of hard-up homesteaders, who killed every antelope, mule deer and sage grouse they could find, often as the only meat to feed their families during desperately dry years.

For Anne and I, that summer was our first close encounter with antelope and we were excited to see them, fascinated that they seemed as curious about us as we of them.

One morning we rode in a new section of our range, on high rocky ground with a smattering of pine and cedar trees, keeping a lookout for the homestead Orlando Olson said we'd find in that area.

"Look at those gorgeous buttes!" I said, breathing deeply of the clear, fresh air with its scent of pine as we checked the draws for cows.

"Yes, see the colors!" Anne pointed, "Purple, red, rose-pink, grey, blue, black…"

> **Brand inspector** (*cont'd*)
>
> could ask about him. We carefully averted our eyes from the men perched on bar stools. Since we were strangers, all eyes followed us.
>
> Terry is a small town and everyone seemed to know where he was.
>
> "Next door," was often the response, to our dismay.
>
> Next door was the dread Yellowstone Bar with its reputation for catering to hard drinkers. To us this was an immoral establishment that we didn't have the courage to enter, even to linger in the doorway.
>
> As we stood stunned, looking at the floor, one of the men at Young's soon realized the dilemma we faced, and volunteered to go get the brand inspector
>
> What a relief to finally have the brand release and get on our way home!

*We discover antelope, as curious about us as we are of them.*

Washed clean by an early morning rain shower, each vein of color gleamed bright, set off by layers of slick gumbo. Small slices of glass-like mica glittered across the face of the nearest butte in the sun.

A meadowlark serenaded us with his piercingly lovely melody from the tiptop of the tallest pine as a porcupine scuttled behind a juniper tree. A stray breeze swished through the long-needled pine branches and touched our faces.

We rode in nearly inaccessible range country, unpopulated, with no ranches and almost no roads.

Still, we knew that some thirty or forty years before, homesteaders claimed much of this land in 160-acre quarter-section plots, too small to be of any use in range country. They plowed the required acreage, planted fields, scraped together meager harvests across these hills. They worked and struggled, caught up in one calamity after another.

Homesteading here—a tragic experiment in an unforgiving land.

*What a disappointment for mothers and fathers who came with bright hope in their hearts and hit years of drought.*

Even as small children we felt the sorrowful tug of such places, scattered here and there through eastern Montana cattle ranges. Knowing hard times and dry years ourselves, we wondered how any family could survive. In the homestead ruins untold families left their scattered records of courage against impossible odds, trying to comply with a law conceived in eastern states that never fit our drier lands.

"There's an old building down here," Anne called to me from across the draw.

Usually this was a signal for us to slide off our horses and explore another forlorn old homestead. Deserted during the dry years of the depression, or long before, a small house would be tumbled into heaps of splintered gray boards crookedly nailed together. Tangles of chicken wire twisted with rusty barb wire clung to a sagging fencepost. Straggly trees attested to someone's long-ago care and attention. Invariably, dingy cotton batting burst from an old mattress, half-fallen into a root cellar. Wild rose bushes grew through a rusty bedspring.

All of it overgrown with ragweed and sagebrush.

If we poked around we found the human clues that interested us most: a child's broken toy, a tiny chair, a battered homemade wagon, the ghost of a teddy bear, a granite tea kettle filled with holes, broken crockery.

Anne and I rode down there, but stayed on our horses by un-

*Francie and Star climb steep hill in search of missing cows.*

spoken agreement.

We knew what we were seeing.

"A log house in the pines," Orlando had told us.

Almost too painful to take in all at once, his story hovered over the place like a shroud.

An elderly couple died here some twenty-five years before. The one-room house was built of logs, more solidly than most, now tumbled. Pines clustered around its stone foundations.

They worked hard for many years, scratching out a garden, doing without—living in this beautiful location near a cedar draw. Then one bitterly cold winter the old man died. His wife dragged his body into a lean-to adjoining the log house where it froze stiff and stayed preserved until spring—when she expected someone would come by and help her bury him. Or so people surmised.

After a late winter thaw, when a rider did show up, he found her dead too, a frozen body in bed, piled with quilts.

Anne and I circled the house, gazing into the fallen logs. We imagined what it was like for the woman living there alone—with the specter of her dead husband nearby.

*Did she die from cold? Or sickness? Or unbearable loneliness and a broken heart?*

We rode on in silence. We'd save exploring for another day.

*We rode in nearly inaccessible range country, unpopulated, with no ranches and almost no roads.*

## Stallion Attack

Riding horses strange to us almost proved our undoing one early fall day when Jeanie and I drove out to Harold Cole's. Dad wanted us to pick up our horses and check our cattle a last time before trailing them back home.  [ANNE]

*"Maybe our horses will be in the corral or the small pasture behind the barn."* I hoped. But no such luck—and no one was home to help.

There were no telephones in that remote Sheep Mountain country, so the Cole's weren't expecting us, and our saddle horses were turned out in a big pasture mixed in with theirs. Ours liked to run when they were with other horses, so we figured we couldn't catch them with a bucket of oats as we did in our own pastures. We'd have to run them into the corrals. The only way to do that was to borrow a couple of Cole's corralled horses.

*But which horses?*

About a dozen of Cole's horses milled around in the largest corral. They raised horses for sale, and were breaking the ones kept in the corral—mostly green-broke—with a few trained saddle horses

*I flung myself violently past the horse's legs. His hooves missed my head by inches.*

mixed in. Jeanie had been away all summer and Francie and I did the riding in this rugged country.

"Do any of these look familiar to you?" Jeanie asked me. "We've got to have a saddle horse that can chase horses. I'd sure hate to catch a green bronc."

In the corral stood a number of pintos and two looked like the kid pony Coles' nine-year-old son rode. Both had saddle marks showing they'd often been ridden with a saddle. No others bore saddle marks and as we didn't want to risk getting a green bronc, we decided if we chose both those pintos to ride, we'd have a better chance of riding at least one gentle, well-trained horse.

We could hear the bay stallion stamping and nickering, locked in his barn, but didn't venture near. Mr. Cole often warned us he could be dangerous.

We drew straws to decide which horse each of us would ride.

Lucky me. My straw claimed the pinto with darker brown spots. He seemed calmer and I was relieved to think this tall horse might be the one referred to as 'the kid pony.'

Then I remembered: I had seen Mr. Cole ride a pinto, and his sons said that was a 'one-man-horse' who wouldn't let anyone else ride him.

*Did we have that pinto?*

Tough luck, Jeanie. She drew the straw for the other pinto. For her sake, I hoped he wasn't Mr. Cole's one-man horse. With only one of our saddles and bridles along, we put them on the horse Jeanie was riding, since he seemed less tame. The bridle had a curb bit, which helps in controlling a horse.

For my seemingly gentle horse we used one of Cole's bridles with a straight bar bit and a saddle that belonged to one of their sons. It had a tackleberry, quick-release buckle cinch, which I'd hardly used before.

My horse stood still as I saddled him, and when I hooked the tackleberry buckle and pulled the latigo strap tight he didn't flinch as many horses do. In retrospect, I think he was probably in shock that I was this close to him.

"Great!" I said to Jeanie. "You're already on your horse. Could we have lucked out and both picked gentle saddle horses?"

I swung into my saddle and we started up the trail. My horse began dancing sideways, shaking his head and though I kept a tight rein on him, became very hard to manage.

"Let's stop and change bridles," Jeanie said. "I think I must have the kid pony and don't need this curb bit. This bridle might be easier for you to manage. Looks like you have the one-man horse!"

So it seemed—*lucky me!*

*Lucky Jeanie riding the kid's horse.*

We found our horses and easily chased them with others back within sight of the corrals. Actually Jeanie did all the work, since it was all I could do to control the horse who wanted only Mr. Cole to touch him. By now we were sure I was riding the one-man horse and the only reason he hadn't dumped me, he was probably still in shock from having this brash teenage girl on his back.

The stallion, locked inside the barn, heard us coming as our horse herd approached the corrals. He set up a terrible squealing and kicking the side of the barn. All our horses flicked their ears forward, milled around nervously and tried to turn back. They were afraid of the stallion making this terrible screaming and whinnying noise even though he was inside a barn, enclosed by corrals.

"Get around those horses and don't let them double back," Jeanie yelled to me.

My horse and I were closing up at a dead run when I noticed a lone bay horse galloping from the barn straight at our herd.

*The stallion!*

He had kicked and jumped his way out of the barn and corrals and thundered right at us.

The horses spooked and dashed away from this ferocious demon.

I urged my horse even faster to turn them back.

This was the last straw for my pinto. Confront the stallion? No way. Time to get rid of this bossy girl on his back. Mr. Cole was a big man, probably twice my weight, and an expert bronc rider. This horse wouldn't buck for Mr. Cole, but figured I was an easy off.

He twisted around at the sounds of the horse herd answering the calls of the stallion as we galloped to the top of a rocky knoll.

I pulled him in tight and he started bucking violently. I couldn't control him. The first few jumps were okay, but as he drew his breath in and out quickly the tackleberry buckle came unhooked from the cinch ring. The latigo no longer held the saddle in place, and I felt it shift. I wondered if I hadn't been able to pull the cinch tight enough. But that was a fleeting thought—the next moment I was desperately trying to clear the stirrups of my boots as the loosened saddle slipped under the horse's belly.

"Jump clear! Jump clear!" In my mind I could hear Dad's often repeated advice.

I flung myself violently past the horse's legs, hooves missing my head by inches.

Even though I occasionally got dumped from a horse, the ground seemed harder than usual. My head came down with a 'whump.'

*Riding at Cole's, Jeanie first selects a horse for running in our horses.*

*I heard the stallion screaming and over my shoulder I saw he was coming for me again.*

It seemed like I was floating, floating, farther and farther into the ground. I grabbed the earth with both arms, but still it floated away. I remembered nothing more.

Meanwhile, the stallion dashed up to Jeanie on the other side. Terrified, her horse tried to get away, but the stallion reared, lashing out with his front hoofs. Her horse jumped out of the way, but as the stallion kept coming after her screaming with bared teeth and rearing, she feared he'd strike her out of the saddle and trample her.

I tried to keep the stallion away from our horses but he charged past me, trying to separate mares out of the bunch and run off geldings. He didn't want me around either. I was desperate because he kept rearing up and trying to stamp me out of the saddle. The other horses ran across the flat between the house and corrals through another open gate. The only way to corral them was to get that demon secured in a barn with the door closed.

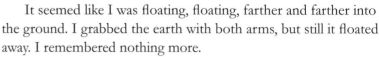

There I was, trying to get the stallion back in the corral and barn. He screamed at us, teeth bared, so close I could count his white teeth. His mane flew out as he tried to smash me every time he reared up snorting and screaming. Just before he came down, my horse and I slid sideways and he missed us again. This pinto kid pony had dodged the stallion before, it seemed. But I realized we couldn't keep it up.

It seemed he had one desire only. *Kill! Kill!* The inside of his wide-open mouth glowed red and the whites of his eyes rolled.

Finally he left to chase the geldings away and I looked over to see if Anne was coming. I couldn't see her. Then her horse came running toward the herd, no saddle, no rider, with bridle reins flying.

I left the horses and dashed to find her.

Anne lay still on top of a rocky knoll. As I rode up she lifted her head.

"Where's my horse?" she asked.

"Over there. Are you all right?"

But she just kept repeating, "Where's my horse?"

Her horse ran with the others, bridle reins flying, circling this small pasture. I thought she must be okay, at least she could talk. So I left her and went to catch up with her horse to take off his bridle before he stepped on the reins and tripped and broke a leg. Her saddle lay behind in a patch of jumping cactus. I'd get it later.

I heard the stallion screaming and glancing over my shoulder saw that he was coming for me again. I looked around for help and saw a fence post nearby. I dashed over, leaned in the saddle and pulled. It came right out of the ground—an old post with no wire attached.

I was so surprised! It was a thin, worn post, dried out from many years of wind and weather, fairly light in my hands. I couldn't believe my good fortune! I raised it up over my head and, each time the stallion reared, I beat at his hooves.

He nearly struck the post out of my hands, but I held on and kept hitting at him. I scored one on his head and he stopped in midstrike.

He turned docile and walked away. Suddenly he was no threat. What a coward! What a relief! I think if it hadn't been for that post being in the right place and loose in the ground, I'd be stomped to hamburger.

I saw Anne getting up from the ground, and decided to corral the horses while I had the upper hand with the stallion.

I must have been unconscious for a while, because the next I remember Jeanie had the horses in the corral and I was walking down a dusty trail toward her.

ANNE

Strangely I didn't hurt anywhere, but everything looked weird. The world seemed yellowish and grayed. A big sage hen with a dozen or more chicks went clucking down the trail ahead of me. These tiny puffs of down darted after her—live balls of brown cotton on two little stick legs.

I puzzled over how to get around her without separating her from her chicks. They were so tiny—too young to get along without her and would die if she didn't find them. I circled off the trail around her. But my head still felt hollow. *Would I get lost if I moved off the trail and never find my way?*

What a relief when I finally reached the corral and Jeanie called to me.

Anne came walking down the trail and the next I knew, there she was in the corral with the horses. They galloped around and she moved right through the middle of them.

JEANIE

*Our cattle graze at reservoir in Little Sheep Mountain community pasture.*

"That's my horse. That's my horse," she said each time the stallion ran by her, stepping toward him plaintively.

She kept following him around the corral saying, "That's my horse."

Then I realized Anne wasn't quite right in the head.

I walked her back to Coles' house, helped her lie down on the couch and covered her with a blanket. By now I realized there was

Western Ranch Life in a Forgotten Era

something drastically wrong and I desperately hoped Coles would come home and help. I was afraid Anne might be bleeding in the brain.

I told her to stay there and went to pick up her saddle.

By this time it was getting late in the evening and quite dark. When I came in to get her, I lit the kerosene lamp and checked her over. She looked all in one piece—but she talked in a jumble of words that made no sense.

If Coles came, maybe they'd take us home and save Anne a jarring ride in our cattle truck. But that might be too late.

I blew out the lamp, helped her in the truck and, leaving all the horses in the corral together, we started for home.

Partway home Anne started to gag. I stopped the truck and she vomited beside the road.

Then I worried that she vomited blood, and got out a flashlight to check. But thankfully she didn't.

> *Although Anne could understand everything we said, her words made no sense.*

She needed to see a doctor as soon as possible, I thought, and when we got to the little town of Terry I took her to the hospital. But that hospital was small with only one doctor, and he couldn't be reached.

So I drove on the thirty five miles home where Mom and Dad could take over.

When I got in the house and told them what happened, Mom put Anne to bed, then worried if she should be kept awake.

Dad took one look and declared she'd be okay.

"Just a hit on the head. It'll wear off," he said.

Fortunately, it did, but took several days.

In the meantime, although Anne could understand everything we said, when she spoke, her words made no sense. Mom tried to keep her in bed, but physically she seemed okay. Just couldn't talk right.

We teased her about the strange things she said and did. But we all were much relieved

## Prairie Fire

"Is that smoke up ahead?" I asked Jeanie as we crested a rise on the narrow dirt road.

"I hope not! Look how long and dry the grass is. If a fire gets started here it will burn forever!"

"There it is!"

As we rounded a curve we saw leaping red flames about a quarter mile off the road.

Jeanie and I were heading home after riding several days for our cattle in the Little Sheep Mountain's big community pasture, sixty miles across the Yellowstone River northeast of our ranch. Usually we took the more traveled route to Terry, but this afternoon we decided to explore this little-used road.

A man was frantically beating at the grass fire with his shirt, his tractor stopped off to the side. It wasn't yet a big fire, but the flames shot high and moved fast. A perfect sixty-foot circle of fire spread outward in all directions from its blackened center. Luckily, no wind picked up the flames.

*Dad feeds our bulls special grain mix through winter and spring in preparation for busy breeding season.*

We scanned the countryside anxiously for signs of habitation in hopes we could alert someone to help put out this fire. At the top of the crest we could see miles of rolling grassland, interspersed with cutbanks and small rocky buttes. Away to the west in the heat of the afternoon danced a mirage at the base of jagged badlands.

But we saw no sign of any ranch, nor had we seen another vehicle since we left our summer range ten miles back.

Prairie fire is one of the worst catastrophes to hit ranching country. Once started, it makes its own wind—roaring rapidly over the countryside through dry grass. Totally unpredictable, it destroys everything in its path. Cattle may burn to death as the fire rolls over them when cornered against a pasture fence.

Jeanie and I looked at each other. We had to help before the fire got larger.

"He's been out here cutting hay," Jeanie said, "A spark from tractor exhaust must have ignited the tall grass. This grass is so dry, a fire can run for miles. He can't put it out alone. It's moving too fast."

She slowed the car and drove off the road down through the shallow barrow-pit and across the grass toward the fire.

"Watch for rocks and washouts!" Jeanie cried.

If a sharp rock punctured our oil pan, we'd be stranded. Glad it wasn't me driving, I hung out the window trying to see jutting rocks through the tall grass and guide Jeanie away from them. I wished for our truck with its higher clearance.

I tried to remember what Dad told us to carry in the trunk for emergency use. Could anything there be used in fighting fire?

"There's a shovel and probably a gunnysack or two in the trunk," I said.

"We've got that can of water in the back seat," Jeanie said.

*One can to put out a grass fire? Could it help,* I wondered?

The young, thirtyish rancher barely looked up from beating the

*One of the worst catastrophes to hit ranching country, a prairie fire makes its own wind.*

*Summering our cattle near the Little Sheep Mountains brought us new challenges.*

Sometimes our bulls fought other bulls in the big Sheep Mountain community pasture. One of our jobs—trying to keep our registered bulls with our own cows.

flames as Jeanie braked the car and we jumped out. She ran to him with our water can while I threw up the trunk lid and grabbed two gunny sacks.

Jeanie thrust the water at him. The gratitude on his face spoke his relief as if looking at a whole tanker full of water. He grabbed the precious water can and, giving her his shirt to beat out smoldering flames, he sprinkled on water, carefully conserving it to put out as many flames as possible.

I ran up with the gunny sacks. Jeanie grabbed one of them and beat at the flames with the shirt in one hand and a gunny sack in the other.

I whipped my sack at the burning grass beside me. *Will we ever get these flames out?*

They advanced more quickly than I could put them out. I glanced back where Jeanie and the man worked side by side. Miraculously, the water he sparingly dribbled on the flames with Jeanie beating frantically beside him, were putting out the flames.

Suddenly both he and Jeanie were helping me. Their part of the fire was out.

We worked together rapidly. Sweat ran down our sooty faces and my eyes stung with smoke.

Then abruptly the flames disappeared. No more fire. The three of us stepped back and surveyed the blackened earth. Not even a wisp of smoke.

Jeanie and I looked at each other and then at the man whose face and bare chest were blackened with soot.

"Thank you! Thank you!" he gasped, his eyes bloodshot and soot-rimmed. "I am so glad you girls stopped. So glad you came over and helped me! That fire would sure have got away from me—I hate to think what would have happened then!"

He was grinning and sweating through his exhaustion, but he didn't tell us his name, and we didn't tell him ours.

Suddenly shy, we blushed under our soot. Here was a strange man we had never met before and he was half dressed. In our embarrassment we nodded to him, dashed over to our car and drove away.

We'd never seen him before, nor he, us. We were total strangers. I bet he always wondered who were those teenage girls who came suddenly out of nowhere to his aid, got all sooty and sweaty beside him, then just as quickly disappeared, never to be seen again.

Summering our cattle near the Little Sheep Mountains presented challenges we'd never experienced before. We didn't know anyone—except for Harold Cole who subleased his share of the community pasture to us. We didn't know the roads—and when driving we felt lost half the time.

Most of all, we didn't know the huge twelve to fifteen section community pasture that was part of a BLM Taylor Grazing allotment with its complex system of creeks, waterholes and gates. And we didn't know where to find our own cattle in that mix of hundreds of cows on that range of some nine or ten thousand acres.

Our job when we rode out there was finding our cows where they hung out together in small groups or sometimes intermingled with others. We'd get a rough count of our hundred or so cows, each with a calf, and make sure all were okay.

The other task: we tried to keep our three to five good, registered bulls moving among our own cows instead of wandering off to breed others.

Dad believed strongly in buying high-quality registered bulls, as this determines half the quality of the calf crop. Bulls make the difference in whether our newborn calves grow big and put on extra weight by sale time, he said. But not everyone in that pasture bought registered bulls, so our job included chasing less desirable bulls away from our cows.

We also took responsibility for the sixty to eighty cows we ran there for the State Industrial School, filling out the Cole allotment.

We branded N-lazy-M on the left ribs, as seen here, after buying a set of cows with that brand.

The Big and Little Sheep Mountains—named for the Bighorn sheep that once ranged those high buttes—consist of rough badlands country. Tall grass sweeps across the horizon, broken by small buttes with jagged rocks and boulders. At the bottom of most draws runs a dry creek bed, with water only in spring or a sudden summer thunderstorm. Pools of water collect in low spots and cattle watered here and at windmills and the reservoir.

Steep cutbanks and drop-offs caused us problems—and not a few heart-stopping rides. Looking for our cattle on water, we rode the creek banks, following narrow trails made by cattle and deer looking for a way down to water. Constantly weathering and changing, such trails often sloughed off into sudden drop-offs.

My heart beat extra fast when a trail narrowed, and I caught my breath, watching ahead for a broken trail.

Riding Dusty one day, I spotted one of our cows flirting with a scrawny bull in a herd of other cattle and rode over to turn her back.

Fast and efficient, Dusty was our best cutting horse. Once he knew which cow we wanted him to cut out, he stuck with her. In fact,

## Cattle brands and earmarks

We branded with two registered brands.

Our original brand *Quarter Circle-W-lazy-A* on the right hip was an old Barrett brand that Mom's dad owned for many years, perhaps when he raised horses in South Dakota before he and grandma married.

Then one day at the sale barn, Dad bought eighty or ninety big Hereford cows in a dispersion sale, all branded *N-Hanging-Lazy-M* on the left ribs. The seller transferred his brand along with the cattle, to save us having to rebrand them all.

An easy brand that didn't blot, like the A sometimes did, Dad branded the NM on more and more of our calves. At a distance it was easy to read when it grew big on the left ribs, but it never looked as classy as the old WA with its quarter circle in the sleeker hair of the right hip.

We also earmarked the left ear with a quarter undercut and cropped half off the right. Not a legally registered mark like a brand, earmarks proved a quick way to spot ownership at a distance. They were a big help in finding our cattle on the open range or when they summered in the big Sheep Mountain community pasture.

One rancher on that range marked with two ear crops. Another used a dewlap or wattle on the neck, which we girls disliked. To us they were disfiguring marks—whether an uppercut, lower-cut or split dewlap on the loose skin of the neck—that insulted the dignity of cows marked this way for life. Undeniably though, they showed up nearly as far away as we could see a cow.

he took control, and there was no turning him until he got that cow where he thought you wanted her. I liked riding him, even though he made my life miserable at times.

He first dispatched that tenacious bull following our cow, then chased her along the trail on a narrow ledge above the creek. When she took off it seemed that she was going right over that high cutbank. But she hugged the narrow trail and Dusty stayed right behind her.

I tried, but couldn't turn him off the trail, sure we'd cave off the edge. I imagined the dirt giving way and all three of us plunging down thirty feet—cow, horse and me. Dusty would surely roll over me, kicking me every direction trying to get up.

Luckily, that didn't happen.

Just before the trail dropped off, the cow dived in the opposite direction. Dusty followed and chased her right back where I wanted her—with a bunch of our own cows and one of our good Hereford bulls.

Thanks to Dusty, that cow gave birth next spring to a big beautiful whiteface calf that grew fast and put on extra weight by sale time.

Riding strange horses always presented a challenge.

One morning Francie and I drove out to Coles' to spend several days riding on the Sheep Mountain range and discovered Mr. Cole had turned our horses out with his range bunch in a large pasture.

"Just ride my horses," he said. "I've got two good cow ponies in the corral and they're well-broke."

Well, okay. But we knew sometimes a rancher's definition of a well-broke horse differed from ours. That day it worked out fine. His horses turned out to be easy riding, willing and eager, just the kind we enjoyed.

Problem was, we didn't know where we were going or even how to get to the Little Sheep Mountain pasture.

Mr. Cole drew us a map of this new country where our cattle were—in the dirt. We knew these maps well. Dad drew them, too. Quickly we tried to memorize the details, but as we rode away we knew we'd lose the location of some gates and trails.

It was a long day, but we found the pasture and all our cows.

Unfortunately, it was nearly dark before we finished. As we turned back to the Coles' ranch where we were spending the night it grew almost pitch black.

We knew the general direction of the ranch but were never sure, criss-crossing numerous trails. Many times Francie and I consulted each other.

*Which trail? What did the map show? Where is the next gate up ahead?*

*Montana Stirrups, Sage and Shenanigans*

Feeling desperate, we rode into the dark without any clue that we headed toward Coles' ranch. If we missed it, there'd be no other ranch for many, many miles in any direction.

Finally the moon broke free of cloud cover and lit the prairie with a shadowy light.

We came to a rutted two track road crossing our path. *Did we come that way?*

Then Francie had an idea.

"Last winter Mom said, 'If you're ever lost in a blizzard and don't know where to go, remember to give your horse his head and he'll take you home.'"

Francie pointed the index finger of her right hand into the darkness as she said it—decisively, like Mom. "Let's turn our horses in a circle, then give them a slack rein and see where they want to go."

Without any hesitation, both horses turned to the right, down the two-track road. Greatly encouraged, we hoped they headed for the barn, not their buddies out on the range. We had no choice but to trust them.

After a long time we came to a gate. Our horses stopped and waited.

"This seems like the gate we came through on our way over," Francie said as she got off and opened then closed the gate behind us.

Our tired horses seemed eager to go. Surely this meant we were getting close. Horses perk up when near home. It's where they get unsaddled and fed grain, which they eat eagerly after a long ride.

Finally we crested a rise and, sure enough, there below us glimmered the lights of kerosene lamps in the Cole's ranch house windows.

Though we'd felt apprehensive about riding Cole's horses that morning, we were mighty thankful we rode them that day. They took us home.

Had we been riding our own horses, this first time out here, they might have pulled their heads in the direction of our own ranch instead—southwest many miles to the banks of the Yellowstone River.

Branding cattle is particularly important in a community pasture such as Little Sheep Mountain where many ranchers run cows together and in considerable numbers. Once we unknowingly left a branded calf out there when we trailed our cattle home.

Like cattlemen everywhere, ranch folks out there knew and respected brands. They treated us well, even when we didn't know them. Contrary to the plots of many western novels, ranchers don't engage in rustling, changing brands or stealing cattle. They watch out for each other—and regard strange vehicles with suspicion.

*"If you're ever lost and don't know where to go, give your horse his head and he'll take you home."*

*Earmarks make it easy to spot our cows when mixed in with others—we crop the right and notch the left as below.*

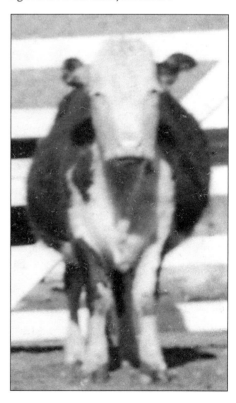

Western Ranch Life in a Forgotten Era

> "Ranchers are the best people in the whole world," Dad said.

Several years after we moved our cattle off that summer range, we received a phone call from a Little Sheep Mountain rancher.

"Your steer is still running in that pasture," he said. "When we round up and sell our calves, we'll bring him in to the sale barn in Miles City. You can come and get him."

To his surprise, when Dad got to the sales barn, there stood a great big four-year-old steer with our *NM* brand on his ribs.

We usually sold our steers as yearlings, for prime meat. This steer was even better. His steaks were tender and marbled with fat, so he brought top price per pound that day—and he weighed far more than any yearling.

Dad had never met the rancher who brought him in, but they parted good friends.

We hadn't missed that calf. Like other ranchers, we expected to lose a calf or two—to injury, illness or coyotes. Likely he had joined calf buddies owned by someone else or taken up with a substitute mom, so got left behind when we trailed our cattle home.

The ranchers out there cared for our big steer over several years. He ate their grass in summer and hay in winter—maybe grain, too. Someone could have butchered him and we'd never have known.

But of course, ranchers wouldn't do that—they were too honest and concerned for their neighbors.

"Ranchers are the best people in the whole world," Dad often said. Later that fall he hauled out a load of alfalfa hay bales to the rancher who last wintered our big steer.

# 6  Hunting—Rancher Style

## Roadkill on the Rez

A car door closed and I glanced out Beverley's front window. There was Grace, one of Bev's closest friends in her battered old car.

ANNE

"The truck in front of me just hit a deer. Even if it's smashed too badly to eat, I can feed it to my dogs," she said. "I haven't any money and my dogs are starving."

I realized then, that Grace probably didn't have any meat—or other food, either.

Life on the Northern Cheyenne Reservation was tough.

It was during this visit with Bev when Grace taught me Native American ways of using the parts of a deer. Primary rule: waste nothing.

Grace BearQuiver was an Indian police judge for many years before taking a job off the reservation. She had a better understanding of white ways than most Cheyennes and yet missed Indian life, so returned to Lame Deer.

I got to know Grace well and appreciated her great Indian values.

Beverley had such interesting friends.

It was the middle of summer when I loaded a few extra groceries, large bags of flour and other staples, and drove down to Lame Deer. Always I took plenty of food with me, because I never knew if Bev's cupboard might be totally bare—after one of her raids on it to supply Indian friends in need.

It seemed they were always in need. With very little employment on the reservation, those who had jobs helped out others. Indian cars were always old and unreliable. Here again, they helped each other.

It's the way Native Americans live. Those who have generously share with those who don't. This was always Beverley's philosophy

*I never knew if Beverley's cupboard would be bare—after one of her raids on it to supply Indian friends in need.*

even before she came here to start one of Indian country's first real newspapers—the *A'tome*. She fit right in.

A fifty pound sack of flour and large bag of sugar didn't last long at Beverley's, since she gave it out in many smaller sacks to her friends even before I returned home after a weekend.

This day, Grace drove up to tell us she'd just seen that deer hit on the road, and wanted Beverley and me to go back and help her load it up in her car.

"It would be so good to have the meat," Grace told Beverley. "But I don't think I can handle it alone and I'm almost out of gas."

Bev looked pointedly at me.

I was skeptical of roadkill, which meant a bloody carcass, not bled out, with mangled, blood-clotted meat to be cut away. Split–open intestines probably were mingled with clean meat, which meant not only a great deal of washing, but also a dreadful smell. I almost gagged at the thought.

I looked at Beverley for inspiration.

There was the 'dreaded look' that Bev used on me as a kid— '*Okay, little sister, you want an adventure, you have to do some of the dirty work!*'

Darn!

"Anne will be glad to take you in her car to go get it," she offered.

*Bring roadkill home in my car!*

One of Bev's friends needed help, so it was unthinkable to refuse.

Earlier, Beverley loaned her car to a man who pleaded he 'needs wheels today,' so guess who had the only vehicle?

Bev conveniently had work to finish the week's *A'tome* newspaper before deadline, so off Grace and I went to retrieve her roadkill.

On Montana reservations, the law said a tribal member could kill and have in possession any wildlife at anytime. They also didn't need a hunting license. But still I felt a little squeamish, worried that a game warden might show up and question my presence with Grace and her deer.

I was surprised to see the deer in such good shape, though somewhat bloody. Hit in the head, its body was intact. We loaded it in my trunk and tied down the lid with four slender legs protruding. Again, I worried a passing game warden might see us, but we got to Bev's okay.

Grace declared she wanted to use the deer for eating, but she didn't know how to butcher it. It seemed to me that historically, Cheyenne men cleaned and skinned the animals; women took the meat and prepared it.

*I was skeptical of roadkill—it meant a bloody carcass and mangled, blood-clotted meat.*

Again the dread look from Beverley.

"Anne will do it," she said. "Anne's been cutting up venison ever since she could hold a gun."

I had no choice.

The hide was covered with clotted blood…. *Ick*.

We pulled the deer behind Bev's house, where no one would see it if they drove up. I gutted the deer and cut out the heart and liver for Grace to soak in salt water in Beverley's kitchen sink.

Grace gathered up the entrails for her dogs to eat while I skinned out the hide.

She very carefully scraped every bit of flesh from the hide, and then we stretched and nailed it to Beverley's wooden fence to dry so it wouldn't shrink. As the day wore on the sun dried it. Then she'd take it to a rough-barked tree, such as a cottonwood, and rub it back and forth to soften it. She didn't use brains, copper sulfate, or any of those other smelly, gooey ingredients the books often suggest, but just this rubbing.

We boned the meat and put it in the refrigerator to cool. As we worked, I asked her questions on how early-day Indians used the deer. She had me pull out the tenderloin instead of just cutting it, then gently pull off the casing surrounding this muscle. She said this sinew, after it dried and was separated into strips, became the very longest sewing thread available from a deer. Unbreakable, it was used to sew garments and moccasins. Needles were made from bone, the antlers used for digging and the hide for clothing.

Indian-made jerky was very thin and came in long sheets about eight inches wide, I'd noticed. Grace showed me how to do this.

She pulled a large chunk of meat from the front shoulder and, holding a long knife almost parallel to the cutting board, started a long thin slice. She continued this slice very thin as she rolled the meat over and over and made it into an almost paper-thin sheet about twenty inches long. Then Grace hung this out on the fence to dry. She didn't use any salt or other type preservative. I was concerned about flies, but surprisingly, a thin film almost immediately formed over the meat, and no flies came.

I stayed outside to clean up our mess, while Grace went into the kitchen to work on the meat. When I went back in the house to start processing it, I saw that another native woman had arrived. She and Grace were cutting and packaging the venison.

I heard the women talking in Cheyenne, one of the most difficult Indian languages. When I was around, they courteously spoke in English, but reverted to their native language when they didn't want me to hear. Naturally this made me curious.

As I washed my knife to help, Grace said in English, "No, no.

*Mule deer, identified by their large ears and black tipped tail, favor the open plains of the Cheyenne reservation, while whitetail deer prefer its wooded draws.*

*Indian-made jerky was very thin and came in long sheets about eight inches wide.*

Western Ranch Life in a Forgotten Era

She won't mind."

So I looked up to see what I wouldn't mind.

The other Indian woman, knife in hand, was slicing and popping raw liver into her mouth. Blood ran down her chin as she slowly chewed this delicate morsel of raw liver.

I fled the kitchen!

Grace BearQuiver brought unique experiences to our lives, yet none as amazing as her final farewell.

When Grace died she returned to tell Beverley and Bev's Cheyenne husband, Jack Badhorse, goodbye.

Bev and Jack were sitting in their living room reading at the exact moment of her death. Things began moving on the mantel; figurines moved—one over to a far shelf and others rearranged.

Beverley and Jack locked eyes.

It was the Indian way to visit loved ones as their dead spirit left the earth. Bev had heard of this, but never experienced it before. Jack nodded to her. Without speaking, they both knew it was Grace, visiting one last time.

Driving back home after dark that evening, Grace had swerved her car.

*Was there a deer in the road? A horse?* We'll never know.

Her car crashed down an embankment and Grace died instantly.

She enriched the lives of all of us who knew her. For Grace, as she had told me, there was no death, only a change of worlds.

*Jeanie and Dad with our elk beside the Yellowstone River. Buck pulled the elk out of the willows and rocks to a place we could load them into the truck.*

Somehow it's comforting to know she is at peace and never hungry.

## Elk Firing Line

We heard many stories of the Yellowstone Park 'firing line'—none of them good—from Dad's hunting buddies.

"Elk hunters will shoot at anything. They line up along the road and fire away. They don't care what's behind the elk," they said. "And if they miss their shot, they'll fight to get their tag on some other hunter's elk!"

JEANIE

Still, Dad sure wanted to go elk hunting. The biggest game in eastern Montana were mule deer and antelope.

The opportunity came when I was a senior in high school

and Francie a sophomore. Heavy snow brought the elk out of the highest mountains and into river bottoms outside the north edge of Yellowstone Park at just the right time for us, during Christmas vacation.

Dad, Francie and I left home in the dark at five, the morning after Christmas, with Buck loaded in the stock truck along with our camping gear and army-green canvas tent.

Buck, a tall bay quarter horse, was calm and dependable as a horse can be. Deep in the chest with the extra lung power needed at higher elevation—just right for his job of dragging out an elk.

Determined to avoid the firing line, Dad saw no reason we couldn't hunt elk as we always hunted—following fresh tracks in the snow, slipping quietly through the brush, listening for the snap of a twig. Stalking game.

Shooting was allowed only in daylight hours and, of course, never in Yellowstone Park. Elk knew this as well as the hunters. So when a deep snowfall and hunger forced them out of the mountains and into river bottoms outside the park for night feeding, they instinctively retreated back inside before dawn.

At the firing line hunters tried to block them. Every year the firing line formed to catch them on the outside and prevent their return. Hundreds of hunters had merely to stand at the edge of the highway and when daybreak came, fire at the elk milling on the bench just outside the park boundary.

Hunters outnumbered the elk. Each hunter had a single tag, and the first to clip his tag on an elk, owned it. That was the rule. It didn't matter if he shot the elk or even carried a gun. Some carried only a tag. Violent arguments erupted. Hunters grew careless in their avid desire to get an elk because it meant a winter's meat for a growing family.

*What if a hunter shoots an elk and runs down to tag him while more elk run by? Do other hunters hold their fire?*

Not always, as we discovered. Dad's hunting buddies told us anger, greed and carelessness precipitate terrible incidents along the firing line, some fatal.

We set up our winter camp north of the Gardiner entrance to the park, as suggested by a rancher Dad knew. Leaving Buck in the rancher's corral and more than enough hay and grain to sustain him, we crossed the bridge over to the western side of the north-flowing Yellowstone River, mostly frozen over, and followed a dirt road to the edge of the park. Our lone camping spot was on a bench of land across the river from the annual firing line.

Over there hunters parked along the highway through the night, waiting for elk to come out of the park—and sunrise.

*Bull elk lose their large spreading antlers in early spring, re-grow them through summer and rub off the velvet and harden them before fall rutting season.*

*What if a hunter shoots an elk and runs down to tag him while more elk run by? Do other hunters hold their fire?*
*Not always, we discovered.*

Dad said we'd ignore all that and hunt higher up on our side, along the park boundary as the fence disappeared up and over the mountain. Francie and I wondered between ourselves how were we going to do this—deep snow and mountains and where were the elk?

This was our first experience snow camping. To pitch our tent, Dad set us to work shoveling the foot-deep snow completely off to bare ground. All the mountains around were deep white, more snow and more mountains than we ever imagined.

We unrolled our beds and lit the gas Coleman lantern. The borrowed tent was as large as a small room with plenty of height and a heavy canvas floor. It grew dark and colder outside but was soon warm and cozy inside from the nice little wood-burning stove. The stove legs sat on an iron plate and a stovepipe went up through the center of a galvanized tin section built into the canvas above us.

We slept snug in our soft, warm bedrolls and woke up in an already heated tent to the smell of bacon cooking. Dad was up early, cooking, and served us hot cocoa with a marshmallow in it before we even got out of bed. Happily, he did that every morning.

For our own safety, Dad said we'd avoid the firing line. We did our best.

> *In the deep fresh snow, our snowshoes caught in the sagebrush and threw us into bottomless snowdrifts.*

The first morning we followed Dad up the steep mountainside along the park fence, our leather-latticed snowshoes and heavy WW II surplus 30.06 rifles slung on our backs. He set a fast pace and sure enough about halfway up found fresh elk tracks leading out of the park into a forested draw. The tracks led us through tall timber and deep into the mountain. There in the heavier snow and more level terrain we strapped on our snowshoes of long wood frames laced with rawhide.

Anticipating what lay ahead, we climbed higher and higher.

*Around the next bend will we see a magnificent bull elk reaching up to nibble pine needles? Or eyeing us from behind that thick stand of brush?*

But inevitably, Jeanie and I were disappointed to find the tracks doubled back into the park—out of range—after slogging hours through deep snow.

Over the next few days we repeatedly climbed that mountain, while Dad ranged higher and wider. Sometimes we needed the snowshoes in deep fresh snow—though often they caught in the tips of sagebrush on the flats and dumped us into bottomless snowdrifts. No matter how deep we fell into the snow, we never reached the ground. In late afternoons we made our way down the mountain to camp.

Trouble was, the fresh elk tracks we followed always circled back into the park at a higher level. Regardless of where they came out, it seemed, their elk wisdom led them to safety before dawn.

Every night fresh snow heaped up around our tent. Still, the larger elk herds did not come out of the park.

Then came the big snowfall—a couple of feet in one night. We were almost buried in it, but Dad said it made for good insulation. As a young man, he simply took a horse and bedroll and laid it in the lee of a snowbank and slept more or less warm all night.

Next morning, long before daybreak, a man gently rattled our tent flap.

"Hello in there." He spoke in a low voice. "A bunch of elk are out here. We need your help. We're lining up along the fence to keep them from getting back into the park before daylight."

Dad went out to talk to him while we dressed. Jeanie and I wondered what this meant. He said about seventy-five elk had gathered on the bench by the river just outside the park.

This was it. The firing line was already forming.

But Dad sent Jeanie and me up the mountain again. "When daylight comes there'll be a war going on down here with bullets flying everywhere. I don't want you girls anywhere near it."

*Elk find good grazing in deep snow inside Gardner entrance to Yellowstone Park, safe from our rifles.*

We didn't want to go up the mountain again; we wanted to stay where the action was.

Then Dad threw us a shred of hope.

"When the firing starts there's a good chance the elk will run up that way to get away from the hunters and back into the park. You girls will be in good position to shoot your elk."

So up the mountain we went, once again, struggling through the darkness and fresh snow, and rested on a granite outcropping. We looked longingly toward hundreds of vehicle lights shining down across the river.

As it turned out, a horrifying pageant was about to unfold below us. High up on our rocky point, we had the best seats in the house.

Gradually the sky lightened. The herd of elk milled around, uneasy that their way back into the park was blocked. Big as cows but with long necks, they nervously held their heads high and horizontal, noses up, like camels.

People got out of their cars on the other side of the Yellowstone River and moved around with rifles ready. Some without rifles knelt, ready to run and clip on a tag before the shooter got to it.

Dad was now in the line of fire—the elk between him and the hunters.

*They can't shoot before sunrise, but when will sunrise come? It's cloudy.*

We watched Dad down there in the early daylight, until he disappeared into a thick stand of willows on our side of the river.

*"When daylight comes there'll be war going on down here with bullets flying everywhere."*

Western Ranch Life in a Forgotten Era

*Appalled and sickened, we couldn't believe what we were seeing.*

Suddenly, in the cold, clear, sweetness of the morning a single shot rang out. Then a volley of shots.

The elk started running north away from the hunters stationed along the park fence. They crossed the Yellowstone and stampeded along the meadow that paralleled the river and the highway over there.

Hunters fired rapidly, loud as artillery fire in the war movies we saw, and at such close range that about half the elk fell.

The rest of the herd ran on farther north while hunters ran to tag their elk. Sometimes two claimed one elk and we could see an angry confrontation. Every elk had one or two people trying to tag it. It looked like the war was over almost as quickly as it began, except for those arguments.

Then suddenly, without warning, the elk herd doubled back and ran straight into the thirty or forty downed elk where hunters knelt, tagging their game.

Immediately, withering rifle fire rained on hunters and elk alike. Men dropped flat behind their game. Shots spurted the snow and we could see drifts of it flying—like smoke. More elk fell. People on the firing line kept shooting into the hunters and elk until every last elk was down.

Appalled and sickened, we couldn't believe what we were seeing.

*Why don't the hunters hold their fire until everyone is safe? Where is their common sense? Their sportsmanship? Their humanity?*

Finally the gunfire stopped. Hunters who crouched behind their elk edged up and peered around. All the elk were dead—not one had escaped back into the park. Jeanie and I wondered if any hunters were killed or wounded, but we never found out.

We were stunned by what we'd seen. Not by the numbers of elk killed, though that was slaughter enough, but by knowing that hunters recklessly fired with all those other hunters mixed right in with the elk they were shooting.

We started down the mountain worried about Dad.

*Was he even alive after all that gun fire?*

Then we saw a man coming out from the far end of the willow trees looking our way and waving his hat.

It was Dad, trying to get our attention in all that distance, and we plowed down to him.

"Give me your tag, Jeanie! Quick! We gotta get it on an elk I shot back here—"

Three dead elk lay deep in the willows. A young man bent over one of them attaching his metal elk tag. Another elk carried Dad's tag and as he clipped Jeanie's on the third elk, he introduced us to the man, a World War II veteran.

"You girls stay with the elk," he told us. "I'll take the truck and go get Buck."

Later he told us the story of what happened.

"I saw three elk run into the willows on our side of the river, but just as I got there the shooting started. Bullets hit the brush and whistled all around me.

"This young fellow called out. 'Hey! Get over here in the ditch—quick!'

"He was in a shallow trench by a broken down barbed-wire fence, with hardly room for me. I didn't need to be told twice. I hit the ground and slid in as best I could."

Dad chuckled at his dilemma.

"A small bunch of elk stood on the river ice between us and the firing line. We were in the crossfire—directly in line with the targets.

"Seemed like everyone was shooting at them—and us. Dirt fell on our heads from shots hitting the bank above us. Bullets twanged against the fence wires in front of us.

"Finally they all lay dead but one cow elk. She fell to her knees injured on the ice and struggled to rise just in front of us. A shower of bullets hit on all sides.

"I was disgusted with the shooting. 'Why can't they hit her?' I said.

"I crawled onto my knees and lifted my rifle. 'I'm going to finish her off before they kill us.'

"'No! Git down, you fool!' My companion grabbed my jacket and jerked me back into the trench.

*"Just then a bullet zinged the fence wire right where my head had been. We hugged the ground—and another round of shots sprayed around us."*

"Just then a bullet zinged the fence wire right where my head had been. We hugged the ground. The cow went down on her knees and then lurched upward and another round of shots sprayed around us.

"He kept saying, 'Get lower. We gotta get lower. I don't want to get hit.'

"'What're you complaining about? You're on the bottom!'" As he repeated this, Dad laughed heartily each time he told the story.

"'It's worse than anything I went through in the war!' the young man said. 'I was in some hot battles. But never anything like this!'

"Finally the cow fell dead and three or four hunters came running out across the ice arguing about who killed her. That elk had so many bullets in her I can't imagine anyone wanting the meat.

"As we crawled out of the ditch, surprised that we were still alive and not even bleeding, I pointed to the thick stand of willows on our right.

"'Come on,' I said. 'Let's check this out.'

"Sure enough, the three elk were still in there, hiding out, just

Western Ranch Life in a Forgotten Era

like we were."

What an adventure! Elk hunting in the mountains, following fresh tracks in the snow, snowshoeing deep into the forests, our cozy winter camp…

It was an exciting one-time experience.

But once was enough. None of us wanted to hunt the firing line again.

## Lucky Shots

"You better get up early and check for deer around the spring," Dad told me the night before hunting season began.

"Last week I saw tracks up there by the water trough."

"Umm. Well, I guess."

I'd had a busy week in high school that first year as a freshman and it took time to shift gears. Maybe I sounded less than enthusiastic.

Early meant getting up at 5:30 a.m. during fall's shorter days and walking up in the hills to reach the cedar draw with the springs by dawn, when a deer or two just might come down to water.

Dad always liked us to shoot our deer early on the first day of hunting season, before the trigger-happy hunters, as he called them, began driving through the hills. Deer were still scarce in our locality in those days—more hunters than deer, though later we had big increases in deer population.

Next morning I crept downstairs in the quiet dark. Chilled, I pulled on a warm jacket before reaching into the corner for the deer rifle.

We kept our guns leaning into a corner behind the back door—a couple of shotguns, the .22 and the .250 deer rifle. I felt my way through them in the dark, but couldn't locate the deer rifle. Finding a flashlight, I looked again. The .250 was gone.

I'd been keeping as quiet as possible not to disturb anyone, but now burst into Mom and Dad's bedroom.

"I can't find the deer rifle. Where is it?"

"Uh…mmm," Dad opened one eye.

"The hired man wanted to take it. I didn't think you wanted to hunt this morning."

*Darn!* Maybe I didn't sound excited about hunting the night before, but now I was all ready and definitely not going back to bed.

I went out and pondered briefly.

*Can I kill a deer with a .22? I don't think so. Probably can't find a deer*

*Suddenly from the top of a hill I heard the scatter of gravel. There, silhouetted against the sunrise, stood a big buck.*

*up there anyway.*

I picked up the .22 rifle, dropped a few shells in my pocket and set off for the spring.

Silently I crept close, keeping under cover of the juniper trees until light enough to see clearly. Nothing.

Cautiously, I scouted nearby draws and then, disappointed at finding no deer, turned toward home and breakfast.

Suddenly from the top of a hill some distance behind me I heard the scatter of gravel in the still air. There, silhouetted against the sunrise, stood a big buck.

Instinctively, I raised the rifle, fired, and he disappeared. Too late I remembered I held only a .22 in my hands.

*Why didn't I wait for a close shot? Why didn't I circle around the hill to get in several quick shots—the only way I'd have a chance to kill a big buck at a distance with these tiny bullets?*

He was long gone and I regretted that quick shot.

Now I had to climb the hill over there and search for blood on the rocks. If I wounded that buck, I had to find him. In our family, a responsible hunter followed up on the shot and did the right thing. You put a severely wounded or crippled deer out of its misery rather than leaving it in pain or helpless against predators.

*What a surprise!*

When I reached the crest of the rocky hill there he lay—a nice buck—dead, with a small, neat .22 bullet hole between the eyes. The small bullet must have pierced a tough skull even at that distance and connected with a vulnerable spot in the brain.

A lucky shot for sure—for me, at least.

Hunting deer with a small-bore .22 was not that unusual for some—Anne, for instance. Good at stalking deer, she shot several that way, until Dad told her it was not a good idea. Not that she missed or wounded deer—but just that it could happen, since a .22 isn't powerful enough for big game or long shots.

Anne became our best and most avid hunter. She cleaned her first deer at age eight, after all. Soon she excelled at silently sneaking up on game and was expert at long shots, too.

For me, lucky meant I got a nice surprise when an unlikely bullet connected. I considered it lucky, too if I didn't have to spend time hunting an injured deer.

That day I had my buck and even made it home to join the family for breakfast.

Along with Dad, I fired another lucky shot several years later—the first and only antelope I ever killed. That time might even have been lucky for Dad.

By then there were many more deer and even the antelope were

*Fleet of foot and keen of eye, antelope move rapidly out of range.*

*Good at stalking deer, Anne shot several with the .22.*

## Shooting with open sights

Like Dad, we girls held steady aim in those days.

We usually shot off-hand, standing up.

*Just lift the rifle, line up the deer in open sights, allow for lead time and fire.*

We often aimed for a head shot, most saving of the meat—or if the animal was too far away, the heart and lung area.

In hunting, it was not our usual style to steady the gun by resting the arm or rifle on a prop as some hunters did. Nor did we use a sling, except in target shooting, as we did at the NRA indoor rifle range in Miles City.

Other refinements, such as telescopic sights, just coming into use after the war, were expensive and eluded us. Instead, we girls followed the example of our sharpshooter Dad in using open sights, not scopes.

With our .250 deer rifle and the smaller .22 we learned two versions of open sights. Either we lined up the 'V' notch at eye level with the post on the front tip of the barrel or, using peep sights, we lined up the crosshairs of a small steel circle with the front post.

As we squeezed the trigger we allowed for bullet drop with distance, as well as wind and deer movement.

We learned from Dad to continually scan the horizon with the keen long-distance eyes of the plains dweller—ever alert for movement and the blending shapes and colors of wildlife. The open sights made it easy to swing the rifle up and zero in quickly and accurately on a deer.

staging a come-back. In Montana at that time we could shoot two deer plus an antelope on one license.

Along with a couple of other people that morning we hunted deer near Uncle Chet and Aunt Margaret's ranch at Zero, ten or fifteen miles east of home.

The other two hunters had just shot a deer each. Unfortunately, they lay dead in the playground of the rural school where Aunt Margaret taught.

It was a Saturday and our aunt was nowhere around. But Dad was concerned. He watched uneasily as the two guys gutted their deer, spilling entrails all over the ground.

"Make sure you get this all cleaned up," he said. "Use the shovel. I don't want a trace of it left on the playground when Margaret comes to school Monday morning!"

He pitched them a shovel from the pickup.

As he did this, his sharp eyes scanned the hills as always, looking for anything that moved or didn't quite belong to the scenery.

Suddenly, he broke into a grin.

"Look at that, Francie—antelope!"

At first I couldn't see anything where he pointed toward the distant hillside. They were a long, long way off. The golden color of antelope blended well with the tawny fall grasses and gravel. I caught just a glimpse of white signal rumps 'flashing in the sun like a tin pan,' as people described them.

Seven or eight in the bunch, the antelope were moving and about to disappear around the hill.

"Want to take a shot?"

I shook my head. "No, it's too far. Can we drive over there? And come around the hill from the other side?"

Dad just laughed.

"Let's try it. You take the one in the lead and I'll take the last one."

Already he was sighting down the rifle barrel.

*Well…okay.*

Typically, he was giving me an edge. If I missed the lead antelope, another might run into my bullet. But his animal in the rear would be long gone, with none behind to catch a stray bullet, unless he led his shot with just the right margin.

Good plan, for me.

But as I lifted my rifle and looked through its open sights I knew that idea was hopeless. No way could I pick out the lead animal at that distance. The slim front post on the tip of my gun covered the entire herd of antelope as it lined up in the open 'V' notch near my eye. They were too far away.

"Ready? Shoot!"

Dad fired. So did I.

Two puffs of dust rolled down the hillside and the herd vanished around the hill.

We jumped in the pickup and raced to see if we could find evidence that our shots had done more than raise those two puffs of dust. Blood on the grass or rocks would mean we'd have to start tracking.

Amazingly, we didn't. Our shots had connected.

There on the hillside lay two dead antelope, each with a bullet in the heart.

"Hey, good shooting!" Dad cheered.

I had to confess.

"I couldn't pick out the lead antelope. So I just shot into the middle of the bunch."

Dad chuckled, "Heh, heh, heh, So did I."

*The slim front post on the tip of my gun covered the entire herd of antelope. They were too far away.*

## Game Warden Dilemma

"Oh my goodness, Jeanie. You just missed the game warden."

Mom laughed ruefully when I sneaked in the door carrying Dad's deer rifle.

"Yeah, I know. I've been out in the barn with Dusty for the last half hour waiting for him to leave," I said.

"He was here for breakfast and do you know what he said? 'Elmer, if I ever catch you or your daughters hunting deer out of season, I'll have to arrest you!'"

Mom didn't approve much of out-of-season hunting, even though she lived her entire life on western ranches among hunters.

Yet we all knew there were too many deer in the alfalfa fields. Our game warden, Jack Nicolai, tried to help. He and Dad were good friends. Every summer the Fish and Game crew put out hundreds of half-grown pheasant chicks in our fields that Dad guarded assiduously until fall hunting. Pheasant season was for roosters only. Dad and the game warden both protected the females from errant hunters.

Many times when we came downstairs to breakfast, there sat Mr. Nicolai at the kitchen table drinking coffee and eating bacon and hotcakes with Dad.

An interesting man, he often talked about his commando service during World War II in Europe, behind enemy lines. He spoke with regret about those duties because they were in opposition to his

Western Ranch Life in a Forgotten Era

moral ethics. He took an interest in all our activities. So there he often sat, engaging in good conversation, even when one of us waited to go on an out-of-season hunting jaunt—at that very moment.

Just the week before I heard Dad talking with him.

"There are too many deer in my alfalfa fields every night," Dad said. "They're ruining my alfalfa seed. We have a great crop this year and I hate to lose it."

"Elmer, if you can just wait another three weeks, till hunting season, I'll make sure we get the hunters out here and cut down on those numbers."

"Three weeks and my seed will be destroyed. Each deer eats a bucket of seed every night and they ruin as much as they eat. They lie down, thrash around and break the stalks so it can't be harvested."

"Well, you know what you can do. You have the right to shoot destructive deer—and just let them lay."

Dad didn't reply. I knew he disapproved of wasting meat like that when many people didn't have money to buy it.

It was legal to shoot them and leave them there, but who would do that? Ranchers ate venison anytime during the year, but wouldn't shoot a deer and simply let it go to waste.

Dozens of mule deer lived up in the dry land pastures of our range and that of our neighbors now the coyotes were gone. The problem was they travelled down at night invading our alfalfa fields,

*Jeanie helps friend pack out a big buck.*

eating and trampling the seed. During the day they went back and slept out of sight in the brushy coulees.

In fall as the seed ripened they caused the most damage. Our alfalfa seed crop was exceptionally heavy that year—and market price was good, too. Alfalfa seed was special. Certain years developed heavy, high quality seed. Other years didn't, and the plants went to foliage, rather than setting much seed.

Our irrigated alfalfa grew lush and thick. For hay, we harvested two and sometimes three crops a year.

When Dad judged it a good seed year, he saved our best fields for seed, either the first or second cutting. That was our best cash crop. Alfalfa seed was so valuable to us it hurt to know that several dozen deer roamed in those best fields every night, eating and trampling it down. Our folks needed that income for living and operating expenses, and maybe, putting a bit aside for our college.

Ranchers knew a time when deer were scarce, after early settlers almost totally killed them off during homestead and depression days. When they began to make a comeback, Dad saw deer as the sign of a returning healthy population. Hunting deer was not necessarily recreation in those days when we were younger, although Dad engaged in it with great enthusiasm. We needed the meat.

He didn't mind feeding a small herd through the year and often talked to us about the balance of nature.

"Coyotes and judicious hunting keep the deer under control."

But now the balance was upset.

We were cattle ranchers, but seldom butchered our own beef. Without electricity and before we could rent freezer space in town, we had no good way to preserve all that meat. Home canning hardly made a dent in preserving that much beef. Our winters were not consistently cold enough to keep meat frozen all winter outside as happened farther north. Even in the coldest winter we usually had our week-long January thaw.

Besides, we sold our steers for cash, a scarce enough commodity in our lives. By this time Beverley was in college and Francie and I would be soon.

So instead of beef we usually ate venison, or Dad butchered a sheep or chickens.

Just the other day Mom, the honest, law-abiding mentor she was for us, mentioned that she was out of meat.

"I'm so tired cooking with only canned meat," she told me one morning.

Mom cooked venison well all her married life. She made delicious Swiss steak from our alfalfa-fed deer. It had great flavor with no wild taste. Her roasts and steaks tasted delectable, too, with just the right spices. And the deer that Francie and Anne shot were always tender with no gamey taste.

So that morning I waylaid Dad, all ready to go to the field on the tractor, and told him Mom needed some venison. But Francie and Anne, our usual deer hunters, were gone for a few days.

"Dad, Mom wants you to go hunting. She says she's out of meat."

"Jeanie," he said. "I'm too busy harvesting oats. If you want venison, go shoot a deer yourself."

"Me? But I've never shot a deer! I don't know how."

"Yes, you can. You're a good shot. The best way to learn, you just ride up there and do it. Try back in the school section."

With that he revved up the motor and took off for the field.

I had never shot a deer before and couldn't believe he really thought I should just go out and do it by myself. I was a good shot with the .22, and knew how to stalk animals.

*Coyotes and judicious hunting kept the deer under control. But now the balance was upset.*

> ## Safety first
>
> Our parents instilled in us these gun safety rules:
>
> **Rule 1.** Treat every gun as if loaded. "Someday it *will be loaded when you think it's not, and that's how accidents happen.*" We heard that often.
>
> **Rule 2.** In holding a gun, never let it cross a person's body—even when unloaded. Dad once fired a hired man for swinging a rifle across the hips of a man standing next to him, "Sorry, but I can't have anyone on this place that careless with a gun."
>
> **Rule 3.** Never put a shell in the chamber until you're ready to fire. This also applies to riding horseback with a rifle; never ride with it loaded. Of course we never shot from a horse. It can cause ear damage and panic.
>
> **Rule 4.** Don't shoot until you're sure of the background. What's behind your target? If your high-powered bullet misses or the animal moves, what gets hit instead? We were fortunate learning to hunt in the badlands—most often a butte or cutbank absorbed stray bullets.
>
> *To reinforce our awareness of safe gun handling, we had the example of Uncle Will. Arriving home late one winter afternoon from hunting coyotes in the Missouri River Breaks, half-frozen in thirty-below-zero twilight, he dropped down from his saddle horse. In that moment, his rifle fell out of its scabbard, the butt end hitting the ground and firing a bullet up his right forearm. He lived sixty miles from the nearest doctor in Jordan. In that bitterly cold weather, it took too*
>
> *(continued)*

We girls shot magpies that bothered our cattle. In the fall, when huge flocks of blackbirds swooped into our grain fields, we'd shoot at them, killing a few. In a noisy, violent rush, the others would swarm off for a few days.

Often we practiced target shooting with the .22, nailing up a target on the clothesline post against the gravel hill. Dad taught me that when I was six and first wanted to shoot something.

I'd killed skunks, sneaking around the chicken house. But for some reason I hadn't hunted deer.

Still, if Dad told us to do something, no matter how difficult, it always seemed to work out okay. Sometimes he gave us pretty tough assignments.

He didn't want us up on the range during hunting season. It wasn't safe—too many hunters and bullets flying around, he said. Most hunters were responsible, but some just shot at any movement in the brush. Others got so excited when they saw deer, they didn't watch where they fired.

At four the next morning before daylight, I took the deer rifle, an open-sight .250 Savage, saddled up Dusty and started for the school section hills. Deer hid out in those wooded draws and, if I rode back there early in the morning as they returned to shelter, maybe I'd get an easy shot.

Mom and Dad bought this land adjoining ours from the school district, when it was no longer needed by schools—two sections from every township so designated in homestead days. Ever after we called it 'the school section'—our half of a square mile of good rangeland with its own springs and one long, deep ravine leading down under the railroad trestle into the Yellowstone River.

The sun wasn't up yet. In the semidarkness, the crisp fall morning was just right for a ride up on the range.

After a while Dusty and I approached the draws of the school section. As I neared the top of each small rise, I dismounted, dropped the reins and Dusty stood still while I crawled on my stomach to the top, watchful of jumping cactus, and peeked over to spot deer. I stalked them this way, with great caution, because I needed every advantage at this strange new game.

Suddenly, there they were, grey shapes through the predawn munching grass, on the opposite slope across the coulee. I waited till the sun shot a single ray directly behind the deer. I counted the shadows—seven. I lay down on my stomach and sighted carefully on the closest one.

It was too dark to judge its age, so I waited. Dad always said not to shoot the deer with the biggest rack—the meat will be tough and besides, the largest bucks need to be spared to sire future big, strong deer.

As I waited, I took stock of what to do next after I shot the deer. Bleed it out, of course. In his numerous hunting stories, Dad pointed out the importance of quickly cutting a deer's throat and bleeding it out to drain blood from the meat.

*Whoops!* Suddenly I realized I hadn't thought to bring a hunting knife. I had to decide, while lying there with the bead on what became a younger buck in the dawning light, whether I should shoot it or just go home. Without proper bleeding, the meat would be ruined.

My buck turned broadside as he ate and in the new light, I saw that if I carefully placed my shot, the bullet might, just might, with luck, cut the jugular vein. I found the place on his neck.

I hesitated and checked his antlers again. Yes, a nice young buck. It's great to stalk your game and have plenty of time to shoot. I gently pulled the trigger and the buck dropped in his tracks while the others ran over the hill. A lucky shot, indeed.

When I got there, his eyes shone bright, shiny black with the gold morning sky reflected in them. They focused on me, then gradually lost their luster and in a few seconds I saw his life go out, just like a flame that dies down into black embers. As his eyes went slowly dull, I felt sad.

Fortunately he lay with his head downhill. Gushes of blood ran in little rivulets past his chin on down the hill, through the dried grass and coagulated in a pool at the base of a clinging sagebrush just below. Hunting was supposed to be heroic. He didn't have a chance. It didn't seem fair.

I'd done all that was asked of me. I jumped on Dusty and galloped hard to the ranch to get Dad to come back and dress him out. I burst in the door of the house with my rifle in my hand.

"I shot a deer! He's up in the school section. He needs to be dressed out!"

Dad was on his way out for morning chores before breakfast. I just knew he'd go up and gut the deer.

But no! My job.

He explained how to dress out the deer. With his explicit descriptions, I could picture just how to do it, but hadn't believed he'd insist on it. Dad only explained things once, so I listened carefully.

I rode fast back to the deer, dressed him out carefully, hearing his voice in my ears.

*"Do not cut into the gall bladder! It's green."*

I was careful, but after I cut open the deer from throat to tail, his stomach gaping open and all his entrails piling out, I couldn't lift his shoulders high enough to empty him. I slashed farther and ac-

> **Safety first** (*cont'd*)
> long before Uncle Will received medical attention and he never recovered full use of his right arm and hand.
>
> There were times, Anne recalls, when she wished for a bullet in the rifle chamber. "I'd sneak up on a deer and then, 'Snick' went the shell, as I slipped it into the barrel. Deer always seemed to hear that little snick."
>
> A good rule, though. One morning Francie got bucked off when her horse spooked from the sting of a loose wire in a barbwire tangle. She was riding with the deer rifle in her scabbard and as she fell, the horse bucked the rifle free. It bounced once and landed, pointing straight at her. Francie thought of Uncle Will and was mighty glad it was unloaded.

*I crawled on my stomach to the top, watchful of jumping cactus, and peeked over to check for deer.*

*Respect meat hunting as an honorable frontier tradition.*

We hunted according to an unwritten code—respect the game you hunt, don't waste meat.

cidentally cut into the gall bladder. A horrible smell rushed out, and I stumbled and fell backward to the hard ground as green bubbles formed.

Realizing the meat would be ruined if any of the syrupy green gall touched the meat anywhere, I scooped out the entrails with my hands and got everything out neatly. The deer was cleaned, thank goodness, but I was a mess.

I wiped my hands and arms as best I could on the stubbly grass and with great difficulty hoisted the deer carcass up behind the saddle, fighting Dusty all the way. He was skittish and kept sidestepping away from me. He skittered around me in a circle as I tied on the deer. Curving his head around, he saw it and snorted.

All this time I was planning a circuitous route back to the ranch in order to avoid the game warden.

He probably rose from his bed in Miles City this morning and sniffed the cool fall air.

*"Yep, today is the right day to be at the Brink ranch."*

When I threw my leg across the saddle, before I caught the other stirrup with my boot, Dusty bucked me off in a patch of jumping cactus.

He wanted that deer off his back and *right now!*

But I already had an alternate plan—hang the deer up by his hind legs.

I dragged the deer into the coulee to a cedar tree, tied the rope around his hind legs and finally got him strung up—head down. The buck was certainly lighter without his insides, but still awkward to handle.

Even though probably no one else would come around to find the deer, the game warden might. I climbed the slope several times to look around and check how well it was hidden in the branches, high enough so a wild animal couldn't get it but low enough so lots of branches hid it. I rearranged it several times. My deer was safe and no insects around to ruin the meat; they disappeared with the recent frost.

I wished I had a towel to clean myself, but again wiped with handfuls of the stickery grass and cleaned off my boots. The smell stuck with me.

Without the deer on his back, Dusty happily took me home. Stopping just behind the last hill above the ranch, I walked him slowly forward and sighted over the crest.

Alas, it couldn't be. Surely not today! That certain pickup was parked in front of our house.

I rode back and approached the buildings from behind the corrals. Darn Mr. Nicolai's friendship, probably relaxing at the table

# Rangeland Hunting Ethics

During our growing-up years we hunted according to a general code of ethics followed by ranchers and most western people. Mom and Dad believed in the balance of nature and its human interconnectedness. They practiced conservation every day. We learned we had a share in nature's abundance, but with that came a sense of responsibility to treat the land and wildlife in ways to sustain them for generations to come.

■ Respect the game you hunt. Shoot to kill quickly. Don't let animals suffer. If we severely wound a deer and it disappears, it's our responsibility to find and put it out of its misery.

■ Don't waste meat. When you shoot a deer, take care of the venison and don't let it go to waste.

■ Don't be too quick to shoot the largest buck with the biggest antlers. The longer he lives the more he passes on his superior genes to future generations—he's probably tough eating anyway.

■ Respect meat hunting as an honorable frontier tradition. Ranchers naturally take offense at boastful 'trophy hunters' who denigrate 'meat hunters'—as if killing the buck with the biggest rack is somehow more honorable than hunting to feed one's family. Ranchers are generous, yet sometimes harbor a streak of antipathy toward hunters who make it evident they deserve only the most magnificent buck on the rancher's land and seem to relish killing him, yet disdain eating the venison.

■ Be discreet. As a wise rancher, use good judgment. Landowners furnish all the grazing, hay, grain and protection for deer throughout every day, every year, so it's reasonable to shoot a deer out of season now and then. But don't talk about it. Don't boast. Keep it quiet. (Not that ranchers don't enjoy a good hunting story—they tell their hunting tales with relish, often with a self-depreciating twist.) Of course the rancher buys a hunting license. But he might be too busy to hunt, along with the extra work of protecting cows, horses and gates from trigger-happy folk who drive through on a hunting spree—and helping them out of whatever trouble they land in.

■ Work with the local Game Warden and game officials. Keep on good terms. Responsible landowners and hunters have the same goals: conserve and protect game and habitat and improve the outdoors experience for everyone. All need to work together. However, relationships with the game warden are subtle and can be complicated. Respect his need to enforce the law. Don't expect to be pals. Don't force his hand. When the law needs to be bent, simply do it quietly. A game warden with integrity and experience understands the needs of landowners and works with ranchers who are good stewards of the land.

*Note: Waste was not an issue in our early hunting years. Deer were scarce and we hunted because we needed the meat; our beef was necessary for income. Later on, coyotes disappeared under a government eradication program and deer populations exploded, especially into our alfalfa seed fields on early fall nights. We watched them filing down cedar draws to the irrigated alfalfa before sundown and trailing back up into the hills at dawn. As many as two hundred might feed there overnight, eating and trampling a great deal of alfalfa seed, another much-needed source of income. Fish and Game officials advised us to shoot them and let them lay if they destroyed property. It was the law. But we couldn't do that, especially when we knew people who needed the meat. It violated ethical principles.*

drinking his third cup of coffee. I washed up behind the granary in the lower corral water tank, but still could smell deer. I rinsed my arms and face again in the cold, clear water, but the smell permeated my clothes. So I hung out in the barn with my horse until Mr. Nicolai left. Then I eased into the house, careful lest he return.

Only years later did I understand Mr. Nicolai's dilemma. He was a good, honest game warden who needed to work with the ranchers because, after all, they are the ones who raise, feed and care for the game all year around and, in season, allow strangers to come onto their property for hunting.

*Only years later did I understand the game warden's dilemma.*

He was concerned about destruction of crops by deer, but also could not authorize out-of-season hunting by ranchers or any others. He needed to uphold the law and arrest any violators if he caught them. But maybe he also believed that ranchers might need some extra venison at times and deserved a little more than the average hunter. He may have noted that ranchers in hard times couldn't afford to continually eat their major cash crop—beef.

When he said, "If I ever catch you hunting deer out of season, I'll have to arrest you!"—I think he was telling Dad to go ahead with what he needed to do, but not to let him know about it. It was unfair to put him in a position where he had to compromise his own principles or do harm to a helpful rancher.

Dad and Mom understood, but continued to deal with deer in practical, traditional ways.

We girls didn't know any of this. So we were worried he was trying to catch us. Actually, this was probably all right because it meant we didn't talk about it much and kept the evidence out of sight.

That first hunt of mine was not the easiest morning of my life, but the eating was delicious.

## Crow Rock Antelope Flight

Stones and gravel scattered everywhere and gray dust trailed off to the east as the pickup slammed to a stop in front of my isolated schoolhouse. My third and last student just went home in another pickup, disappearing on the road to the north spreading its own trail of dust across the short-grass prairie.

JEANIE

Breathing quickly and almost out of breath from excitement, my new friend, Shirley, jumped out of the cab and ran to meet me.

"So glad you can spend the weekend with us! We're going antelope hunting Saturday. You'll just love coming along," she said.

I liked to hunt and was interested in antelope. We were out in the wide Cohagen country where the ranches were miles and miles apart—plenty of undisturbed grass for the beautiful animals.

Methods used to hunt game depend a great deal on what kind of game you're hunting, and with whom you're hunting. All the time Dad was teaching us to shoot, he emphasized safety, safety, safety.

My worst fears were realized early that Saturday morning. Five or six cowboys showed up at sunrise at the ranch, almost as eager to meet the new school teacher as to get started hunting. They stood in a tight group by their pickups, pretending not to notice me until introduced, but taking peeks anyway.

Shirley, her husband and I walked over to them. I asked Shirley if she brought her rifle.

"Oh no, I don't shoot. I'm just going along for the fun."

No one offered me a rifle, but always there were several extras at every ranch. I assumed I'd be shooting, too, because I was invited to hunt.

"Does anyone around here have an extra rifle?" I asked her.

Hearing that, a cowboy grinned shyly at the others.

"Nope. Hell, no! No woman knows for nuthin' about shootin' a gun, nohow."

He grinned at the other cowboys who, equally shy, grinned back. A cool joke among themselves… Someone muttered something back as they jumped into the pickup bed. I said nothing.

I heard of another rural teacher the next year who shot her abusive husband right there at the school. When the news reached me, I thought of those trigger-happy guys in the back of the pickup who didn't think women can shoot. I guess she knew how to handle a gun.

Shirley jumped into the cab of their pickup, slid over next to her driver husband and motioned me to climb into the seat beside her. I got in, slammed the door and rolled down the window.

So there we were—cruising along, she and I supposedly hunting, but without rifles. A little strange, but who's to argue.

Behind me, the load of guys stomped around the truck bed. The pickup sped along.

Through the rear window I saw them loading their rifles and filling their clips as we turned off the relatively smooth dirt road and flew across the lumps and bumps of the raw prairie. We raised dust in long rolling clouds behind us.

On a ridge we stopped to scan for antelope. In the early morning there was hardly a breeze. Nothing moved. The grass, still standing high from the summer rains and the rising sun turned the prairie golden for as far as I could see.

"There they are! The big herd I saw yesterday," one of the cow-

As a country school teacher, Jeanie keeps handy a .22, with scope she purchased from her first paycheck.

boys yelled. "Let's go!"

A mile away about thirty antelope grazed with only their white rears showing. Antelope are the same caramel color as the dried grass, except for their bellies and rumps that are a shocking white. Even at this great distance the sharp-eyed animals that can see grass move three miles away, suddenly took off to the west.

"There they go!"

"Step on it!"

"Floor it! Yippee!"

The guys and the rifles they held banged and bumped together back there. Down the coulee we went. It seemed to me a useless chase—they were so far away and almost out of sight over a low ridge. But our driver didn't hesitate. He shoved the gas pedal to the floor.

Shirley and I struggled to keep our seats. Bouncing at high speeds over the prairie was not safe.

I was alarmed, too, that Shirley's husband would punish his vehicle so, but he was caught up in the spirit of the chase and nothing else, not even the welfare of his beloved pickup, mattered to him now.

It's a wonder no one was accidentally shot, because in the truck bed the rifles were held with abandon as the hunters tried to keep their balance.

I don't know why we didn't break an axle.

"There they are!" one cowboy yelled.

Someone fired, then another.

"Stop that shooting! We're too far away!"

The driver swept down off the ridge and across the flat. The antelope fled, their white rears bobbing as they ran.

That was the worst part—catching them. We hit scary high speeds. Never in my life had I travelled so fast over rough country.

Gunning it over sagebrush, coulees and washout holes we gained slightly on the herd. Shirley and I clung to each other and grabbed at anything in the cab we could.

Running down antelope herds—seemingly a popular event out

*The men bounced all around back there, rifle butts slamming into the floor. Those rifles were loaded.*

Antelope flourished in the big open Crow Rock country north of the Yellowstone.

in that wide country—was not for me. As Shirley's guest for the weekend, how could I have refused this debacle? To them my being invited was an honor.

On our ranch we stalked our game. Who wants to eat meat heated up from running? Makes it taste gamey. Yuck! Venison is great when the animal is shot while quietly grazing.

We gained on the herd, collectively a soft carpet sweeping over the prairie, low and flat to the ground, almost without motion, yet rapidly swallowing up great distances.

We suddenly hit a hole that slammed Shirley and me against the roof. The seemingly flat prairie held sharp rocks and prairie dog holes hidden in the dry grass and we hit them all.

Going at such a speed made it difficult for the shooters to stay in the back of the pickup while they crouched down and gripped the steel sides. They bounced all around and thumped and bumped, their rifle butts slamming into the floor. I knew those rifles were loaded. I saw them put in the bullets.

The driver hung on to the steering wheel, but Shirley and I had nothing—and bless her heart—she was enjoying this immensely.

The caramel brown of the pronghorn antelope and the distinctive white patch that covered their rear flared out as they ran—dozens of marshmallows bounding up and down on a dun colored prairie ahead of us. As they reached top speed the bouncing stopped and they seemingly flowed along smoothly.

When we pulled up to the herd, they stretched out low to the ground with their legs extended far to the front and back of their elongated bodies. Normally three or four feet tall, now they flattened to the ground. All in sync at about sixty miles an hour and not one hoof appeared to touch dirt.

Our herd was in a tight group, smoothing out like a flight of low-flying birds. At that speed, not one bounced or stumbled although there were rocks and rough spots on the prairie. I think they could have gone on tirelessly for miles. We were the ones who hit all the bumps.

A three-strand barbed wire fence cut across our paths. We were now on a dirt road and had a cattle guard to cross so we didn't slow down.

I was sure the herd would turn at the fence because antelope don't jump. But no, without slowing down they flowed under the lowest wire unchecked—seemingly all in step, a poetic sight.

During this time I couldn't say a word, but the yelling from the back continued. The animals were so beautifully choreographed and this deliberate action on our part, so unspeakably crude.

Dad would not have been pleased.

## Pronghorn antelope

Once almost numerous as bison, pronghorn antelope nearly became extinct as the west was settled. But with good conservation practices by stockmen, the herds made a strong comeback in short-grass rangelands.

Antelope have the largest eyes for their size and the keenest eyesight in the animal world. The eyes are located far back on the sides of their heads so they can keep watch even while feeding.

The antelope's great speed and remarkable vision are adapted for life on the plains, giving them the ability to spot predators such as coyotes and in former years, wolves, miles away. In short-grass country their very survival depends on their eyes and their exceptional speed. They are faster than any land animal except the cheetah, and can maintain their speed far longer.

They are not really antelope, but pronghorn, a species with no known living relatives. African antelope are not even distantly related.

*The last man tripped and his rifle went off accidentally. No one even noticed.*

I tried to see the speedometer but it was impossible; the needle was jerking around the dial too fast. As we wildly closed in on the herd, I felt blessed indeed, not to have a rifle. I didn't want to shoot these defenseless creatures.

When we pulled alongside and hung parallel to them, the shooting began. Wild shots peppered the antelope and the hills around. No one could possibly have used their rifle sights. Dust spurted all over the prairie and bullets zinged off rocks from misguided shots.

After several antelope dropped, that was it. Our mad race was over.

The driver jammed on the brakes. Before he bucked to a stop, the hunters leaped off the side of the pickup to get their game.

The last one tripped and his rifle went off accidentally. No one even noticed.

## Lively Skunks

"A skunk just went around the back of the granary! Hurry, before he gets in the chicken house!"

Here came Dad, excitement on his face and delight in his eyes. This should have forewarned me; but how could I resist the eagerness with which he presented the .22? How could I resist such confidence in my marksmanship? How could I resist the sudden responsibility of joining the ranks of my hunting older sisters?

ANNE

Briefly I wondered, *Was there a reason my older sisters didn't shoot skunks?*

This was my first hunting experience and, like my next older sister Francie, I was pretty young. Our family had a target set up on the clothesline post, and once we were strong enough to hold up a .22, we were encouraged to 'hit the target.' Jeanie was only six years old when she started shooting; I was a little older.

Stealthily, with rifle in hand, I peered around the granary.

Sure enough, there was the skunk, slowly lumbering toward the chicken house. Someone had forgotten to close the door after the hens went in for the night, and that skunk was bent on using its advantage.

First I ran to the chicken house door and stood there challenging the skunk.

*Lesson learned: always stay upwind from the skunk…*

The skunk slowed, raised his head, nearsighted eyes squinting as he puzzled out my existence.

*Shoot!* I told myself. *Quick, before this big guy turns around, lifts his tail, and blasts me.*

Skunks usually move slowly, but this one managed to reverse directions as I took aim and fired. The yellow squirt of skunk aroma flew by me just as I saw him fall over, and too late the breezes blew this fine mist all around me.

Dad was delighted when I sought him out after killing the skunk, though he did mention that I ought to have Mom 'help me clean up.' Mom, needless to say, was not excited about my killing the skunk, did *not* want to help me clean up, and in fact, did not even let me come in the house.

"Here's a bar of soap," she said, with her face scrunched up and holding her nose. "Go out to the water tank. I'll bring you some clean clothes."

*Second lesson learned: if you fail to stay upwind, don't expect any sympathy, or help in ridding yourself of the smell…*

This was a year of many, many skunks. Some carry rabies which spread to domestic animals. It was a wet spring, which seems to bring more mice, which brings more skunks, who are excellent 'mousers.'

"There's a skunk in the chicken house!"

This announcement brought our entire family to red alert.

There's nothing like the smell of skunk. It sticks to your clothing, clings to your skin no matter how hard you scrub, and goes before you, announcing your presence to all in your path.

Growing up youngest in a family of four girls taught me many things I would have otherwise not known. For instance, I always knew skunks were bad news. They smelled; they sprayed when startled—on the dogs, in the barnyard, on the car, on anything they could get close enough to, often including my older sister Jeanie, and therefore possibly me.

My mother hated skunks for a practical reason. Besides mice, skunks also loved to eat eggs. Any hen that stole her nest out away from the hen house could expect to have it raided by a skunk. And since our chickens rambled all over the barnyard, a few hid nests there, good bait for skunks.

Mom had lots of chickens. During and immediately after the Depression, raising and selling poultry and eggs was a steady part of our ranch income. So from my earliest childhood I remember my older sisters in the evening, running back to the house after closing the hen house door.

"There's a skunk in the chicken house," one of them would loudly yell to Mom.

Mom dropped whatever she was doing and dashed to the rescue.

*Stealthily, rifle in hand, I peered around the granary. The skunk slowly lumbered toward the chicken house.*

Western Ranch Life in a Forgotten Era

*Anne perfects her marksmanship by shooting targets on the clothesline post.*

> One day
> the inevitable happened—
> I got a direct hit.

She seemed to know how to avoid that awful smell. But my older sisters, when they tried her methods, were not always so successful. Often they came back reeking of skunk. Unless it was a direct hit, soap and water or tomato juice did the trick.

When old enough to graduate to the chore of closing the hen house, I had already observed this proficiency in the art of getting the skunk out without being sprayed. Of course the ideal method was simply to close the door early in the evening before the skunks, nocturnal animals, started prowling. But that meant not too early, or some wayward hens got locked out, not having gone inside to roost. If left outside, an owl or coyote might get them.

Once inside, the skunk climbed up to the nests, a neat double row of connected wooden boxes, and crawled inside to get the eggs. But if we gathered too early in the evening, hens that laid late in the day added their eggs for the skunk to eat, leaving fewer for sale. Everything had to be timed just right. Gather eggs after all hens laid, close the door after all chickens were on the roosts—but not so late skunks got in.

Ceramic eggs in nests encouraged the hens to lay there rather than on the floor or to steal a nest outside somewhere. I often wondered what the skunks thought of those white china nest eggs.

After my first skunk experience, the responsibility of ridding the ranch of skunks was mine. They say that 'skunk scent' is used in making the most expensive perfumes, but I didn't believe it.

We had several years of skunk overpopulation, so it was important to keep down their numbers. Problems grew worse in the fall when they moved in close around buildings and the young were full-grown and hungry.

Reducing the skunk population every fall was my job and I became quite adept at their extermination.

It wasn't so bad when I went to the county school. Other kids had the same skunk duty, and odor, that I had. Our teacher, the only odd smelling one in school, kept the windows open in defense.

Unfortunately, our small country school closed, so I had to attend town school in my fifth grade. Town schools are filled with kids who grew up together and knew each other well, so it's hard for a new student to join and make friends. Having known many in 4-H and Sunday school, this wasn't a real problem for me until later in the fall when skunks became active.

This was the fall that I lost count of the number of skunks I killed after my tally reached twenty seven.

One day the inevitable happened: I got a direct hit.

Dad had large wooden calf feeders that he pulled in and out of the corrals, and these had skids on the bottom that skunks could hide under. I'd peek under one end and see a furry ball, then wait until the skunk came out. Unluckily, this one skunk was very close to the end where I went to peek, and he was all ready when I bent down... We used some of Mom's homemade lye soap on me, tomato juice, lemon juice, and anything else we could think of.

But when I left for school the next morning, there was still that particular aroma preceding me.

Our fifth grade room teacher seemed to think this odor was coming in the windows, and quickly closed them. But still the odor persisted.

"Yah, yah! Knew you were coming before we saw you," teased some of my friends during recess. I knew they were wishing they could hunt skunks, so I just grinned.

But the real insult was when the kid who sat behind me came running up.

"Anne's wearing her special perfume again," he taunted.

We country kids had a lot more muscle from ranch chores than the town kids did, so I promptly cold-conked him.

"Girls aren't supposed to be so strong," he mumbled as he staggered back to his feet.

However this was a good education I gave him, as later in life he became a highly successful Chief of Police in one of Montana's larger cities, encouraging women to join the classes he started in strength building and self defense. Even so, it's unrecorded how many women there defended themselves, instead, with skunk perfume.

## Two-Fifty Savage

The big buck with a huge set of antlers slowly and steadily climbed the ridge. Quickly I bent down and crept to the top of the small gravel hill, making my profile smaller and less like a human on the off chance he turned back and saw me.

It was a long shot. Thankfully there was no wind, and avoiding prickly pear cactus, I wiggled myself into a prone position. He angled up the hill away from me as I waited, hoping for a side shot to get a sure bead on him at that distance.

Sometimes a mule deer will turn and look back down the hill as it briefly pauses after a steep climb. This could give me a still shot and a perfect angle.

I decided on a neck shot. At that distance, if I aimed almost at the top of his neck, the bullet's trajectory dropped it down four or

*I knew they were wishing they could hunt skunks, so I just grinned.*

*Antlers on a whitetail buck grow with prongs coming from the main branch, unlike those of mule deer.*

five inches. No head shot this time; he was too far away. Getting him in the neck would be good—it meant less meat messed up and lost when I butchered him.

Too many times I'd listened to Dad's admonitions and aimed for the heart. First of all, this meant ruining the heart, which was my favorite meat. If the bullet entered the deer a few inches too high or to the side it meant cutting away blood clots and slimy, bloody-shredded meat from the shoulder.

Disgusting.

I followed this buck with the two-fifty lined up along his neck as he purposefully climbed the gravel and yucca strewn steep hillside. Not the best shot. I waited, hoping he'd turn and look back my way.

"Get out there early, the big bucks leave the field before dawn and you can intercept them just before they reach the badlands," Dad told me the night before.

Usually I wanted younger animals—fat and tender spike or three-pronged mule deer were the best eating. But for two years, friends of a relative from back east came to our ranch on the first day of hunting season and I was delegated to take them out to the deep coulees where deer hid out in daytime. They wanted only trophy deer, big-antlered bucks they could hang over a fireplace and brag about their hunting prowess.

Guess I listened too well, and decided I should have one of those hanging on our wall.

So there I was, flat on my stomach in the dirt, tracking this big one with the two-fifty three thousand Savage.

This was also the first time I was using the .250 for a year or more. Dad bought me a brand new thirty-ought-six two years back. In our family we never bought expensive gifts, so it was a total surprise when he handed me this beautiful rifle.

"Now you can fit right in with those Eastern greenhorns you take hunting," Dad said with irony in his voice. "You can take just as long shots as they do and learn to brag about it."

At our home we never bragged. It just wasn't done.

Dad expected us to know our limitations, and not shoot unless it was a sure shot. He always asked us about our shot when we came home with a deer: how far was the shot, was it uphill or downhill, was the wind blowing, was the deer running or standing still?

Then he proudly told his friends about the skilled shots of his daughters. But while we could talk about each others' exceptional shots, we knew never to pat ourselves on the back.

So when he talked about the Eastern trophy hunters' shots, I just grinned. We both knew I outshot them every time with the little .250 open sights—and them with their expensive high-powered scopes.

The .30-06 was shiny dark mahogany—not a scratch on it—and amazingly there was a big scope mounted on top of the barrel.

My heart fell. It was too big and fancy.

But Dad was so excited—his eyes sparkled and he grinned from ear to ear. How could I hide my disappointment? But I did. This was a huge gift. I was in awe of this magnificent rifle and handled it reverently.

I pulled back the bolt and looked into the barrel. No bullet of course. Then I stepped out onto the porch and raising the rifle to my shoulder sighted at Mom's clothesline pole where magpies often perched.

The rifle was heavy, made for a big man. It fit awkwardly into my shoulder and felt unbalanced. I had to take a much shorter grip on the barrel with my left hand, and then couldn't hold it steady with my hand so close to the magazine. The slick finish on the stock slipped against my hands as I held it—not like the worn, easily gripped wood on the two-fifty.

*How can I ever hit anything—let alone a running deer?*

I was glad Dad was behind me and couldn't see my face. By the time I put the rifle down and turned around, I was ready to smile, grateful for the gift.

*But if we needed another high-powered rifle on the ranch, why couldn't Dad have given me the old trusty .250 that I loved so dearly, and kept the new .30-06 for himself?*

This was a very special gift, and I thanked him in the manner in which it was given.

Inside, I hurt. Now I'd have to use it. So ever after when I went hunting I took the shiny new .30-06.

Hunting became a chore. No skill needed to sneak up on a deer—with the scope I could place my shot at any distance, leading the animal if it was running, or aiming a little higher if it was far away. No longer a need to outwit my target. It was like shooting a yearling steer docilely grazing in the pasture. I had become like those Eastern hunters.

But this day, with the .30-06 loaned out, I held my old friend the .250 in my arms. And now I was on a 'real' deer hunt.

While I waited for the deer to reach the ridge top and stop to look back, I thought back lovingly on the many hunts this gun and I had together.

At ten I shot my first deer. Not with the .250, but with the .22. After killing a number of deer with the .22, Dad realized I was a serious hunter and insisted I use the .250.

"The two-fifty will shoot straight in any wind, and you can lead

> *At that distance, if I aimed near the top of his neck, the bullet's trajectory dropped it down four or five inches.*

your shots if the deer is running because it's much faster. Uphill or downhill it will do a better job for you. I know it's heavier, but from now on I want you to take the .250," he told me.

There was no way of knowing how many deer I killed with that gun. It became my trusted friend, accompanying me on many exciting hunting adventures—stalking deer far back in the hills, following hoof prints in the snow, riding out on a brisk spring morning and scouring the alfalfa fields at dusk.

I entered the hunting arena just at the time when our deer were multiplying fast, without coyotes to keep their numbers down. They became a terrible nuisance and ruined fields and overran pastures. Dad pleaded with all his friends to come out and shoot deer on our land.

"If they're destructive, kill the deer and leave them lay," the Game Warden said.

But we couldn't waste meat like that. Instead Dad sent me out to bring in the venison and we ate every deer I killed. I always got a license during hunting season and those deer I could talk about. The others quietly filled the freezer. All shot with the .250.

So I was glad to have the feel of this trusty gun in my arms.

I lay there watching—the 'V' of open sights filled with the steel front post as I followed the heavily antlered buck picking his way to the top of the ridge.

He stopped and slowly turned his body sideways to me as he surveyed the steep coulee behind him.

I held my breath and slowly squeezed the trigger.

A loud bang. I watched as the buck ran over the ridge and out of sight.

This could not be! I didn't believe it! That deer didn't even break stride.

*How could I possibly have missed him?*

Thinking he'd be lying dead on the other side, I ran across the brushy coulee and up the opposite ridge. Nearing the top, I found his tracks in the gravel and at the very top saw his footprints where he turned sideways to me. Next, the tracks were farther apart as he bounded away from my shot.

Cresting the ridge I expected to find him lying in a heap a few leaps away—but he was long gone.

I was stunned. *The shot was perfect. I must have hit him.*

Tracking him further, I spent an hour following him until I could see where he went deeper into the badlands. Not even a spot of blood.

I finally had to admit that I'd missed the deer, and headed back home.

*"If they're destructive, kill the deer and leave them lay," the Game Warden said. But we couldn't waste meat.*

*Why? Why had I missed that perfect shot?*

Walking the long way home, I finally decided I was no longer able to hunt. Something—maybe I had become shaky—ruined my ability. Since I had no idea where my bullets went, maybe I was not safe with a rifle.

Dad, Mom and the hired men were at the table eating their noon meal when I walked in.

"What happened?" Dad asked, noticing the dejected look on my face, and seeing my hands were not bloody from field dressing. As usual he quickly sized up the situation.

"I don't know," I sadly replied. "It was a perfect shot."

I explained my hunt and ended by telling him I just couldn't shoot anymore.

There was silence around the table.

"No Anne," Dad said. "It has to be the rifle. That two-fifty has shot many, many deer. I used that rifle to bring home the meat for years before you were even born. The barrel must be burnt out."

It should have made me feel better knowing Dad had such trust in my ability, but somehow it seemed even worse. My friend, the .250, couldn't just quit working like that. I felt it wouldn't let me down.

"You set up a target and we'll sight it in after we finish eating," he said.

We walked out a short distance beyond Mom's garden to our usual shooting range, where I'd set up a paper target in front of a cutbank backstop, 200 yards away.

Carefully I took a fine bead on the small bull's-eye, held my breath and gently squeezed the trigger.

I fired the usual three times. Dust blew up in the cutbank each time I fired—wide of the target.

"No, no, Anne. That's just terrible. You've shot a foot or two wide all around the target." Dad shook his head. "Try again."

My second try was no better.

"You do it," I said.

"No. You would have put those three bullets into the middle of that bull's-eye. I can't do any better."

"Please," I said. "It can't be the rifle. Just fire three shots."

Dad took the rifle from me and fired.

As he laid the rifle down I ran to the target.

Triumphantly I held it aloft. Three small holes pierced the bull's-eye, making a perfect triangle.

The rifle was okay! The barrel was not burnt out!

But my elation was short-lived as I realized my hunting days might be over.

*Discouraged at missing shot, Anne wonders if she can trust herself to hunt again.*

"It's okay, Anne," Dad said. "You've hunted a lot of deer and had a lot of fun. Sometimes we have to turn to other things."

Mom took a different view when we returned to the house.

"You've been shooting your new rifle with a scope—makes things bigger and easier to hit," she said. "Time to get an eye exam. I'll call Dr. Rowan."

As usual, Mom was right.

A short time later, wearing my new glasses, I held the .250 to my shoulder at our firing range.

*Bang! Bang! Bang!*

And there, right in the middle of the bull's-eye was the perfect triangle pattern of three bullet holes!

## Seven Deer in Seven Days

Once more I saddled up, took the .250 and headed for the hills.

I had returned from a week-long high school trip to find Dad hospitalized with pneumonia while Mom taught at the distant Cottonwood rural school. She filled in for a teacher who abruptly abandoned her job.

It was a busy time at the ranch, with fall harvest in full swing. Dad asked me to carry messages back and forth between him and the hired men, besides cooking for them and the neighbors who were helping.

The first two requests were easy to fill; shuttle messages to the men and cook for them. But that's when a big problem appeared: the meat in the freezer ran out. I was alarmed—the big chest freezer was usually well stocked. And how could I cook for working men without meat?

There was no way to contact Mom at her isolated school, and Dad was quite ill so I didn't want to concern him. I certainly didn't have money to buy meat for that crew, but I could rely on my skills.

After school I drove to the hospital and got the men's work orders from Dad for the next day and, at home saddled up, took the .250 rifle and headed for the hills that flanked the ranch.

Hunting on horseback helped to avoid getting the deer spooked; but of course, we *never* shot from horseback. When shooting, we tied the horse to a sagebrush or cedar tree and crept over the hill for a still shot. If there wasn't a suitable place to tie the horse, often he ran off and I had to walk home, or found him waiting at the nearest gate. Sort of a game our horses liked to play, I think.

Then while hunting, I had to deal with rattlesnakes. They

*The meat in the freezer ran out. How could I cook for working men without meat?*

warmed themselves in a sunny spot those short fall days and coming on them suddenly could give a snorty horse an excuse to buck. Getting bucked off was to be avoided: these badlands were covered with jumping cactus. And besides, how could I land on the hard ground without damaging the rifle?

It was just before the opening of deer season, so the deer weren't easily spooked. Simple enough to find the deer, circle them and get close enough for a sure shot.

I was lucky the first day, and shot two small, young bucks not far from the ranch. Good eating!

I had long ago learned to use a head shot, since I cleaned and butchered the venison myself. Much easier when the innards weren't all shot up. Dad chided me for not taking the easier heart or lung shot, but my favorite meat was the heart and I didn't want it ruined.

I field-dressed the two bucks, then dragged them to a place easy to load when I returned with the pickup. With big deer I cut them in quarters, since they were too heavy for me to drag and lift into the pickup. This time I only had the weight of small deer to manage and loaded them with ease.

Once home, I kept an eye peeled for the hired men, since I didn't want them to know of my pre-season hunting.

Then there was the Game Warden to watch out for—when Mom was home he might be at the ranch eating breakfast when I came in from an early morning hunt. I learned to stop at the water tank and clean up a bit—wash off the blood, and wipe as many deer hairs off my clothes as possible. Still, I often wondered if he suspected.

Game wardens appreciated our family since Dad welcomed their policies and made the ranch available for planting young pheasants and fall hunting.

I lived in horror of being caught, but not worried enough to quit hunting.

The hired men never went in our family garage, so this was where I hung the deer and cut them up. Built into a hill, with three sides against banks, the large double garage stayed cool.

I soon discovered with all the errands Dad had me doing, and going to school, there wasn't time to butcher the deer. So before cooking, I ran out to the garage and sliced off a roast or steaks for that meal.

Mom was a great venison cook—and certainly she cooked plenty of it over the years. Her delicious venison Swiss steak rolled in seasoned flour and browned, then simmered with canned tomatoes and onions, was a favorite at our table. She served her roasts, crusty on the outside and moist inside, with boiled potatoes and flavorful

*Whitetail deer flash their tails as they dash for cover. We said they wave their long white tails back and forth like arms.*

venison gravy.

Steaks, like those I cut from those young bucks just naturally turned out tender and tasty.

I had helped enough to know Mom's techniques and the men seemed to like my cooking. Anyway they kept lavishly praising the food—and eating more. Evenings I prepared large meals so there'd be plenty for them to heat up next day at noon.

I was awed at how much meat they ate and by the third day it became evident that I needed to go hunting again.

This time I found three tasty young bucks and repeated the process of getting them safely into the garage and finally into the oven and frying pan.

By this time, Dad showed signs of improving—but the men kept eating. The last three deer dwindled fast.

So it was back to the horse and the .250.

This time I stalked a big dry doe and another young buck.

Surely this meat would keep the men filled for awhile—seven days in all. Plenty to last over the weekend when Mom would be home.

A last day of frantic activities: message carrying—hunting—cooking—and school. Dad was released from the hospital and I was there to bring him home.

As we pulled away from the hospital he asked in a casual voice, "Did you get the beef I took in to get butchered at the meat packing plant?"

*Whaaaatt?*

Some days are like that—blood and guts all over—fighting with a horse that wants to dump me—watching for rattlesnakes and jumping cactus—lugging floppy, bloody, dead deer up and onto the back of a pickup—hiding out from hired men and the game warden.

All the while butchered beef, frozen in neat packages, was waiting in the town locker for me to take home.

Dad saw my stunned look and asked, "Well, what did you do for meat?"

*Seven deer in seven days—that's all—and no one the wiser!*

> *By this time, Dad was improving—but the men kept eating. The last three deer dwindled fast.*

# 7 On the Home Front

## Amazing Cherry Pie

"Jeanie, I need you in the corral," Dad called as he burst through the back door.

"But I'm making a pie for dinner." Rising quickly to the situation, Jeanie turned her dimples on him. "Uncle Chet and Aunt Margaret will be here."

Jeanie loved dabbling in the kitchen. Not ordinary things like potatoes, roast and beans, but more enticing foods like stuffed cabbage, doughnuts, divinity or latticed pies. Mom was a meat and potatoes cook, so if Jeanie wanted something more exotic, it was up to her to fix it.

On the ranch many tasks required help from us girls. At these times, Dad enthusiastically swept through the back door announcing Jeanie just had to help him with a certain ranch project that needed two more hands.

Jeanie was always more than ready for him.

"But I just got the fire built up. I'm starting a batch of doughnuts,"

Mom, who wearied of cooking three big meals every day, usually chimed in, saying that she had her hands full with other chores, and that Jeanie was making the meal special with her added delicacies. Dad had a notorious sweet tooth, and couldn't resist the appeal of Jeanie's desserts.

Along about the time of the June rise of the Yellowstone River, when it had been raining for several days, Jeanie gathered together all the fat she could find for frying raised doughnuts, and spent all afternoon in the kitchen. She spread flour all over the table, and on it, circles of doughnuts covered with several dishtowels while she waited for them to double in size to fry them.

So the extra help Dad needed fell to Francie or I, or whatever hired man he could pull from his field work.

*There is nothing like coming in from the cold to the tempting aromas of fresh fried doughnuts! But not all Jeanie's concoctions fulfilled their promise.*

Western Ranch Life in a Forgotten Era

If we were fishing in the Yellowstone or outside working in the cold rain it was such a delight to come into the warm kitchen where Jeanie's roaring fire heated her doughnut frying fat.

And the smell! There is nothing like coming in from the cold to the smell of fresh fried doughnuts bringing with it the promise of crunchy sweetness.

But not all concoctions fulfilled their promise.

That Sunday, Dad decided to vaccinate a few calves that got missed, which meant he needed help. Beverley was away for the day. And what was Jeanie doing? She was making a cherry pie for Sunday dinner.

Cherry pie takes a lot of time to make. First she made the crust, then started with canned cherries. She thickened the juice, adding just the exact amount of sugar, cornstarch and real butter. This she poured into the bottom pie crust, and set it just under the pantry window ledge. Spring breezes floated through the window, cooling her pie filling.

*On another day, Jeanie wields the vaccine gun, her usual job when we work calves.*

Normally she then started the task of weaving the lattice crust on top, but noticing the kitchen stove was not hot enough, she momentarily left her pie to add more wood to the fire.

While she was gone, Dad finished vaccinating the calves, washed his syringe and placed it on the pantry window ledge to dry.

Sometime later Jeanie returned, rolled and sliced the top crust into narrow strips to weave across her pie. This was a painstaking job, weaving carefully over and under. Through the years, she became a master at this—the finished, uncooked pie was perfect.

She slid it into the oven, keeping careful watch on the stove as it baked. Wood kitchen ranges, with no thermostat, have minds of their own. Skill is needed to control the heat, to bake a pie to golden brown perfection. Cherry pie would be the ideal ending to Mom's roast beef meal.

Sunday dinner came and the table was filled with good food and fellowship. The big moment arrived for dessert. Dishes were cleared from the table and Jeanie brought in her masterpiece. All eyes were on the pie.

The honor of cutting it was given to Aunt Margaret.

"Dear me. I surely hate to cut this pie," she exclaimed. "I surely hate to mess up this beautiful crust by cutting it!"

We all held our breath as Aunt Margaret poised the knife over the pie. With a decisive swipe, she cut into it.

"Klink!" There came a loud sound of metal on glass.

She carefully moved her knife around. "Klink!" it went again.

"Oh my, dear me," she cried.

Jeanie grabbed the pie and leaped from the table.

Mom quickly followed and they went into the pantry and shut the door.

Soon the door opened and Mom and Jeanie returned to the table with a plate of cookies.

What happened to the gorgeous pie? We didn't often get cherry pie for dessert, but looking at their faces, we knew better than to ask.

As soon as permitted to leave the table, we dashed for the pantry to see what had happened to the cherry pie. There on the counter sat the pie with its peeled-back crust exposing Dad's six-inch vaccinating gun, smeared with cherries and cherry juice.

The syringe had fallen off the window ledge—landed in the middle of Jeanie's open pie—then was covered with juice and eventually, her lovely lattice crust.

Around Christmas time, Jeanie made fudge and all sorts of other good candies. Dad's parents were both from Sweden, and like most of that ancestry, he had a sweet tooth for candies and pastries of all kinds. We kids all inherited that trait. One type of candy that Jeanie never perfected was divinity, though it took a disaster for her to admit defeat.

Baking at Christmas time was also a Swedish tradition, and family and friends expected to be treated to a lavish array of goodies when they stopped by. Jeanie knew many people could make good fudge, taffy, and caramels—few could make good divinity. Good divinity was stiff enough to drop from a spoon on a cookie sheet. Inside the outer crust was an oh-so-smooth, melt-in-your-mouth creamy mixture. This became a challenge to her culinary skills.

So each Christmas, and a few times in-between, Jeanie cooked up a batch of divinity. It was difficult to get it to set, so we scooped it out of a big cake pan and ate it with a spoon. Then Jeanie would seek out another recipe and gleefully begin the extended ritual of trying for perfection again.

This demanded her total attention and concentration. Ranch activities carried on without Jeanie. She was making a special treat for us all.

Out in the corrals, we dodged mad cows and devious calves, slipped in new manure, and choked on dust kicked up by running cattle. But all the time we thought of the divinity Jeanie was making, and how we'd pop a morsel into our mouths.

Late in the afternoon we finished, and we all met in the kitchen for coffee or a glass of milk. Where was the divinity?

"It seems not to have set," Jeanie mournfully announced. "I put it out in the back room to cool." And then hopefully, "Maybe it's

*What happened to the gorgeous pie? Looking at their faces, we knew better than to ask.*

hard by now."

"I'll get it!" I jumped up from the table, thinking how I could quickly slip in a finger and get an extra taste.

There it was—the cake pan placed high on a shelf above Mom's Maytag washing machine in the back room. Carefully I raised both arms and gently lowered the heavy cake pan.

"Yiiii! Yuck!" I almost dropped the pan!

There in the middle of this pan full of smooth soft white divinity was a little gray mouse. He looked up at me with his beady black eyes, front feet struggling to escape the divinity. A faint path behind the mouse showed where he fought to get through the sticky goo.

"Jeanie, Jeanie," I called.

Jeanie came rushing in—I thrust the pan at her. She looked in horror at the little gray mouse.

Quietly, empty-handed, I returned to the table and let her deal with it.

Soon Jeanie came walking into the kitchen with a plate of cookies in her hand.

Once again, cookies saved the day.

*We could hardly shirk our duty, when others gave so much. Gold stars hanging in people's windows attested to that.*

## For the War Effort

"Jeanie, Francie and I will hoe beets. Sure we will!" Beverley turned toward us. "Won't we?"

Well…

It didn't sound like much fun. We hoed weeds in the garden; a hard, never-ending job.

"We'll get paid. Just like the Mexican beet workers!"

Jeanie and I looked from Beverley to Dad, both of them smiling enthusiastically. We exchanged glances. No match for their enthusiasm myself, I looked to Jeanie for help.

*How can we hoe an entire field?*

"I've got a small field just right for you girls, just thirteen acres and pretty clean of weeds," Dad said. "I'll go sharpen the hoes."

It was too late to protest. Of course. We already knew it. Beverley volunteered us and was off and running. Just a matter of working out the details.

And it would be nice to earn some money. We each received a small weekly allowance, but for most of the work we did at home we were not paid, of course. We didn't expect it.

*We're all in this together—for our family and the war effort. We need to pitch in and do what we can to help—whether with the cattle, horses, garden, canning or housework. Alas.*

But was this one of those jobs you get stuck with and can never escape again? Probably. Once started we couldn't quit and we knew it.

An element of patriotism kept us on track, as well. We could hardly shirk our duty, when others gave so much. The gold stars in town windows attested to that. And blue stars, too, for sons in the service. We had no brothers, but Mom's brother Chet was serving in the Seabees on the Aleutian Islands.

At the time, we didn't know we'd win that terror-filled war. The Nazis swallowed up one country after the other, goose-stepping viciously across Europe. One dreadful event followed the next and Americans responded by coming together to meet the terrible needs of war.

Sugar, rationed and severely restricted for domestic use, was urgently needed in the war effort. It was an important ingredient in all explosives, from bullets to bombs. During wartime, Dad, like many others with irrigated land up and down our valley, plowed up his crops and planted sugar beets.

"Let's go. We'll need our big hats. And a jug of water," Beverley said.

Dad took us out to the field. We set our water jugs in shade at the end of the rows. That field looked awfully long.

"This is the beet plant. See, with just two light green leaves? They're coming up all along here, so it's easy to tell them from the darker weeds. We use a short-handled hoe for thinning, like this."

He leaned over and chopped a six-inch bite right out of the row, whacking out beet plants along with weeds.

"Take out one hoe width between plants. That gives room for the beet to grow and spread out."

"What if two are growing together, like this?" asked Beverley, taking a swipe with her hoe.

"Don't leave any doubles. That's important. Pull out the smaller, weaker one with your fingers. Leave the strongest."

*How will I decide which one lives and which dies, when they look so much alike? What if I decide wrong?*

"Okay, girls. Each of you take a row and let's get started here."

Dad watched while we hoed a short distance. Then satisfied, he drove off and left us there.

"Oh, look how far we've come in such a short time!" Beverley waved cheerfully back to where we'd started.

*Sugar—an important ingredient in all explosives from bullets to bombs—was urgently needed in the war effort.*

From left, Francie, Jeanie, Anne and Beverley hoe sugar beets, needed in manufacturing ammunition during war years

Western Ranch Life in a Forgotten Era

# Garden Bountiful

Mom and Dad raised a big irrigated garden every year. We all spent lots of time planting, pulling weeds and harvesting the great quantity of vegetables and fruits that grew there. Everything flourished under irrigation and Mom and Dad often marveled over their bounteous harvests, in stark contrast to results in their earlier struggles gardening on dryland hillsides during the years of drought and depression.

All summer we ate fresh radishes, lettuce, green onions, peas, beans and carrots, along with fruits like strawberries and raspberries. One of Mom's specialties was creamed new potatoes with fresh peas. *Mmm—simply delicious!*

When ripe, we filled overflowing buckets and tubs with gorgeous red tomatoes. We picked heaps of cucumbers and cabbage, also pumpkins, squash—and the muskmelons and watermelons that tasted even better after first frost.

Summers, it seemed that Mom—and often we girls, too—worked all morning in the garden and all afternoon canning and caring for the produce. All that became markedly easier when we got electricity and a freezer in 1951, and Mom shifted to easier, faster freezing methods.

We grew many new plant varieties selected from Wills' Seed Catalog while snow still covered the ground. And to brighten house and garden, Mom lined the garden patch with rows of flowers. Zinnias and her favorites, delicate cosmos.

Mom loved asparagus, but didn't need to raise it. It grew wild along our main ditch bank—spread by seed from somewhere upstream. Each spring she sent us out to cut the tender spears poking through green grass.

Two rows of plum trees furnished us and our neighbors with lots of good eating and more fruit than Mom could can or make into jam.

If Dad had time, we sold strawberry plants in spring when the strawberry patch sent out thousands of vigorous runners in every direction. To prep them for sale, on the day he went to town or the night before we laid out ripped-open burlap bags, counted fifty strawberry plants onto each, laid half each direction on the burlap, added wet sawdust and rolled them into tight bundles. With thirty bundles, at fifty cents each, they brought a tidy sum.

Their abundant garden was critical to the family food self-sufficiency that Mom and Dad always practiced. But by no means was it the only one. They raised and preserved most of our own food: beef, venison, pork, chicken, eggs, milk and cream.

Our family had always plenty to eat and a generous, well-balanced diet.

In early years our garden grew below the ditch, which meant easier irrigation and productive soil, but less convenience since our house was above the ditch. In fact, we were fortunate that our house, barns and corrals were built above the ditch, unlike many other ranches in the valley. We benefited from a drier, better drained location, with fewer weeds and mosquitoes and plenty of room to spread out.

So one year Mom and Dad decided to move the main garden above the ditch in a handy spot near the house. Dad and the hired man hauled in truckloads of rich topsoil and manure, fenced out the cows and piped in ditch water with a gas pump set in the ditch.

From then on they could direct water precisely where and when needed.

Best of all, it was a quick run out to the garden for a handful of lettuce or a cucumber for salad. And on frosty fall mornings Dad swung by the melon patch on his way in to breakfast, bringing a couple of cold and delicious muskmelons to the table.

*Mom and Dad hand cultivate their large garden, providing us vegetables for the entire year.*

Amazingly, we had covered a lot of ground with our short-handled chopping. Because of Dad's careful cultivation, the weeds were in a very thin line with the little beet plants, so it went rapidly.

Thinning beets took about three weeks. Up three rows and down the next three; up three more and down again, across the entire field. And those rows were long. We rested a short time at noon, then back to the field.

When I debated over which double to pull out Beverley came over to help.

"It really doesn't matter, they're so close."

She laughed. "Besides, if you accidentally take out the bigger plant, it makes the other one stronger. It'll grow all the faster—so all is not lost!"

At the end of the first day we were dead tired. I wondered how we could continue, but was determined to keep up with my older sisters, and they kept going. So I did, too.

The second day, they admitted to a few aches and revealed a blister here and there on hands and heels.

The third morning we all felt so stiff and sore we could hardly get out of bed and by evening Jeanie and I weren't so sure we wanted to do this. But Beverley remained cheerful and kept up our spirits.

"The third day is always hardest," Mom told us, with sympathy, at suppertime. "Tomorrow will be easier."

She was right.

The fourth day went quickly and easily, much to our surprise. By then we were in the swing of it and hoeing went well after that. We didn't ache and could work all day long.

Dad sharpened our hoes every night—and they were sharp. An expert at this, he ground, filed and even heated them in the forge to pound out a thinner edge.

When we finished thinning, that big field looked awfully skimpy—only those flimsy little plants with their two light-green leaves spaced out every six inches all the way across it. But Dad said it looked good.

We hoed each field three times. First, thinning and weeding; then weeding twice more. All this had to be done at just the right time.

After thinning beets in two or three fields we had a week or so of respite, then came the second weeding.

Dad cultivated close to the plants and kept the fields clear of weeds between rows. We again used a short-handled hoe with a lot of bending over because the plants were still small. Short as we were, bending and hoeing hour after hour could still get painful.

Irrigation enables us to grow large vegetables such as this squash Francie is showing.

## Raising potatoes

Planting potatoes took a whole Saturday in early May, usually in the most wet, cold and miserable weather possible. But Beverley and Jeanie didn't seem to mind. Maybe they didn't have plans for their Saturday. And always they had a few jokes up their sleeves.

Some people planted just enough potatoes in their gardens for their own use. But our folks raised potatoes for sale—in great quantity. We planted a couple of acres—or even more—in one of our most fertile and weed-free fields, and they produced abundantly under irrigation.

Down in our dark basement we cut the old withered left-over potatoes into sections. On every piece we left two or three eyes, from which sprouts emerged.

We loaded these seed potatoes into large buckets and burlap bags and hauled them out to the field in the Bug, our old cut-down and wheel-base-widened truck. We always had to laugh, loading our stuff into that Bug and hanging out a leg or two over the side.

Then taking one row each, we planted the potatoes about five or six inches apart in shallow furrows that Dad plowed. When we finished, he harrowed the soil back to cover them.

If weeds grew, we hoed.

If potato bugs infested, we plucked them off and dropped them into a can of kerosene, walking up and down the endless rows. Or, though they didn't let us, Mom and Dad dusted them with deadly toxins, Paris Green powder or DDT.

Then, worst of all, came harvest. 'Picking potatoes,' we called it.

*(continued)*

We got so tired hoeing. We rested some at noon, but never quite enough. Still, there were fewer weeds this time.

In other fields, we saw our Mexican workers bending over much farther than we did, moving rapidly up their rows.

*Do their backs ache more than ours because they are taller and work faster?*

When the third hoeing came along, the beets were bigger and we used long handled hoes. Finally we stood upright. So did the Mexicans, we were glad to see.

Sometimes, on a hot afternoon we turned silly, and the more tired, the sillier we got, giggling helplessly over nothing much. Still, always aware of the need to keep one beet plant every six inches.

"Oh, gosh, we need a rest," Beverley might say, wiping sweat from her eyes. "When we get up to the end, let's go wade in the ditch."

We took off our shoes and ran in the water, splashing and laughing. Then, much refreshed, we returned to work and the rest of the afternoon went quickly.

By fall, large white sugar beets burst from the ground, completely filling the spaces we left for them.

Sugar beets grew well under irrigation, but were a tremendous lot of handwork, so the government brought us a series of workers. Mainly ours were Texas Mexicans, Mexican nationals and German prisoners of war. They swept in, worked hard a few weeks, then disappeared, to be replaced by others when needed.

Each year Dad saved out two or three small fields for us—less weedy and easier to work—of five, ten or fifteen acres.

Beverley was about twelve when we started hoeing beets. She was our organizer and taskmaster, and always a cheerful one. With Beverley, everything was possible and everything could be an adventure.

Jeanie a year and a half younger and I, two years younger than that, made up her somewhat-flimsy team. But Jeanie was tenacious, too.

Anne at age five was much too small, but as always, ready and eager to join us. She loved being out in the field with us. Mom would bring her out for awhile and let her work beside us.

Later she got her own row.

"Jeanie was so good," Anne said. "She was the fastest of all. When she finished her row she came down mine. I always felt so relieved when I saw her working on my row—it was such a good feeling. I didn't want to be left behind and miss out on the interesting things my sisters were talking about."

And Beverley was right about the pay at the end—it was pretty good. Dad paid us by the acre, the same as the Mexican workers.

And although we finished at the same time they did, each acre took us much longer.

We pored over the Sears and Montgomery Ward catalogs, tallying up our probable earnings.

The last year Jeanie and I sent for Victory bicycles. At that time, near the war's end, they didn't make the greatest bicycles with the best rubber. Like sugar, rubber was rationed and substitutions made. The bikes we yearned for boasted wide balloon tires, but when ours arrived they ran on hard, narrow tires, not easy to ride—but we had good times with them and rode several years to school.

With help from friends, Dad butchers three hogs. Meat is canned in glass jars, smoked as ham and bacon, eaten fresh—and some taken to a meat locker in town. Mom renders the lard in oven.

We always wanted to earn money and Beverley, our enterprising older sister, was ever eager to launch a new money-making job for herself and us.

Gardening offered another opportunity, since our parents always raised a large, productive garden. Beverley grew and sold sweet corn from about age twelve, while Mom steered Jeanie and I into cucumbers.

We learned to pick our cucumbers in their prime—and then keep them picked so new little cukes kept coming and none grew large and pithy.

When Mom planned a trip to town she'd call for us to get our cucumbers and corn ready. Jeanie and I hurried out to our cucumber patch with the little red wagon, filled it with our best specimens, carefully clipping their stems and cleaning them with a damp cloth where needed. Mom delivered them to the grocery, where she traded eggs for family groceries and collected payment for our vegetables, always at their freshest.

We earned some cash, too, from our many 4-H projects—mainly clothing, foods, garden, home décor, and once, my two yearling steers—from ribbons and awards at the Custer County fair. Jeanie's breadmaking won top awards, and selling my steers added money to my college account.

We saved money for special items and college, recording our small deposits and withdrawals in little brown bank passbooks.

From the time we were young, Mom told us about her good times at the University of Montana in the early 1920s, so we looked

*Francie's 4-H Hereford, freshly washed and brushed, with a neighbor who stopped by to inspect the steer.*

forward to our own college years and saved for them. She put herself through four years of college entirely by her own efforts, mostly in low-paying service work, like washing dishes and waiting tables—where she refused tips because of her Scottish pride and independence.

When we collected enough savings we bought war bonds, costing $18.75 each, to be cashed out at $25 in ten years. We felt good about those bonds; that we could lend our money where so urgently needed—to support the Allied cause on both the European and Pacific fronts.

As in hoeing beets, with each money-making job, Beverley set us an example of making fun out of the worst drudgery jobs.

The most challenging work of all was hoeing beets. We liked earning money, but it was hard, tedious work and we surely could not have continued but for Beverley. She cheered us on and made even that dull job adventurous. We stuck with it three or four years, until the war ended and Dad stopped raising sugar beets.

Jeanie and I had our amazing Victory bikes to show for it, plus money in the bank.

## Frog Egg Breakfast

"Look! Frogs' eggs." Francie pointed, crouching down at the pool's edge.

There in the shady, spring-fed pool, the frogs lay motionless; their eyes and snouts just breaking the surface. We never minded the frogs that shared our springs and swam around in our drinking water.

ANNE

We poked the puffy masses at the far edge of the pool just under some low bushes.

Sure enough, upon closer examination we noticed the grayish jelly mass had little black dots in the center of round, tiny balls. They all clung together in a sort of gelatinous goo.

It was exciting to find frogs eggs. To Francie and me it meant that soon there would be little tadpoles—or as we called them—pollywogs, swimming through the water. At first they have no legs; just a big fish-like belly with two eyes. The front legs appear, the back end still looking like the tail of a fish. Finally, little bumps emerge and grow into long hind legs.

In the past when we found frog eggs, Francie and I frequently

### Potatoes *(cont'd)*

Again it took a perfectly-good Saturday and sometimes the entire weekend, unless we had help from a hired man or State School boy. Dad let the potatoes mature till after frost and then he chose the worst cold, rainy or snowy fall day—weather too miserable to do anything else. He plowed ahead of us, carefully turning the furrows, while we carried along our buckets, bent and picked up all but the smallest potatoes, digging with cold fingers through mud or snow to get those hidden gems not quite exposed by the plow.

*Ugghh, potatoes! What a miserable, muddy job.*

But Mom and Dad invariably exclaimed over the prolific harvest of large, sound red potatoes to take us through another year.

detoured by the spring to eagerly watch the metamorphosis of the eggs into pollywogs and then into tiny frogs.

To Jeanie, frogs' eggs meant something entirely different.

"Hummm. What do you suppose we can do with this?" Jeanie said with that particular gleam in her eyes, as she scooped up a large gelatinous mass in both hands.

We were a little shocked that she picked it up in her bare hands, but then we realized the wheels were turning in Jeanie's ingenious mind, and we eagerly awaited her proclamation.

"It looks a little bit like cooked oatmeal," Francie said.

"Yesss!" Jeanie cried. "Let's put this in the hired men's oatmeal! Won't it be a good joke when we tell them what they've eaten!"

"What do you mean?" asked our cousin Patty with horror.

Our two cousins from Oregon were visiting. A rare event, so we had brought the girls, Patty and Bobbie, on a ride up to Spring Coulee—one of our favorite cool spots on the open range above our ranch.

Bobbie and Patty, brought up to be proper small-town ladies, were instantly alarmed. However, they were learning we had no such silly restrictions. After all, how could Francie and I resist Jeanie's excitement and enthusiasm?

Jeanie's brain simmered with ideas.

One time when the hired men and the rest of us were at the dinner table, Jeanie discovered a new straw hat one of the hired men had just purchased.

What a temptation! It didn't even have any dirt on it. Just returning from hunting prairie dogs with the .22, she set his hat on a yard post and shot a hole through the crown—smack in the middle of where his head would be. A new hat had to be initiated!

That hired man was a little testy when he found his new hat.

Or the time we stuffed the toes of another hired man's boots with cockleburs—or the time we put a dead garter snake in the hired man's bed.

He slept there two nights before he found it and was pretty mad about that.

So Francie and I watched Jeanie eagerly as she grabbed an old can beside the spring and let the ball of frogs' eggs slither in. Triumphantly she mounted her horse and led the way back home, clutching the can close to her side.

As soon as we rode into the corral, Jeanie took her can with its precious contents and placed it in the far end of the horse trough, behind the winter heater where Dad wouldn't notice. The cool water coming into the trough ought to preserve the frogs' eggs as well as the spring water where the mother frog laid them.

*Jeanie delighted in thinking up possible pranks and shenanigans.*

Western Ranch Life in a Forgotten Era

# Whooping Cranes Arrive

Every fall, ducks, geese and other migrating water birds stopped by our swamp, ate leftover grain and stayed awhile, if hungry and the weather favorable.

One day around 1941 a flock of a dozen or more whooping cranes glided down, landed and spent several days resting up in our swamp by the big curve in the highway. They grazed in leftover millet in a harvested field nearby and slept with heads tucked under cocked wings, looking like white sails floating in the swamp.

They would have made a tasty lunch for coyotes, so Dad watched them by day as he worked, a rifle close by. People traveling on the highway were astounded to see such great birds so close. If a car stopped, Dad went right over to talk about the beautiful birds' vulnerability and need for protection, and encouraged them to move on. At night the Game Warden came to watch them. No one who wanted a trophy got the chance.

We girls walked home from school the long way on the highway to see them while they stayed, with instructions to move quietly and not talk. To us girls they seemed huge; whooping cranes stand as tall as five and a half feet with a wingspan of six feet across. It seemed unusual, the game warden said, to have that many whooping cranes in one migrating flock. Now, with increased population, these cranes appear to migrate in smaller, family groups.

At that time it was thought there were only fifteen or sixteen whooping cranes left in existence. Most of them might have been in our swamp that fall.

Next morning Jeanie got up early. This alone should have alerted Dad that something was up. Jeanie was not a morning person. Mom was pleased to have the extra help getting breakfast for the family, which on this day included our visitors and two hired men. Some of the men Dad hired were real dopes, and these two were at the head of Jeanie's list. She had no tolerance.

Dad insisted on cooked cereal as breakfast's first course for him and the men.

"Oatmeal sticks to the ribs and it's a long time until noon."

They finished the meal with all the pancakes, eggs and bacon they could eat.

After cooking oatmeal, Jeanie and Patty managed to slip a quantity of the uncooked frogs' eggs into the bottom of the men's bowls. They blended in well. Even Francie and I couldn't see anything extra in the bowl.

The men ate their oatmeal in silence, then went on to Mom's sourdough pancakes, eggs and bacon.

The five of us were a little disappointed that no one noticed our efforts, so we decided to have frogs' eggs again tomorrow.

The second day of oatmeal and frogs' eggs went just the same. We were perplexed. It wasn't a joke on the men if they didn't know about it. However, we decided to try again the next day. Surely someone would notice the little jellied balls with the black dots in the center.

It seems that either Mom or Dad might have wondered why Jeanie appeared so early each morning—three days in a row.

Unfortunately, on the third day, Mom announced we were out of oatmeal, so Jeanie needed to make Cream of Wheat instead. No matter, we thought. Frogs' eggs and Cream of Wheat works well, too.

Jeanie dropped some frogs' eggs into the bottom of each bowl and cousin Patty spooned in white Cream of Wheat. Jeanie added another dollop of frogs' eggs, covered by another scoop of white mush.

The resulting Cream of Wheat had a strangely gray tint. We held our breath and watched covertly as the hired men took several bites, then quickly put their spoons down. Dad looked into his cereal bowl with suspicion.

We girls burst out laughing.

"OK girls, what did you do to this cereal?" Dad had only to look at us to realize we had pulled a prank.

We gleefully explained.

The hired men looked a little sick, but we hurriedly passed them Mom's good pancakes and syrup.

"Oh, my," Mom said, turning back to the stove.

Dad chuckled.

Why didn't our parents get upset with our pranks?

Secretly they may have enjoyed them. When they lived in the Breaks on the Missouri they seldom saw their friends. When possible they spent evenings around the supper table, repeating tales for everyone's enjoyment. They made fun when they could. There was little to laugh at in a day's work, they said, so pranks on each other seemed uproariously funny. Dad's best stories were ones he told on himself, related with much enjoyment and laughter.

## Saved by a Whirling Dervish

Flames surrounding Donny lit up the terror in his eyes as he crouched behind the stove in shock—red gas can thrust toward the fire.

Just waking up, I heard from the kitchen a soft *'BOOM'* followed by a crackling *'Whoosh'* of rushing air.

Although not loud, the terrible, unknown sound brought me leaping out of bed and dashing toward the kitchen—unaware of all my glory in my bright red underwear.

*The terrible, unknown sound brought me leaping out of bed and dashing into the kitchen.*

There in the kitchen, red flames shot to the ceiling and leapt along a trail on the kitchen floor into the pantry. Backed in the corner behind stove and sink, trapped behind the flames stood Donnie.

He held the red can of high-test gas in his hand, gas dribbling from the can onto the trail of flames. This most volatile gas that we had on the ranch, and *never ever* used to start a fire, was dribbling from the can onto the trail of flames.

On winter mornings Dad rose early and we'd hear him shaking down the ashes in the kitchen cook stove. If the fire was out, he started it anew with wood and a little concoction, kerosene mixed with ashes kept in a metal can with a lid. If not, he added wood, then went out to the corrals and barns to check on the livestock. When the rest of us got up the rooms were warm, the kitchen range hot and ready to cook breakfast.

Dad banked the fires in our three coal-burning stoves every night so they gave out only a small amount of heat. Placing large chunks of coal on the glowing embers when he went to bed and closing down the damper, or draft, meant they burned slowly and saved fuel through the night.

The kitchen stove, built for cooking, was not as efficient as the heating stoves. Often the embers burned out during the night and a new fire had to be started. But not always.

On this morning, Mom was gone substitute teaching at a rural school, and the job of getting breakfast fell to Anne and me. Something alerted Dad to the barns, so he left the house before getting the fire going in the kitchen stove. This was not a problem, since he knew Anne and I would do it if he didn't return before we got up.

This time, though, Dad had hired a boy from the State Industrial School in Miles City to work on the ranch. He was eager to help, but with little life experience.

The State School was a place for boys in trouble, juvenile offenders, but sometimes they were kept there because they had no place else to go. Donnie was one of those boys. That day he got up early and decided to start the kitchen fire, which he'd never done before.

Anne and I abandoned our unheated bedrooms upstairs, as we usually did in winter and slept in what was originally the dining room, set off from the living room by double doors and located behind the kitchen. It was warmer here because stoves burned in rooms close by, but our downstairs bedroom could still get cold.

On frigid nights Anne and I slept in thick flannel pajamas or long underwear. I had just received a brilliant red two-piece set of long underwear for Christmas. I'm sure Mom envisioned me wearing this under my jeans as we rode after the cattle in times of blizzards and bad weather. But during the school week I had other plans for this soft, warm, comfortable suit—it was perfect for crawling into icy bed sheets on a cold winter night.

Anne said she envied me this elegant fashion statement!

So there we were, sound asleep in our warm double bed, with wool blankets and a thick wool quilt pulled up around our heads when I heard the soft explosion and flames crackling from the kitchen.

The kitchen range and sink were across one end of our large kitchen, close to the pantry—a large hallway leading to the cream separator room and back door with cupboards on one side and a long tin-covered counter on the other.

Donny crouched in the corner beside the stove.

*Beat out a fire with a rug,* we were often told.

Grabbing up a throw rug with both hands I hit hard at the flames and, surprisingly, they went out in that spot.

It worked! I beat harder.

"Get out! Come on through!" I shouted at Donnie, with a path momentarily clear.

As he escaped and ran by me, to hover near the door, I saw he still held the red can.

*Backed into the corner stood Donnie, trapped, holding the can of high-test gas. Red flames shot to the ceiling.*

"Yank a coat from behind the door and help me beat the flames!"

But he just stood there stupefied.

Flailing frantically, I attacked the leaping flames again with the rug. Watching uneasily over my shoulder as I pounded the flames, I half expected another explosion from the can of high-test gas, not yet empty. Luckily it didn't blow up. Donnie just stood there, helplessly, looking stunned. Finally he set down the can. I expected him to come help, but it seemed he couldn't move.

"Anne, help!" I called, hoping she'd hear me and rouse from her deep sleep.

No time to waste; I turned back to fight fire. All along the floor in front of the cook stove leapt a high, thin line of orange flames where Donnie had spilled or thrown the gas. I swiped at them with the rug. Looking beyond, I saw most of the fire was now in the pantry hallway, and so high and intense I could not see past it.

Turning my attention there, I began beating out the even bigger and hotter fire.

Anne appeared in the doorway and instantly took it all in. She grabbed coats for herself and Donnie.

"Come on!" she commanded, thrusting an old coat at him.

Released from his helplessness, Donnie followed her. Anne directed him to beat out the stray flames in the kitchen and rushed to help me.

Together we attacked the raging fire in the hall pantry. Flames ate into the floor-to-ceiling storage doors, bubbling the paint. Cardboard boxes on the long tin counter were starting to burn, which would soon catch the walls.

So far, it seemed that mostly gas was burning, with dancing flames and the pungent smell of gas everywhere. We knew we had to work fast before the woodwork and floors caught.

Then, as suddenly as the raging fire began, the leaping flames disappeared; the fire was out.

Anne and I stood back and looked at each other with relief, astonished at what we had done. With no thought of calling the fire department or running for safety, even with half the kitchen in flames, we succeeded in putting out the fire by ourselves!

Eyes filled with horror, Donnie related his story.

"The gas exploded and suddenly everything went up in flames!"

When he came downstairs he found the stove cold and figured the fire was out. He thought to help by starting it with kerosene like he had seen Dad do it.

But the fire was not dead. And the can he grabbed was not kerosene, but the white, high-test gas we used only in our Coleman

*Flames bubbled the paint. We had to work fast before the woodwork and floors caught fire.*

lamps, stored in a red can to remind everyone of its danger.

When he poured gas onto the red-hot coals, it blew up. Donnie in his terror still held the can, letting it splash gas across the floor as he retreated in horror into the corner behind the stove. He was just lucky the gas can didn't explode all over him; it still held gas when the fire was out and Anne set it outside.

Donnie didn't even get his eyelashes singed.

Anne gave me an odd look and broke into a little giggle.

*What was so funny? The fire was out, but black soot covered the ceiling. Mom would not be pleased.*

Then Anne pointed at my outfit, and I realized I was still attired in my long red underwear.

Quickly I departed the kitchen while Donnie was still in shock.

Donnie was not laughing.

"I was trapped back there," he gasped, with his dark eyes big and round. "Then through the fire came this red devil—a whirling dervish, dancing, swirling and eating up the flames as they came closer. The devil saved me!"

How embarrassing. *Why did I ever wear my lovely red underwear to bed?*

> *Quickly I departed the kitchen while Donnie was still in shock.*

*Jeanie hangs out clothes on washday*

## Echoes in the Kitchen

"Think you can do some mowing for me this morning, Beverley?" Dad said as he came through the door. "The alfalfa in that field by the bee hives is ready to go."

FRANCIE

"Do I have to do it right now? I just got my bobbin wound and the tension tightened on the sewing machine so it's finally just right."

Beverley protested, but at the same time she sighed and put away the skirt she was sewing. It never failed. Dad came in for help just as we got started on what we hoped was a morning of sewing. No one else was using the sewing machine and our thread was on the bobbin—probably wound on top of someone else's thread, because we only had two of the long narrow shuttle bobbins for our treadle Minnesota sewing machine.

Changing thread could screw up the always-touchy tension big time.

We each began sewing 4-H projects at age ten, old enough to join the Blue Ribbon 4-H Club. From then on we girls designed and sewed many of our own clothes—dresses, blouses, skirts and jackets—and exhibited them at the Custer County fair. Jeanie once made a lined green wool suit, earning a Best-of-Class purple ribbon, the overall winner of all the clothing projects.

We modeled our winning clothing in a style show on the grandstand at the beginning of the big evening show. Special and scary, too.

But we learned to sew much earlier.

As a six-year-old, I sat on a stool beside Grandma while she cut small, dainty quilt blocks for a doll quilt we made together. She threaded my needle and tied the knot, then guided my stitches as I sewed two blocks together, and then two more and finally a square.

An excellent seamstress, Mom every year or so received a big box of clothing from her aunts. We girls had fun dressing in the grown-up clothes, parading around with maybe a purse and hat or two from the box.

Mom looked it all over and planned what garments to refit for herself or make anew for us. Then she started ripping seams apart.

We girls helped Mom bake from the time we were able to cut sugar cookies and make 'snowmen' from little balls of bread dough. Over time we each developed our specialties.

Beverley's specialty, lemon meringue pie—delighted us all. Mmm—it was so good! She spent one summer perfecting it for the 4-H demonstration contest at the fair.

She fed us several versions and we eagerly ate them all. Even those she considered failures—a slightly less tender crust or an imperfection in baking the golden meringue. Our wood-fired oven, ever temperamental, could be roaring hot or way too slow or heat unevenly, not browning her meringue uniformly. An electric stove was still in the future, waiting for the electric lines to come.

However, the luscious, lemony tang of her fillings, smooth texture with hint of grated rind, never failed. Anne and I hung close with spoons as she filled the baked shell—to forestall her scraping bare the saucepan.

Jeanie often made bread, winning a first in the State 4-H bread baking contest. But, sadly, she didn't get a state or national trip because of war and post-war economies, as Anne and I did later.

One hot summer day, while Mom was shopping in Miles City and I was home alone, I mixed up my usual large batch of bread—enough for five loaves and a pan of cinnamon rolls. Then I had an inspiration. There sat several flats of freshly picked strawberries, plump and ripe.

*How can they not make a delightful addition to my loaves? After all, raisin bread is good—strawberry bread will be even better with its bright red color and delicate flavor!*

Unfortunately it was bad idea. The dough didn't look that great

### Ripping seams

Using her special method for ripping seams in garments she remade, Mom took one side of the fabric in her left hand and pulled it taut against the seam, while she wrapped the other side over her right hand, also pulling tight. In her right thumb and forefinger she held a razor blade, which she sliced against the stitches. Every little while she pulled with both hands to rip additional stitches. Then again a deft razor stroke or two, and another rip. We heard her in the night, ripping seams by dim kerosene lamplight; often it was hard-to-see black thread on dark fabric.

*Beverley fed us several versions of lemon pie and we eagerly ate them all— even what she called failures.*

as I kneaded in my mashed and stringy strawberries. As it began to rise in the warm pantry, it looked even worse. The strawberries turned into a mass of terrible-looking gray splotches throughout the bread dough.

It didn't taste good either. Yuck! What a mess! Dismayed, I realized it looked like I stirred in gumbo dirt.

Six loaves of bread spoiled.

*I have to get rid of this or I'll never hear the last of it! One look by the hired men or my sisters and I'll be teased forever.*

I didn't want to waste it, but feeding rising dough to the pigs might make them sick. Stuffing it down the drain—not good for the plumbing.

I punched it down into a tight ball, grabbed a shovel, went out to the garden, dug a hole and plopped in the disgusting mess, covering it over with garden soil.

Problem solved.

What I didn't notice, as the sun beat down, heating up the already warm earth, was that my dough began to rise there in the garden, breaking through and lifting the soil that covered it.

Later Mom came home from town, but didn't get out of the car. Instead of hurrying in with sacks of groceries as usual, she just sat there.

Finally I went out and opened the car door.

"What's the matter? Are you all right?"

Mom gazed out toward the garden, never moving her eyes or changing expression.

"Oh, my *stars,* Jeanie!" she said. "That is the biggest mushroom I have ever seen in my life!"

*Anne and Francie share a magazine after dinner, while taking a break from pulling weeds in the garden.*

**Jeanie punched down the bread dough into a tight ball and buried it in the garden, not noticing the sun beating down.**

"Stop," Jeanie commanded the hired man. "The butter's coming! Stop churning!"

*Slosh—slosh. Swish—swish.*

"Stop turning. Don't go past the butter!"

The weekly chore of churning butter seemed to take forever. On a hot day, it took even longer. Everyone took a turn—even the hired man coming in for dinner. After washing up, he'd reach for the two-gallon green metal churn, turn the handle ten minutes or so, then pass it on to the next person.

We listened for key sounds, as when the usual deep sound of the wooden paddle churning thick cream suddenly gave way to thin sloshing buttermilk. Lifting off the lid, we saw the separation—bits of butter, floating golden islands, in the thin white buttermilk.

Mom took over at that point. Gently she pressed the yellow butter particles together and drained off the buttermilk. With a wooden paddle she worked the butter in a special wooden bowl, added salt and squeezed out extra liquid until she had a solid block of butter ready to store in our ice box.

The buttermilk did not go to waste. An avid milk-drinker, Dad liked buttermilk even better. We girls drank it too.

"You live thirty years longer if you drink buttermilk," our Swedish Dad declared with a chuckle as he drained his glass.

Maybe. We knew Swedish people lived long lives and were famous dairymen. Indeed, he lived to be nearly ninety-four.

Mom traded eggs for groceries every week at her favorite grocery store. We girls helped her gather, clean and pack eggs in large wooden thirty-dozen egg crates.

Every March she bought hundreds of Leghorn chicks. About half were pullets that replenished her flock of laying hens. The other half, cockerels, reached one-and-a-half pounds by the Fourth of July and sold as tasty young fryers to the Met (Metropolitan) Cafe.

Mom and Dad settled the cute little bouncing yellow chicks in our small brooder house. It had no insulation except for hay bales set a couple high around the outside. With the cold, and often snow and blizzards of early spring, keeping a warm, even temperature for tiny chicks was a real challenge—but critical. That brooder house must have leaked heat everywhere.

With a fiery kerosene brooder stove, the small chicks cuddled together by the roaring heat under a circular metal hood or brooder. Danger of fire was high, so either Mom or Dad slept out there until the chicks feathered out and didn't require so much steady warmth.

The feathering-out process disappointed us girls anew every spring. We wondered how those cute yellow peeping puffballs that cuddled so lovingly into our necks could change so swiftly. As sleek white, long-legged teenagers they squawked and scattered when we tried to catch them.

Pleased that her young roosters were ready to sell as prime fryers by the Fourth of July, just as the restaurant requested, Mom butchered them a dozen or so at a time. Only one problem—we had to pluck the feathers off before we took those fryers to town. A smelly job after their wet-feather plunge into hot water.

On the other hand, in the belief it kept the meat more flavorful, the cafe owners wanted the insides left in until ready to cook.

*Yay!* We didn't have to clean entrails out of the chickens! They were sold intact. Stark naked bodies, but with long yellow legs and sharp yellow beaks still attached.

Tasty fryers, I'm sure.

Mom canned the old hens—and beef and pork in the fall when the men butchered—until the locker plant took over cutting and

## Buzz

Grandma liked to play games as she sewed with us gathered around her.

In *Buzz* we counted quilt pieces, saying *"Buzz"* instead of seven or multiples of seven. One of us began counting, but if we did it wrong and said *"Seven,"* someone caught the mistake and shouted, *"Buzz."* The winner started the game again.

Before marriage, Grandma Barrett was a dedicated teacher and likely identified well with the getting-bored pupil.

As we worked, she rocked gently in her small oak rocking chair with its unique sliding sewing drawer under the seat. She opened the drawer by pushing it to one side or the other to reach exactly what she needed in the five compartments and spool holders—scissors, thread, pins needles, quilt patterns and scraps of cloth.

Sometimes we stopped sewing to play *I Spy* with her silver thimble.

"I spy!" she'd cry, clapping her hands gleefully when she found where we hid it.

storing meat in freezer lockers for us. As soon as electricity came, we bought our own large chest freezer. Freezing foods made life much easier.

We girls helped can peaches and apricots bought by the crate, and made jams and jellies from wild chokecherries and buffalo berries we picked in the hills.

We helped decorate rooms and painted walls and woodwork, as Mom worked to strip layers of old-fashioned wallpaper from the ten-foot-high walls of our large downstairs rooms.

Then stitched curtains and draperies. We took these 4-H projects and colorful illustrated reports, with our own garden vegetables and flowers, to the Custer County Fair.

Once while painting the two-story front stairs hallway, her tall ladder propped on the stair landing, Beverley dropped a full gallon of dark blue paint. It hit the landing and clunked down the stairs, spilling paint all the way.

She yelled and we all came running to the foot of the stairs.

A friend, Mrs. Reinie Whittmeyer, visiting us that day, recalled it this way:

"We were alarmed to hear the paint bucket go clunk—clunk—clunk all the way down every step. Beverley screamed and we all rushed into the hall in time to see the bucket thumping down the last of the stairs, spilling paint on every tread to the very bottom.

"Everyone stood there shocked and silent for a moment. I wondered what your Mom was going to say. I couldn't imagine a mother not giving out with a resounding scolding.

"But after seeing Beverley's dismayed expression, scrunched precariously way up there on the tall ladder against the ceiling, suddenly everyone burst out laughing, your Mom and you girls, too. Then you all got rags and cleaned up the mess together. Even I helped. We had so much fun cleaning up that I understood why your mother's reaction was so right. Still, I shook my head in disbelief."

"Putt—Putt," went the gas-powered Maytag washing machine.

"Putt—Putt! Putt—Putt!"

So loud it echoed off the rimrocks if Beverley and I happened to be riding up there on wash day.

In washing clothes, someone first had to chop wood at the woodpile to heat water. We heard every whack, from up there a half-mile away. Fascinated, we watched as Jeanie chopped, the sound traveling so slowly over that distance that we heard a delayed chop of the ax on the upstroke. No sound at all came when the ax struck wood.

"Whack!" we heard as her ax paused at its apex.

Silence when it came down and split the cottonwood stump.

*Jeanie sweeps the fishpond clean in spring. Her large goldfish winter in the heated livestock water tank and we fish them out when weather warms.*

**Slipped into a wool sock, we even took them to bed with us to keep our feet warm on a cold night.**

Then, "Whack!" Again with ax high in the air.

We laughed, delighted.

"Putt—Putt," the washer called. We knew it was time to get back home and help with the washing.

Mom tripped the starter with her foot.

"Putt—Putt." The explosive sound meant the motor was running well and the agitator swishing.

She fed clean clothes through the wringer from the soapy water into the rinse, anxiously warning little Anne away, as the clothes squeezed between two white rubber rollers. Even when old enough to use it ourselves, we were leery of getting our fingers caught between them. We heard stories of kids getting an arm caught in the wringer; they still carried the scar,

Still, it was hypnotic. *What little kid can resist trying to run a sock through those rollers?*

The drain hose, too, attracted kids. At that certain young age, I guess each of us unhooked the drain hose, just once to see what would happen and it gushed soapy water all over the floor. Who knew that lifting the hose high would stop the flood?

Clothes dried quickly on hot sunny summer days; in winter not so fast. Sometimes they froze dry on the line and Mom brought in long, white underwear, frozen stiff. We danced around the room with these 'ghosts' till they slumped, limp enough to finish drying on a rack by the stove.

After that came ironing—but first, shirts, blouses and dresses needed to be starched, sprinkled and wrapped till evenly damp.

Oh, temperamental irons! At first ours were sad irons, left all week on the back of the stove, accumulating dust, smoke and soot. They locked onto a grab handle for ironing, usually too hot or too-quickly cool. Slipped into a wool sock, we took them to bed with us to keep our feet warm on a cold night.

Mom eventually replaced them with a gas-powered iron—a miracle in that we could control the temperature somewhat. But not necessarily cleaner. Gas irons could spit soot on a white blouse.

After the war ended in August 1945 many things changed—copper for electric lines changed our lives most.

Still, electric lines weren't built across our ranch until 1951, the year I was in junior college and Anne already in high school. That fall and winter we wired our large two-story house. What a project that was! It seemed to take forever.

Dad hired John Grauman, our mastermind electrician, to super-

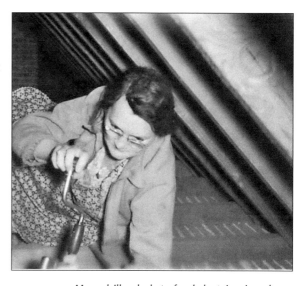

*Mom drills a hole to feed electric wires down from the attic, that winter when we all wired our big two-story house. Not till 1951 was war-rationed copper available for electric lines to our ranch.*

## Canine friends

Our dogs played a special part in our lives. Jimmy, part collie and shepherd, eagerly herded sheep. Dad brought home Rex our Springer spaniel as a puppy who grew into an exceptional hunting dog. He was also a fierce watchdog—unless the visitor carried a gun. Then he wiggled all over with excitement, expecting to go hunting with them. One day he treed a trucker while he loaded up our steers.

*Our loveable dog, Jimmy, with Anne.*

Earlier we had Jack, a mostly-black shepherd-cross that Dad found hit by a car and starving along the highway. We nursed him back to health and, with his natural affinity for handling livestock, he was just the dog we needed. Jack's one dastardly deed was killing our pet magpie Kaw-kaw that escaped the house one day and fluttered into the garden below. Beverley, a sixth grader, wrote a poem about both pets for our Tusler school newspaper that she and the teacher ran off with blurring purple hectograph ink.

She closed her poem, "Then our dog Jack bit poor Kaw-kaw in the back. The end."

Dogs were our daily pleasure. Sometimes they expected us to throw sticks for them. Other times we just sat in the sun petting and talking to them, while they smiled into our faces. We felt sure our dogs loved us as much as we loved them. They followed us everywhere, found the rattlesnakes before they found us, and were our constant faithful companions.

vise. We all helped him—Mom, Dad, Anne and I, every spare minute. Mr. Grauman's wartime crash on a B-52 bomber left him with paralysis of both legs and he worked mainly from his wheelchair. Despite this he was agile, energetic and congenial.

Anne and I crawled up the attic hole time and again to laughingly snake wires down between the walls to Mr. Grauman, who called merrily up to us through hollow outlet openings. And we wired countless light switches and outlets under his direction and sharp eyes.

Then, at last it happened! Job finished.

We flipped the switches and our large two-story house flooded with light, just like in daytime.

What a miracle it was when the lights came on and the new refrigerator and chest freezer began humming!

Surely this was progress. Yet even so, we missed the soft kerosene lamps and the way our family gathered together around the kitchen table—sharing that warm light as we read, studied and played board games.

## Three-Legged Kitchen Help

Dad appreciated a good joke, but he could be stern, especially if we girls were having a loud argument over nothing. He did not abide arguing and knew how much it upset Mom. [FRANCIE]

With creativity highly valued in our family, we girls found ways to spice up our most tedious chores—such as washing dishes. After every meal, *every meal*, we had to wash dishes. There we stood, side by side at the kitchen cook stove where we kept the dishpan and rinse water, so the water stayed hot.

"I know what let's do," I said one day as Jeanie and I tackled the job. "Let's tie our legs together and go three-legged."

"Okay," she agreed. "We can't go anywhere, anyway, till we're done."

Then she giggled and dashed off to the rag drawer for one of Mom's old stockings. We wrapped it around our ankles several times, my right and her left, tying the ends together in a tight knot. Away we went, hopping three-legged around the kitchen, just for fun.

I was washing dishes, rinsing them from the boiling hot tea kettle, and Jeanie was drying. As we each moved through our jobs—I brought more dishes from the table to dishpan and Jeanie carried dry dishes into the pantry—the other hopped along.

Then we began arguing. In our house, dishwashing arguments usually happened because the person drying discovered a smear on a plate, plunked it back in the dishpan none too gently, complaining about the poor job the washer was doing. After a time or two, the washer grew irritated; because maybe it was a very slight—or even imaginary—smear and maybe the dish slammed back too sharply into the dishpan.

In retaliation the washer slammed it back into the drying rack. Grounds for an escalating fight.

"There! Is that better!" I swiped an offending plate with the dishrag.

"No, it's not. Look at that. It's still dirty!"

"Where? I can't even see it!"

Our argument heated up. We shouted, standing close together with our ankles tied and our voices getting louder and louder.

Mom folded socks at the dining table, trying to ignore us. Dad was reading in the living room with the door shut.

"Be quiet in there, you girls," he shouted through the closed door.

We hushed, but it didn't last long. By this time neither of us wanted to lose the argument. We both wanted to prove we were right. Shouting louder seemed a good way to do it.

Now we stood side by side washing and drying dishes at the stove, shouting at each other, our ankles still attached.

Suddenly Dad opened the living room door, his face red with anger, one hand on the doorknob, the other holding his magazine.

"You girls stop arguing right now! I've had enough!"

Then he issued a startling command, in a tone of voice we knew we dared not ignore.

"Francie, you get in the bathroom and Jeanie, get in the pantry." He pointed us in opposite directions.

Stunned, we looked at each other. What to do? We had to obey. When Dad gave us an ultimatum like that, we did it. Immediately!

We hardly noticed that Mom looked up, alarmed, a peculiar expression on her face.

No time to untie our super-tight knot in the old stocking. We had to move.

So away went Jeanie, crawling with one leg toward the pantry, while I crept toward the bathroom, stretching my leg back as far back as I could. Our ankles still attached in front of the stove.

No doubt we looked ridiculous—earnestly trying to follow those impossible orders as quickly as possible.

We heard a hoarse chortle from the living room doorway. Dad disappeared quickly, and the door slammed shut.

*So away went Jeanie, crawling with her free leg toward the pantry, while I stretched toward the bathroom.*

## Games

Evenings, gathered around the kitchen table by the light of Aladdin lamps with their brightly glowing mantels, we played cards and board games. We liked Old Maid, hearts, whist and rummy, checkers and Chinese checkers. Mom loved to play bridge, but didn't teach us, because it was so involved and she really didn't have the time; she only played it at community card parties. And of course, in our bedrooms, we read library books for long hours by the dim flickering light of kerosene lamps. Later, we had a friend who sneaked us comic books with their torrid, violent covers—something Dad abhorred, but we could read the funnies in the newspaper. He confiscated and burned comic books if he found them. That Superman could leap tall buildings in a single bound thrilled us, but not Dad.

As we sat there on the floor, pulling our legs together, sheepishly working at the knot, tightened almost beyond loosening, we heard an odd coughing sound from Mom. She was trying to keep from laughing.

The argument forgotten, we turned back to the stove, loudly clearing our throats, trying to stifle giggles.

When we couldn't stop, we burst into laughter and hopped through the pantry into the separator room. Holding our sides and laughing hysterically, we found some old sheep shears out there and cut ourselves apart.

## The Gutsy Bug

"Let's go!" Bev announced, and once more we girls headed toward a new adventure.

FRANCIE

We hiked up Battleship Rock and Bathtub Rock, exploring the unique formations on the massive crests of those hills and others. We set out on longer hikes into the hills across the highway, checking out deserted homesteads, and the falling-down house built into another rock formation on land we eventually bought from Arkansas Bill and Sugar Babe. Sometimes we hiked all the way to Spring Coulee or rode horseback to visit the Hill kids some three miles away.

Beverley loved exploring and setting off on new adventures. So did Jeanie—especially if they promised the unique opportunities she thrived on, setting up pranks, building a campfire or spontaneous side trips.

Both were natural leaders and it was my good fortune to be just enough younger so if either wanted someone to come along or help out, I was handy and willing. Anne's turn came when she reached age five or so. Our two older sisters both came up with great ideas on how to spend a spare afternoon, separately or together.

Off we'd go bouncing across the hills in what we called 'the Bug'—an old stripped down truck, its roof cut off and rear deck half-loaded with fence posts and equipment being used around the ranch. In late summer and fall, as they ripened, it took us to favorite chokecherry, buffalo berry and current patches that grew wild in our pasture draws and the open range above.

The Bug led us into great adventures, though it was almost cer-

*Our Bug, ready for an adventure or work day.*

tain to give us a challenge or two. Just when we were off the road up some gravel draw far from home, we had motor trouble, a flat tire, ran out of gas, got bogged down in soft gravel, or hit a rock causing oil pan leakage. We learned to park on a side hill in case the battery failed, and the Bug taught us simple mechanical skills and driving expertise through necessity.

"We're stuck!" Jeanie called.

We dreaded those words. She loved the challenge of driving through hilly pastures and gullies or along some gravely trail. Grabbing shovels—emergency equipment always kept in the Bug—we set to work digging out. Sometimes this took an hour or more of hard work, but better than the long walk home for a tractor to pull us out.

"Weren't you bored?" we were once asked by someone who wanted to know what life was like growing up on a ranch.

The question stumped us because we never experienced it. There was so much to do, so many things to investigate and, using our imaginations, we could turn most any outing or chore into fun.

Moving to the ranch east of Miles City on the Yellowstone River added a whole new dimension of excitement and adventure. We 'swam' in the irrigation ditch on hot afternoons—or actually, we splashed and played and paddled up and down. Since our fields were the end of the line, the main irrigation ditch was no longer the big canal it was farther upstream. The moving water was only a foot or two deep at the most, depending on how much water was released from Tongue River Dam many miles upstream and left unused by others.

We made mud beaver slides behind Dad's shop where the gumbo clay in the ditch was right for this. We pulled up hands-full of wet clay from below the water line and plastered it on the slope till we'd built a nice slippery slide. Then, keeping it wet, we took turns sliding down—hitting the water with a huge splash, climbing out, running around and splashing down again.

We enjoyed winter sports and took it for granted we'd be outside in winter, either playing or working, except on the very coldest days.

"It's a full moon—let's go skating. The ice should be perfect!" Dad called in his enthusiastic voice.

"Let's stop and see if the Herzog kids can come," chimed in Beverley.

We were all skaters. Even in a cold snap we went skating when the ice was right. Raised in northern Minnesota with abundant water and ice, Dad was a good skater. One Christmas we girls all

Beverley and Anne enjoying rock climbing.

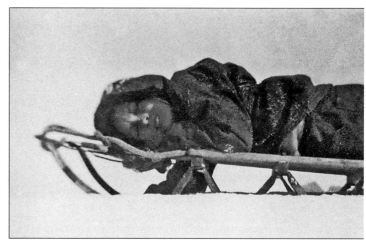

Anne, sledding.

## Paper dolls

One magical summer Jeanie and Francie played paper dolls every day in the storeroom upstairs. Cut from last year's Sears and Montgomery Ward catalogs, our families ranged up and down over all the boxes stored there.

Jeanie made up long entertaining stories that continued day to day, week to week—of our moms and dads, babies, little kids, teenagers, hired men, friends, visiting cousins, aunts, neighbors and older brothers and sisters gone off to college. Lively conversations went on between these characters as they travelled along the boxes and between our people's homes.

We had only two tasks that summer—wash the breakfast dishes and bring in the milk cows for evening milking. As soon as possible after the dishes we escaped to our upstairs hideaway and picked up where we left off the day before. Probably it was Jeanie's job to take care of Francie, at age six, and keep both out of trouble, so others could work undisturbed.

Our paper dolls were mostly cut off at the ankles, no feet, because that's the way the catalogs displayed them. And always in black and white. No matter: they lived active, interesting lives in our imaginations, and were as real to us as our own neighbors and friends.

Our story-telling skills flourished that fanciful summer, interrupted only by Mom's calls in late afternoon that it was time to go get the cows.

got skates and he took us skating on many Sunday afternoons and evenings on glistening ice when reservoirs were free of snow.

When she reached high school, Beverley organized skating parties with friends. Jeanie built up a roaring bonfire and no one objected that Anne and I came along, skating around the reservoir and up the creek that ran smoothly into it. When chilled, we warmed ourselves at the fire. Often someone brought out a guitar and began a sing-along.

When the snow was right for it, partially thawing, we built snowmen and forts.

We spent many hours after school before dark and on weekends sledding down the hill near our house. This ended when a midwinter snowmelt exposed too much gravel and rocks on our favorite slopes. Chinooks were especially exciting—a warm wind that suddenly, in minutes, raised the temperature forty or fifty degrees.

We tried sledding behind a horse, taking turns—with the rider holding the rope. We also experimented with skiing behind a gentle horse, not very successfully. Our horses didn't like it and refused to pull straight, dancing sideways. Sometimes they tripped on the rope so whoever was riding had to let go.

Between farming and ranching jobs our family sometimes camped at a favorite fishing spot where we could both fish and hike. We also took in rodeos, parades, the fair, community picnics and get-togethers with neighbors and friends. Church and 4-H meetings and their activities played a big part in our youth; there were Sunday school, Methodist Youth Fellowship, 4-H camps and fairs. All year we worked on our 4-H projects.

Our parents joined enthusiastically in our activities and, shortly after WWII before the auto industry got rolling again, our church had no means to get the youth to our church camp at Luccock Park in the mountains near Livingston. Priority was given Dad for a new sugar beet truck during the war effort, so he piled the kids' bedrolls into the back, put canvas over the stockrack and loaded in the campers. Everyone talked about that fun trip for years.

Work was a priority on our ranch, but fun and recreation played a large part too. Many times we were isolated from close neighbors or friends, but we never felt lonely or deprived.

As we grew older, we each had special jobs around the ranch, but Mom and Dad made sure there was always time for fun.

## Gathering 'Round the Piano

"And those caissons go rolling along…"

All who happened to be at our house for dinner gathered around our piano singing as Beverley belted out the current popular song.

Mom felt she lacked a good musical background and was determined to give us girls plenty of opportunity to develop our natural musical abilities. When we moved near Miles City, she wanted to start us on piano lessons as soon as possible. Beverley was nine, and it was time to start.

It took a few years for Mom to find a good piano at an affordable price. I remember how she saved for it. The woman selling her Schumann piano—for just $50—lived down near the Yellowstone on what Miles Citians call 'the island,' in the northeast part of town. We went there together to look at it. Dad came, too, in the truck. As they loaded it up, the woman looked us over somewhat wistfully as if hoping that giving up her piano would open new vistas and bring much pleasure to this family of four young girls.

It certainly did.

That wonderful piano was almost never silent again through our growing up years.

Beverley took to the piano like a lark and shared generously with everyone within listening range. She'd come in from the field, wash up for the noon meal, then sit down at the piano and play and sing at least a few minutes before dinner. Always smiling and welcoming, she'd call us over, hired men and all, to sing around the piano. Again at suppertime she'd sit down at the piano, singing as she played, or experimenting with a new piece.

Jeanie played well, too, although I think she preferred playing for herself, when others weren't around.

I had a good start, also, and loved joining in with Beverley's joyous voice and those of various hired men gathered around the piano. Eventually I could play and sing fine by myself. It wasn't anyone's fault but my own that my playing was tortured at times, or that I much preferred running around outside—to the drudgery of that required half-hour-a-day practice.

And of course, there were all those recitals at the formal Benson conservatory where we took our weekly lessons. They were scary enough that I did practice and memorize, though I never quite learned to enjoy it.

We saved our money to buy sheet music, and lingered many hours in the music store, debating how to spend it.

Mom definitely had more urgent demands for her cream and

*We often sang from sheet music around the piano that Mom bought with her egg money. We practiced piano lessons under the eye of the big mule buck Francie mounted.*

*Ready for church in our Easter Sunday outfits—organdy dresses and hats, Beverley, Francie, Anne and Jeanie.*

*In the summer of 1941 we took a long-anticipated trip to Yellowstone Park. In awe, we experienced its amazing natural wonders. On the porch of our rustic cabin, Francie, Anne, Beverley and Jeanie watch, fascinated, the antics of the ever-present playful bear cubs, while Dad fly-fishes a nearby stream.*

egg money, but she didn't flinch from buying Community Concert tickets for all of us through our growing up years.

At one point Beverley wanted to play violin and Mom rented an instrument and arranged lessons for her. The teacher even lent us a small violin that fit perfectly for little six-year-old Anne, and Beverley taught her to play.

A few years later Anne also learned to play piano, but she yearned for a guitar, so she talked me into sharing the cost with her. We pored over the Sears Roebuck catalog and finally sent in our order. It cost fourteen dollars with instruction book.

We both learned to chord on the guitar, but I still preferred the piano. Anne felt confident in performing on the guitar as long as I played melody on piano. When she felt ready, which was probably too young to know this, at age ten or eleven, she volunteered us for a musical number at the 4-H achievement night. Even though our first performance could not be called a raging success, we were often asked to fill in for musical numbers. Some turned out well—others we tried to forget. Still, 4-H brought us many musical opportunities, and soon we branched out to playing at Farm Bureau and other community functions.

That heavy old Schumann upright Mom saved for years to buy continued to hold a fine tune over a half century later in Anne's home. We often wondered if the lady who looked so wistful when Mom bought it knew how much we enjoyed and cared for her piano.

## Rangeland Hospitality Code

Mom looked out the window that hot summer afternoon, amazed by what she saw.

Groups of children and teens, led by a few adults, straggled down through the hills toward us, cutting across from the highway toward our house.

JEANIE

We were the last place in the Yellowstone Valley heading east out of Miles City on US Highway 10. Our ranch extended up into the bluffs and rangeland with the highway winding through the hills for ten miles before the appearance of another human habitation.

If people ran into car trouble or needed help, they might leave their vehicle and start walking toward Miles City. Because our buildings were in sight of the highway, they eventually came to our ranch, unless they caught a ride first.

Mom, watching out the window, couldn't see a vehicle on the highway above, so she knew the kids, struggling red-faced and weary, had walked a long way in the heat.

Some staggered and were being helped by adults.

"I wonder, Dad, if you should go to meet those young people," Mom said, as he joined her at the window. "They look so tired."

Dad took one look.

"Oh, Mom! Better get some water ready—they'll be thirsty. Look how red their faces are!"

He hurried out the door to meet them.

Mom dashed to the pantry and cleaned the cupboards of all the drinking glasses, cups and pint canning jars she could find. At the sink she filled them with our sparkling, cold spring water, lined them up on the table and went back for more.

Our kitchen filled with exhausted, overheated kids.

On their way home from church camp near Glendive, mechanical trouble stalled their bus up in the hills beyond our ranch. One of the leaders thought he knew how to cut across to our ranch and brought them down some rugged coulees.

It was close to a hundred degrees outside.

*I knew hard-up, but this looked truly desperate. The car spilled kids of various sizes.*

They stayed with us several days until their bus was fixed—parts had to be ordered over the weekend and it was late Saturday. They came from the western part of the state, Helena or Missoula, too far for their parents to come for them.

Our big old Texas-style ranch house had six large bedrooms—eight in a pinch—and they filled them all, about thirty kids plus adults. We girls were all at 4-H camp that week and the house must have seemed empty.

No longer.

"They had so much fun!" Mom said. "They held sing-alongs, played games and made up skits and performed them in the living room. Everyone was so happy."

"Where did they all sleep?" I asked.

"Our beds were filled for once. Because they came from camp, they had sleeping bags. Their leaders said to put them on top and not muss up the beds. Others slept on the floor."

"And you fed them all?" I asked.

"We had beef in the locker and our full vegetable garden—plums and ripe strawberries and canned goods. They pitched in with the cooking and food preparation. One leader and some kids made six plum pies for supper two nights. They cleaned up and washed dishes," she said.

"We enjoyed having them."

One time a car broke down on the highway near a field where Dad and I were mowing and raking hay.

A tall, gaunt man came walking through the field toward us, a frayed bib overall hanging off his lanky shoulders.

FRANCIE

He didn't quite meet our eyes.

"Car just quit. It'll start—I think. If I can just get a pull."

"Sure, we can help you out," Dad said.

The man took off back across the field.

"Seems kind of down on his luck," Dad remarked as we drove the truck over there.

I knew hard-up, but this looked truly desperate.

The car spilled kids of various sizes out the doors, all looking hungry and bedraggled—but silent. In the front seat a woman held a fussing baby.

The car had seen better days. A rusty black, it was an old, boxy style with hard straight lines. Ropes held down a bulging trunk; boxes clung to the running boards. It sat there at the edge of the road, tired and despairing as its occupants, as if it lacked the energy to go on. I wondered if they'd ever get it running again.

Dad hooked up the log chain. He looked at me.

"Francie, pull them a ways to help them get started."

*Me?* I was fourteen that summer, never before faced with a problem like this.

"Git back in the car!" The man growled at the kids.

So off we went down the highway toward Miles City, with me trying to drive smoothly and steadily so as not to snap the log chain.

After I felt him jerking at the chain, grinding, trying to get the motor going, I slowed, hoping the car would take up the slack.

But no. When I stopped, so did the car. One mile, two miles. More jerking and grinding.

"What do you think?" I asked, getting out.

"A little farther—I guess," he said.

I wasn't sure what to do, but got back in the truck.

*Dad needs me back in the field. He didn't expect me to go so far. But how can I leave this helpless family beside the road?"*

More miles. After awhile he stopped trying and so did I. By this time we were half way to Miles City and I knew just what to do.

I pulled in at Krumpe's machine shop and turned them over to Mr. Krumpe and his cheerful mechanics—Leo, who was deaf, and one-legged Mike who got around just fine on his welded peg leg, likely designed and built right there. These were Dad's hunting buddies and the shop a place where we often met Dad on our way home from town.

"I'll get 'em going," Mr. Krumpe said, when I explained what

*I glanced at the kids, with their stringy hair and haunted eyes and the weary, defeated-looking mother. Her eyes shifted away.*

happened.

"How much do I owe you?" the man asked me with his haggard look.

I glanced at the kids, with their stringy hair and haunted eyes, and the weary, defeated-looking mother. Her eyes shifted away.

"Nothing. It's okay." I shook my head, embarrassed that he asked.

Of course, nothing. We didn't take money for helping out.

Mr. Krumpe unhooked the log chain and laid it on the cab floor. Then he grinned at me and patted the hood.

"Okay, Francie, I'm sure your dad wants you home. I'll take care of this. Now, get going."

I didn't need to hear more. I was glad to go and leave this desperate family in his kind and capable hands.

Dad and Mom never turned away anyone who needed help. Helping people is a rangeland tradition. Feeding them, expected. Mom told us that Grandma Barrett always set an extra place at the table during the 1920s when they lived near the Jordan road. Frequently a traveler, lone cowboy or neighbor stopped in to sit at the place and eat a meal with their family.

JEANIE

If anyone came to our ranch near mealtime, we invited them to eat.

Even if it wasn't mealtime, Dad asked, "Have you eaten?"

If they hadn't, however inconvenient it might be, Mom stopped what she was doing, prepared a meal and made people feel welcome as they ate.

Neighbors, hunters, tramps, hitchhikers, tourists and other strangers all ate at our table.

Some were especially enjoyable, as the time the Director of Music for the Montana State Office of Public Instruction had car trouble in the hills above the ranch. Of course Dad helped him bring his car down and into Miles City to get fixed.

Repairs took several days because it had overheated and done serious damage.

Where did he stay? At our house, of course.

Mom rarely had opportunity or leisure to visit with others in the teaching field. Having taught in high school for many years, she was delighted with the conversations they had.

Dad and Mom both hated to see the man go, because he had such an engaging personality, was realms of information and a great conversationalist.

Right after WWII, veterans hitchhiking nearby Highway 10 often stopped at mealtime. Often they were too proud to accept a meal

*Francie pulls a hapless family and their car to Krumpe's repair shop.*

*Mom didn't notice that the veteran who came up the road had an empty sleeve—with only one arm he went out to chop wood.*

Western Ranch Life in a Forgotten Era

without giving something in return. One day Mom didn't notice the man who came had only one arm and an empty sleeve—some men tried to hide their injuries. With only one arm he went out and tried to chop wood.

When Mom looked out and saw his difficulties, she called him in without comment, according him the respect and dignity he deserved; told him the meal was almost ready and if he liked he could wash up.

Another time, Dad picked up a boy of about eight or nine walking along the highway. He was sick and unable to eat dinner. Mom thought he had the flu and put him to bed with his dinner on a tray nearby. In the morning he was gone.

Dad called the sheriff and they looked for him without success.

One day a trucker lost a tire, coming down out of the hills. Negotiating the steep grade and sharp curves of Highway 10, he felt a jolt and noticed in his mirror one of his rear dual tires running free of the wheel. As he came around the last curve, the tire caught up and passed him.

He came into our ranch and spent the night while we all looked for his tire. No luck. Off and on all winter we looked, and every time he came by on his trucking business he stopped and searched.

Then next spring, Dad rode out to look at a strange object—he thought it must be a calf caught in a washout hole. There, jammed under big sagebrush, lay the tire. After it landed, the tall green grass apparently snapped back upright, hiding it. During winter, the grass turned brown and snow matted it flat, exposing the tire.

Dad called the trucker and he gladly came to pick up his tire. By this time he was a family friend, so he ate supper and stayed the night. Occasionally when he drove through he stopped by to visit.

And of course, there were the bad apples.

One night a rancher's son from the Jordan country, driving by our ranch, stole two saddles from our barn. Dad reported the loss next morning and the same day the young man was arrested and jailed in Terry, our saddles and other stolen goods in the back of his pickup. The sheriff from Terry phoned Dad and said he'd return the saddles in a few days on his way to Miles City.

"Don't bring the saddles," Dad said. "I need to go up and have a talk with that boy. He should face the people he stole from and understand how he's hurt them."

Knowing those low-voiced serious talks, we girls thought to ourselves, "If Dad has a talk with him, he might think twice next time."

When he returned he said, "The kid was embarrassed and apologetic. Said it was his first time and whether it was or not, let's

*I intended to stop and ask if they needed help, but then I had a bad feeling about them.*

hope it will be his last."

One summer morning, returning from an errand in Miles City, I slowed down to enter our driveway when, right at our turn, were two men walking along the highway hitchhiking toward town and carefully watching my approach.

*Must have had car trouble up in the hills.*

I intended to stop and ask if they needed help, but as I got closer I had a bad feeling about them. So I didn't stop—just kept going.

At the top of the hill, I turned around, parked and watched them continue down the highway beyond our ranch. Then drove home and told Mom about those sinister looking men.

An hour later a hunter stopped in and reported a burning car he came upon, back on a neighbor's range. It was engulfed in flames that he couldn't put out.

"Somebody burned a perfectly good car on purpose. Can you think of anybody who would do that?" he asked.

We couldn't.

"Oh! Those two men." Suddenly I remembered.

He phoned the sheriff who found them still hitchhiking. Though they held their thumbs out, nobody but the sheriff wanted to give those tough looking men a ride.

Wanted in Chicago, a criminal investigation revealed they had robbed a bank, shot someone, stolen a getaway car and escaped pursuing cops. Since they knew they could be traced by the stolen car, they turned up a gravel road a short way into our hills and torched it.

If it hadn't been for that hunter, and the vibes I got from those sinister-looking men, they could have gotten away, leaving their crimes behind.

We honored the western tradition of helping people, expecting the best from them, trusting unless it proved undeserved. To be paid for this, or even offered payment, was seen in rangeland code as almost an affront to hospitality. In a sense, it put a price on something that was truly beyond dollars and cents.

We certainly knew and appreciated this when we were on the receiving end of rangeland kindness and generosity.

As Dad said, "Ranchers are the best people in the whole world."

> **Despite the bad apples, we honored the western tradition of helping people, expecting the best from them.**

*Mom and Dad tell wonderful stories around the supper table. We learn from their stories to think positively, and not to gossip because they don't.*

## Pioneer Storytelling

"I never saw a ghost, but one night just before dark I came close to thinking I did. Every two or three days I rode a 60-mile circle, checking my coyote traps in the Missouri River Breaks. This was a long way from home through a lonely stretch of broken prairie and buttes."

FRANCIE

Finished with supper, Dad pushed back his plate and surveyed his audience with a twinkle in his eye.

"Suddenly I heard a blood-curdling scream and saw a flash of white off to my right, just for an instant out of the corner of my eye. When I turned to look it was gone.

"It sent chills up and down my spine!"

As he watched, the white apparition rose up again, dancing in the dim light, screaming. It whirled and leaped in the air for a moment, then vanished.

*Was it a ghost? Sure looked like it.*

"I was scared at the way my horse acted. He spooked and snorted, rolled the whites of his eyes, threw his head, jumped around and refused to go any closer. *What did this horse know that I didn't?*

"He sure wanted to get out of there and so did I. It was getting late and I needed to find a place to sleep. For sure I didn't want to sleep anywhere close to that!"

"A third time I saw it and again heard the shriek. Was it a human cry? Some crazy person waving a white sheet? Yet, no one lived within ten miles.

"I thought, if I don't go over there and get a closer look, I'll always wonder. And maybe I'll decide that it really was a ghost. I didn't want to do that." He chuckled.

Dad worked his terrified horse in closer. Then in the gathering darkness he saw what it was.

"A white horse! A white horse trapped in a deep washout hole, so deep he could see out only by rearing up. He leaped, pawed his front hooves into the air, beat them against the gumbo turf, trying to jump and claw his way out. Then he'd fall back out of sight.

"And those blood-curdling screams… I've heard a lot of horses scream and squeal, but never quite like that. Even up close it made me shiver."

Dad explained how he freed the white horse. With a spare coyote trap, he dug a slanted opening in the gumbo, dropped a rope around the horse's neck and pulled hard and steady while the trapped horse made a mighty lunge and broke free from his prison.

"He shook himself, nickered his thanks—and followed me to the nearest water hole."

Growing up, we girls were fortunate that both Mom and Dad were storytellers, usually around the supper table at night. They told fascinating tales of family, community, good times, hard times and hilarious incidents. They accepted and respected each individual in the tolerant way of the west, and never gossiped about other people's business. Instead, their stories recalled epics of courage, endurance, hardship and individual enterprise. Details of homestead and pioneering days were vividly described, so real we might have lived through them ourselves.

After all, they made it through those tough times, and there were wonderful stories about how they and the neighbors did it, how they socialized and helped each other, their amusing stumbles and triumphs and, above all, optimism and survival.

Dad told his stories with great gusto and humor. He'd finish eating, lay knife and fork across his plate and push it back, chuckle and begin.

"I remember one time …."

Then he might tell about the time a hard-bucking bronc threw him off on a stump, breaking his nose so it bent far to one side. Then a year later it happened again, different bronc—only this time he reached up and pushed his broken nose back straight again and held it there while the blood flowed. And there it was, with only a slight bump on top.

He enjoyed poking fun at himself and related stories like this with laughter and a hearty punch line.

Other times he told of hunting wolves, running wild horses out of the Breaks, horse racing, rodeoing, swimming cattle across the Missouri, watching the Wright Brothers' brief flight at the Minnesota State Fair, and practical jokes he and his brothers pulled growing up.

Afterwards, when we'd all had a good laugh, he got up, went in the living room and lay down on the couch with the Saturday Evening Post.

Then, if we stayed on at the table talking, Mom might tell a story. Hers were quieter stories, usually told after the men had left the table—though often a hired man lingered, too, just in case he'd hear another story. She usually took more time to eat than others, since she'd been up and down serving the meal for our crew, six of us and a hired man or two.

*The white apparition rose up again, dancing, screaming. It leaped in the air for a moment, then vanished.*

*The Sioux Indians were fascinated with little Chet because of his bright red curly hair.*

Sometimes we begged for a favorite: "Tell us about the time…."

Mom's stories recalled serious and sometimes painful events, full of precise details about life and hardships she'd seen. Still, her basic optimism prevailed and validated the triumph of the human spirit.

She told of her family's four-hundred-mile covered wagon trip from southeastern South Dakota to a Shade Hill homestead near Lemmon in 1906, when she was five.

Uncle Chet was just a toddler. They spent three days crossing the Sioux reservation west of the Missouri.

"The Sioux Indians were fascinated with Chet because of his bright red curly hair. Every evening more of them came to hunker down around our campfire and exclaim over the little boy with curly red hair. They called him 'Little Fire Hair,' crooning words we couldn't understand and gently touching his curls.

"One evening, Chet tagged along after us, while Will and I played running games with the Indian children.

"Then suddenly he cut through the campfire circle, trying to catch up, and toddled right into the fire's red hot coals.

"He screamed. But before our dad could get to him, two Indian men grabbed him up and ran with him to where their horses stood tethered. Scooping up a handful of wet, steaming horse manure, they held it against his feet.

"Chet stopped crying.

"He never had any pain or scars from those red-hot coals. Somehow that warm horse manure pulled the burn damage right out of his poor little feet."

Another story from Mom's covered wagon trip testified to her mother's courage in the face of a big herd of stampeding Texas longhorns.

As they travelled across the open range, they often saw herds of the near-wild Texas cattle in the distance.

"We were warned to stay away from them. 'Watch out for the longhorns. They're unpredictable,' said a cowboy we met on the trail. 'If they stampede, you can't stop 'em. They'll charge right over your camp, trample whatever's in the way—and toss it high, whether it's a dog, child or someone's bedroll.'

"That evening when we stopped to make camp, Dad and John went down to the river for water. As they returned, their shiny water buckets attracted the attention of a herd of longhorn cattle coming over the hill for water.

"The lead cows shot up their heads, snorted and began running toward the men with their bright tin buckets. In an instant, here came the whole herd, a hundred or more, stampeding faster and faster, in a mad charge down the hillside."

Five years old at the time, Mom remembered well her terror. John's wife screamed, grabbed her children and hid in a wagon bed.

The men, still too far off, ran for the wagons, while the big herd of bellering cattle bore down on them with the sound of thunder and clacking horns.

"My mother acted fast. She grabbed a big white dishtowel that was drying on a sagebrush, snapped it and ran right at the longhorns, flapping the dish towel in their faces and shouting at them.

"The lead cows skidded to a stop and stared at her, snorting, heads high and eyes wild.

"I was so scared," Mom said. "I can still see the whites of their eyes. I can hear their bellowing and almost feel their panting, slobbering breath! I thought they'd gore my mother to death before the men reached the safety of the wagons.

"Then, *'Whoosh!'* The whole herd stopped dead, blowing hard and pawing the ground. Mother kept moving toward them, threateningly, snapping her towel.

"Suddenly they all spooked, threw up their heads, turned and stampeded back the way they came."

Dad grew up in a Swedish family of six boys with one older sister. Full of exuberance and energy, the brothers hunted, fished, trapped and hiked in Minnesota's north woods around St. Hilaire. Their grandpa, an immigrant, was a sturdy blacksmith, and their dad followed in his footsteps. He also raised bees and collected a string of racehorses to keep his sons busy. The brothers raced their father's horses at county fairs and wore out several motorcycles, said Dad.

As a boy of thirteen, he ran away to Montana and took a job with a rancher rounding up wild horses from the Missouri River Breaks. He also helped with breaking the wild horses.

Later his parents and brothers came and proved up homesteads along the Missouri River bottomlands north of Jordan. They ranched, raised cattle and planted alfalfa on the rich river bottoms. When some of his brothers returned to Minnesota, Dad stayed to become part of the ranching West.

He told us of the time he and his brothers rode the sixty miles into Jordan and decided it would be fun to ride their horses into the Foster Drug store.

They rode up the aisles and scattered customers, tipping their hats to the ladies and greeting the druggist.

"Mr. Foster took it well. He laughed—since luckily, none of our horses misbehaved.

"'Good to see you boys. Now, how about you tie your horses

*Grandma grabbed a white dishtowel and snapped it, running toward the herd of stampeding longhorns.*

outside?'

"Our oldest brother Moritz, stayed out by the door. He said to Mr. Foster, 'Sorry to cause a ruckus—my brothers haven't been to town in awhile. Can you fix us up with milkshakes and grilled cheese sandwiches? We're mighty hungry.'

"Before we got done eating, it seemed like half the people in town dropped by the soda fountain to see us and have a good laugh. Turned out to be quite an evening!"

As a teacher, Mom appreciated readers, whatever their education or unique interests. For herself, she cherished her literature books from college days. A favorite was her set of complete works of Shakespeare.

One day the books disappeared, she told us. She and Dad had gone to town on a two-day trip and when they returned the books were gone, the only thing missing from their home. She mourned their loss, but could be philosophical about it. In that lonesome country during hard times few homes had any books at all.

"Maybe whoever took my Shakespeare still reads them and still enjoys those plays." she said.

One day, in late afternoon, two drifting cowboys stopped at their ranch and were, of course, invited to eat supper and stay the night.

*Mom and her sister Pearl drive through a Giant Sequoia tree at the Mariposa Grove. In the early 1920s, when few women dared undertake a cross-country trip, they drove from Lewistown, Montana, to Berkeley, California, for summer college classes.*

"While I fixed supper, Pete volunteered to set the table.

"Jess picked up a book of children's Bible stories and was soon engrossed in reading, his elbow propped on the table and his long legs sprawled out to the side."

As he set out the plates and silverware, Pete complained about having to walk out around the long-legged cowboy's boots each time he went by.

"Jess—get your nose out of that kid's book and help instead of sticking your legs out in the way, like that!"

Jess didn't budge. "Leave me alone. These are darn good stories! I never got a chance to read Bible stories when I was a kid. Now I'm gonna read 'em!"

And he did, all evening by kerosene lamp.

Mom's quiet humor enlivened many of her stories.

She told us of when she coaxed her younger sister Pearl to drive with her to California to enroll in summer classes at the University of California, Berkeley. This was around 1924 when few women would attempt such a trip, She had saved every penny when she started teaching—after graduating from the University of Montana, earning her way there by waiting tables—and bought a Model T Ford.

For Mom the summer session was a dream come true—to travel and continue her education at a noted university. She registered for a full load in her favorite studies of literature, Spanish and math. Aunt Pearl, hoping for a more relaxed summer with plenty of time to enjoy the ocean and San Francisco sights, signed up for just one class—creative writing.

"But Pearl didn't enjoy that class," Mom said. Every day her instructor stood in front of the students and ridiculed their papers as the worst of creative writing. He seemed to delight in humiliating them.

"Pearl lived in terror of the day he might read one of her manuscripts aloud."

In the teacher's eyes, the worst sin was to write without knowledge of the subject.

"Write about what you know!" he shouted at them.

Aunt Pearl took that message to heart. Mom tried to help. He didn't sound like a teacher to nourish the creative muse, but Mom encouraged her to write about life back home in Garfield County.

So for her next assignment she chose a setting she knew well: a wild west rodeo in her home town of Jordan, Montana. Rodeos there happened somewhat casually when some rancher rounded up wild horses from the open range and brought them to town for sale. Local cowboys, eager to test their skill and daring, lined up for bronc

*"Sounds like the bucking horse is lying on the ground," the professor jeered. "Three men holding him down—biting his ear!"*

riding on a Sunday afternoon.

Aunt Pearl spent many days writing and re-writing her rodeo story, agonizing over dramatic phrases she feared might draw the professor's scorn.

"Oh, dear. I hope he doesn't read it out loud," she worried.

"It's a good story. You have vivid word pictures." Mom reassured her. "It describes a rodeo exactly. Your professor's going to like it."

Unfortunately, he didn't.

Hers was the next story read aloud—and ridiculed.

"You can tell Miss Barrett has never seen a rodeo! Listen to this, 'Joe held the rope and eared down the wild black stallion with his teeth. Two other cowboys ran to hold the squealing bronc by kneeling on its shoulder and hind quarters.'"

The professor jeered. "Sounds like the bucking horse is lying on the ground. With three men holding him there. One biting his ear? Ridiculous! So how does the imaginative Miss Barrett think the rider gets on a horse in this position?"

The class giggled and leered. Flushed with embarrassment, Aunt Pearl shrank deeper into her seat.

"She goes on inventing. Listen, 'The stallion bucked and reared and kicked ferociously while the rider fought to stay on. He stuck with it a minute or two, then flew into the air and came down hard on the ground. With the wind knocked out of him, he lay still for a time, then slowly picked himself up and retrieved his hat as the black horse galloped off into the sunset.'

"Galloped off into the sunset, indeed! What nonsense." He hooted.

The other students burst into laughter and turned to stare at her with their knowing West Coast faces.

Aunt Pearl couldn't meet their eyes and escaped quickly when class ended.

She never told the professor or any of her classmates that hers was indeed an accurate portrayal of rodeo at its grassroots in the American West—before the arrival of bucking chutes, arenas or the

*Frontier rodeos, Jordan style, were held in the wide open spaces and bound by riders on horseback, not arenas or fences. Other times, as below, cars formed a partial fence*

ten-second ride.

"That was real rodeo just as it happened in cattle country at the time—the only kind of rodeo Pearl had ever seen."

With four small children in the middle of the depression, times were so tough Dad decided to return to trapping one winter. By then, most beaver and mink with their high-quality furs were trapped out of the rivers and backwaters of eastern Montana and only muskrats remained. Muskrats were small, but with enough pelts he might get by.

He had no money to buy the fur-bearer's trapping license and without it was not allowed to sell furs.

"A trapper friend let me sell pelts on his license until I earned enough to pay for mine. Then I sent word to the sheriff.

"In a week or so, the sheriff came by with the license. He signed it. I paid and we stood talking on the porch."

"Where do you figure to trap?" he asked.

"Well, I thought I'd try the Redwater."

"Naw, don't do that. They ain't no mushrats on the Redwater!" said the sheriff. "Don't waste your time there."

For once Dad was at a loss for words. In the corner of the porch was a large canvas-covered pile. He avoided looking that direction.

"I sure wanted to flip back the tarp and show him that big pile of muskrat pelts!" Dad laughed heartily as he told the story.

"All of them caught on the Redwater!"

## Running the Tumbleweeds

"Your horses! There go your horses!" Mom called as she and Beverley dashed out the door.

Jeanie and I ran to the window and, sure enough, there went our tumbleweeds bouncing across the prairie in the wind, their twine ropes dragging behind. Tears ran down my four-year-old cheeks. *Will they ever catch them?*

Beverley, Jeanie and I each had our favorite steed captured from some fencerow, where tumbleweeds piled up, often as high as the fence. When the wind blew hard, as it often did that summer, we ran with them across the prairie. It was great fun when at the end of a long twine our wildly bucking tumbling-tumbleweed ponies pulled us along in a

*Swimming in Yellowstone River backwaters near Savage with friends. From left, Beverley, Francie, Jeanie.*

FRANCIE

Western Ranch Life in a Forgotten Era

stiff wind, fast as my short legs could run, leaping and plunging and kicking. The bigger the tumbleweed, the harder it pulled.

We kept them tethered to the fence on the sheltered side of the house, far enough apart so the twines didn't twist together in a sudden gust.

As very small children living in the desperate years of the 1930s in the most desolate part of Garfield County dryland, we didn't have many toys. We lived south of Fort Peck Dam among the Missouri River Breaks. This Hell Creek area, well-named for its rugged terrain, was already well known in museums throughout the world for its dinosaur fossils.

Most families left Hell Creek and the Missouri Breaks before we did—in 1937, just before Anne was born.

We cherished our little red wagon, pulling it around daily, filled with various loads, building treasures and live kittens. Sometimes a bleating lamb warmed itself in the wood box behind the kitchen stove and we took it for rides too.

One evening when Mom went out to hunt the milk cow, she came over a rise to see in the distance a small figure pulling the red wagon up a hillside. Leaving the cow she ran to catch up.

"Jeanie! Whatever are you doing out here? Why did you leave the house?"

"We lookin' for Dim's snake!"

Jim Roy, a bachelor neighbor, riding by that morning, had talked of a rattlesnake that killed his three-year-old niece. With Mom gone, Jeanie put me, a toddler, in the wagon and set off to find the lost snake, while Beverley stayed home to care for the house as, of course, we were all instructed to do.

Mom shuddered, wondering what Jeanie would do with a coiled rattler if she found one.

Yet, with her teaching background, Mom firmly believed encouraging a child's imagination caused a young mind to expand and grow beyond formal education.

"What could this empty sardine can be?" she asked us as she washed it and carefully bent back the sharp edges.

How special that sardine can was! We built roads and bridges in the ash pile, with ash clinkers for hills and trees from our coal-and-wood-burning stoves.

Our sardine can became a shiny auto with rumble seat, just like the one in which a babysitting neighbor girl and her boyfriend gave us a ride one Sunday afternoon. Luckily, the boyfriend didn't drive his 1929 Model A Ford coupe very fast or far with us aboard. My sisters put me in the middle so I wouldn't fall out. Perched out there in the open air of the rumble seat with nothing to hold us, the wind

*Jeanie, Anne and Francie play in the snow.*

blowing in our hair, it was a once-in-a-lifetime adventure that we loved, but never experienced again.

It seldom rained, so we built our own mud hole and shaped the gumbo clay into people, tables, chairs, cups and saucers. We spent weeks building a mud corral and barns, adding cows, horses and sheep, with a lurking coyote or two in the trees. All of them dried in the sun, with deep fissures and cracks, hard as rocks.

One day Jeanie and I stirred up a special mud pie using a clean bowl from the kitchen. Inspired by the clean bowl, we added an egg from the chicken house to the mud and beat well. Then, since Mom was working in the garden, we spirited sugar and flour out of the kitchen and put that in too. By this time we had invested a lot in our mixture—and it didn't look too bad.

I was eager to eat—even though it had a mud base.

"Hmmm," Jeanie didn't say much, but took a second small taste.

I'd eaten dirt before and knew it would be gritty. Yet there was an interesting sweetness to that mud pie with its genuine cake ingredients.

Our imaginations flourished in those early years, as we made our own fun. We were isolated country girls surrounded by kings and queens and cowboys, all our storybook friends, castles and ranches.

Sometimes in those days our family went down to fish and play on the shore of the Missouri River or Fort Peck Dam. On rare occasions the folks got together with others for a picnic or country school dance. We didn't attend church, as the nearest was sixty miles away in Jordan.

On our rare trips to town we peered out the car windows in amazement at children who had to stay in their yards without any building materials or animals to cuddle.

It's doubtful they envied us, but we wished they had tumbleweed ponies to run with across the wide-open prairie and were thankful we had such a rich and adventurous life.

*We perched in the open air of the rumble seat with nothing to hold us, the wind blowing our hair.*

# 8 Working Livestock

## Wild Heifer

One blistering hot day I spent hours trying to bring back a young heifer from a herd of wild Grieves' cattle up on the open range. Associating with that wild bunch was never good for our cattle.

JEANIE

It was almost time to put his bulls in with his cows and, knowing Dad wanted to use his own bulls, Mr. Grieves stopped in at the ranch and told us one of our heifers was in a bunch of his cattle. However, being a herd animal, that heifer learned to run with those crazy cattle and kind of liked it. She became as wild as they were.

Dusty and I topped a hill. Down below, suddenly, there was a big Grieves' herd, fanned out and grazing. Calm red bodies and white faces cropped grass peacefully. When they saw us they immediately bunched up and faced us, their round, cropped ears standing straight out, eyes rolling in their white faces—as terrified a bunch of cows as I'd ever seen, but normal for Grieves.'

We stopped on the rise with the uncomfortably hot wind blowing directly in our faces. It dried out my eyeballs and I squinted trying to pick out our heifer with one ear cropped and the other notched.

Suddenly the whole herd turned tail and fled down the coulee. Grieves cows had the peculiar habit of all throwing their tails up in the air at once as they ran. Running with Grieves' cattle, our heifer learned to spook easily.

Dusty knew his job—we took after them with a vengeance, as only Dusty could. Since his very favorite job was cutting cattle and he was our best cutting horse, he knew that when we found a bunch of cattle, his job was to find the right animal, tail it out of the herd in the direction we wanted.

But in this case first we had to catch the herd.

As we caught up with them, the grime and prairie dirt fly-

> *When they saw us the open-range cattle bunched up and faced us, their round, cropped ears standing straight out, terrified eyes rolling in their white faces.*

ing, Dusty kept right on dodging through the center of the herd. I couldn't believe it.

*What is Dusty doing? I tried to rein him in, but couldn't.*

*Then I saw the heifer turning in front of Dusty. Apparently he saw her long before I did.*

No matter how quickly she dodged, or swapped ends, trying to lose herself within the galloping herd, Dusty anticipated her. It took all my strength to stay in the saddle when he made his whipping turns.

Finally, after I was almost exhausted, there the heifer went, running out alone on the side, us in pursuit, away from her buddies and trying to get an edge to dodge back in, but not succeeding.

We almost never left gates open, but I learned sometimes when you're running an animal out of Grieves' alone, you just do—and I was glad I did. We got her through the open gate, running hard, and turned her down a coulee the right direction while I closed the gate.

Both Dusty and I galloped some distance with the heifer ahead of us, trying to dodge back. We were in a lather by the time we neared the ranch several miles away on that awful, blistering and windy day. My clothes stuck to me, so sweaty even the wind couldn't billow them out.

The big white corral gate was open waiting for us when we neared the ranch buildings. Getting her in was going to be a cinch.

But oh, no! There was Francie sitting on the top rail of the closest corral, completely engrossed in skinning out a coyote, the brim of her captain's cap shading her eyes while she waited for us to show up. The coyote surely smelled to high heaven in this heat.

It seemed like summer after summer, Francie was there sitting on the top rail of the corral fence, wearing her captain's cap, skinning out a coyote, deer or some little creature in her spare time.

It could be especially harrowing riding into the ranch if my nervous saddle horse caught the scent before I did. Cattle and horses weren't afraid of coyotes, but for some reason, with that awful dead smell hanging in the air, they were.

When Francie saw us coming, she quickly slithered down the corral rails to the ground, trailing her carcass behind her and disappeared out of sight behind the horse barn. Sometimes our cattle or horses didn't even catch a glimpse of her, but her carcasses left such a rank smell that they spooked anyway.

I think the dead animal odor was overwhelming to them, especially the coyotes and bobcats she skinned, maybe because they were predators.

Francie disappeared fast, dragging her coyote behind her.

But too late—the heifer caught the smell. Dusty and I, going

*Jeanie with our favorite cutting horse Dusty.*

**It took all my strength to stay in the saddle when Dusty made his whipping turns.**

Western Ranch Life in a Forgotten Era

slow now to thread her through the open gate into the corral, lost her completely.

She went berserk and headed back as fast as she could into the hills.

I was beat. I pulled Dusty up and caught my breath. Francie came running up.

"Where's your heifer?" she asked.

I didn't answer because we dashed back to turn her. Dusty and

## Working Cattle

Cattle require care throughout the year—from calving in spring, June branding and selling calves and culled cows in fall, to providing nourishing, high-protein food every day summer and winter. For close-up work we appreciated our new homemade chutes that were easier, safer and caused less stress than the old ways of roping and restraining the animal on the ground.

*Francie hazes while Dad ropes and leads a cantankerous heifer that refuses to go in the chute for doctoring.*

*In closer quarters, Francie and Dad half-drag, half-push the ornery cow into the chute holding pen.*

*At left, a range cow that refuses to feed her newborn calf is milked out. Colostrum, her first milk, is critical for the newborn for its antibodies and concentrated nutrients; Francie holds cow in narrow chute with posts placed ahead and behind her.*

I brought the heifer down again and corralled her, but only because she was winded and ready to give up.

We had hard rides on Dusty. Once running fast down a barbed-wire fence cutting out a steer, I tore a leg of my jeans almost completely off without so much as a scratch on my leg. He knew exactly how far away to stay from a line fence.

Although he liked to buck, he never did it cutting cattle. Throughout all the years sorting cows, Dusty never failed us.

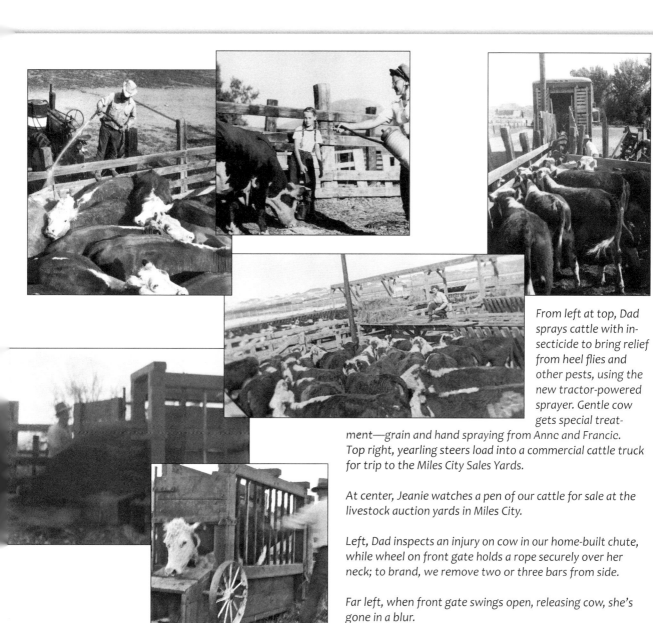

From left at top, Dad sprays cattle with insecticide to bring relief from heel flies and other pests, using the new tractor-powered sprayer. Gentle cow gets special treatment—grain and hand spraying from Anne and Francie. Top right, yearling steers load into a commercial cattle truck for trip to the Miles City Sales Yards.

At center, Jeanie watches a pen of our cattle for sale at the livestock auction yards in Miles City.

Left, Dad inspects an injury on cow in our home-built chute, while wheel on front gate holds a rope securely over her neck; to brand, we remove two or three bars from side.

Far left, when front gate swings open, releasing cow, she's gone in a blur.

Western Ranch Life in a Forgotten Era

## Coyote Attacks

Beverley taught us the sheepherder's lament she heard at a Woolgrowers convention.

FRANCIE

"When the last man dies on this earth, a coyote will be there to pick his bones!" a crusty old sheepman told her.

When we fed-out lambs during fall and winter months, sighting one or two coyotes in broad daylight called for instant action. Many times we girls were sent out to keep skulking coyotes away from the sheep while Dad worked around to get a good shot at them.

One spring morning we were roused at daylight when Dad spotted three coyotes racing into the buffalo berry thicket in Art Hill's pasture corner above our irrigation ditch.

"You girls run through the fields and I'll circle around above the ditch," he yelled up the stairs. "Get to the lower openings and guard the trails out of the berry patch so they don't sneak out down into the fields. They haven't come out there yet, and don't let them!"

*A coyote scouting near the ranch in broad daylight means eminent danger.*

Only half-awake we all four leapt from our beds, pulled on yesterday's clothes, and took off on the run through fields and across ditches to the buffalo berry patch, about a half-mile away.

This was a thicket of thorny trees with grey-green leaves and blackish branches, tightly grown together, broad as a barn. Taller than a person on horseback, you could not ride through it, but only walk and crawl through tunnel-like trails the cattle had made to cool themselves and scratch off insects.

We didn't much like going through these cool tunnels because of the stickers and thorns and the sudden creepy twists and turns of shadowy crisscrossing trails. It was easy to get lost, but even so, we sometimes explored them.

*A sheepherder's lament: "When the last man dies on this earth, a coyote will be there to pick his bones!"*

As we ran toward the thicket we saw Dad circling around on the hills above with his rifle held ready. We knew he'd position himself where he could watch the upper tunnel exits from a guarded distance.

In only a few places did tunnels of the buffalo berry patch emerge on the lower side, and we each rushed toward one of these exits. We took up our stations silently, ready to shout and wave back any coyote that came escaping our way. It was scary because trapped animals sometimes attack.

We heard two quick shots, silence, and another two shots. Then Dad yelled for us to come out. He killed the first two coyotes. The third, far out of range, escaped into the hills.

In October one year, our parents bought a band of a thousand new fall lambs off the range to feed and grow out over winter on beet pulp.

Since the truckers came earlier than expected, Dad sent them to unload the lambs on a flat across the highway where they could fill up on grass. Meanwhile he had to finish work on the corral before bringing them in to the feed bunks.

In the emergency, Beverley stayed home from school to herd them on Eagle, with help from our sheepdog Jack. It was not safe to leave the lambs alone in unfamiliar country. Just taken from their mothers, they could scatter and get lost.

However, a more deadly danger lurked.

When a heavy fog from the river began to roll in, Mom and little Anne, not yet in school, went to help her. As she arrived, Mom saw a shadowy coyote leap at a lamb through the fog. Another coyote streaked by, intent on hitting the other side.

Leading Eagle, Beverley was running, shouting, pushing lambs and waving her arms as coyotes attacked one lamb after another. The dog dived toward them, barking and snapping.

"Hey you! Git—Git away. Git 'em, Jack!"

Attacked and pulled to the ground, one lamb bleated frantically. Another blatted his terror at a ripped and bloody leg.

Jack ran back and forth barking, snapping at the marauders. The coyotes grew bold under cover of fog, gliding away into the mists when she came near, attacking at other points.

Dad came speeding up and jumped out of the car with a rifle.

"We need more help," he said to Mom. "Go get the girls from school."

Jeanie and I were shocked when Mom came rushing into our country school room without even knocking.

"The coyotes are into our sheep and we can't watch all sides in the fog. I have to take the girls home," she said to Miss Wyss.

"Yes, yes, it's all right," the teacher told us.

Mom propelled us toward the cloakroom.

"Hurry! Just grab your coats and put them on in the car. The coyotes are attacking the new lambs."

Beverley was crying and shouting furiously at the leaping coyotes when we got there. This was frightening in itself: Beverley never cried. Just as she ran off one coyote, another charged in. Being the same color as the fog, the lambs faded in and out.

Bev jumped back on Eagle and raced around to the other side.

Dad was trying to get the coyotes in his sights and fired off a shot or two. We heard a yelping and guessed he'd got one. It seemed the coyotes had killed a couple of lambs and dragged them back into the sagebrush.

"We need to get them down off this flat and into the corral,"

*Coyotes prey on helpless lambs.*

*Just as she ran off one coyote, another charged in.*

Western Ranch Life in a Forgotten Era

Mom said.

We spread out around the sheep with Mom and Anne, waving our arms and shouting. The fog was so thick, we could hardly see ten feet. But the lambs bunched tightly in panic, and when we were close enough together the coyotes didn't dare dive between us.

Riding Eagle, Beverley kept the lambs close.

As the fog lifted a bit, we surveyed the damage: several bleeding lambs, two dead, along with two dead coyotes and two or three canine raiders shadowing off into the brush.

On another morning, Beverley and I discovered a coyote den in the high, steep, bluffs above our lower pasture. We were riding above the ditch, at some distance, when one of us suddenly spotted the mother coyote lying way up there on a large rock. The sandstone matched her tawny color exactly.

She didn't move when she saw us, just turned her head and watched. So of course we had to climb up and investigate. She seemed reluctant to leave, then trotted off up a cedar draw, and we didn't see her again.

*Coyotes are tough survivors. Sheepmen say, "When the last man dies on this earth, a coyote will be there to pick his bones."*

We headed straight for her rock, scrambling up the steep cliff, and there found a well-trod path.

Just beyond, from a dark hole, emerged little yips, yaps and quarrelsome growls. We lay down and reached in our arms, but neither of us could reach in far enough to feel pups.

Excited about finding our first coyote den, we toyed with the notion of spending days here playing with the puppies, keeping our find secret. But sobering thoughts soon followed.

Our family battled terrible destruction from coyotes. Coyote pups grow quickly, and full grown they revert to their wildness and turn vicious. We'd been told that often enough.

Mom said an old bachelor neighbor kept a snarling flea-bitten coyote, once a cuddly pup, on a chain in his yard. He had neither the heart to shoot it, nor to let it go.

> *Dad brought home a sack of six frisky and cuddly coyote pups that he'd smoked out of the hole.*

Besides, Montana paid a bounty on coyotes. Six pups were worth money. We had to tell.

Dad lost no time bringing us back to point out the exact position of the den up the face of the cliff. He started a smudge fire that revealed two or three more openings spilling down the cliff. An expert trapper, he understood coyote intelligence. We knew he'd set traps at all openings, camouflage his traps well and there'd be no human scent. The mother coyote didn't stand a chance.

Next morning he brought home a gunny sack full of six coyote pups that he'd dug and smoked out of the hole. He tumbled them out into a washtub—frisky, cuddly and various shades of tawny golden beige. Jeanie, Anne and I played with them all afternoon cuddling and petting. We even taught them to drink milk from a bowl. Playful as a litter of puppies, they were unafraid, wrestling with us and each other, growling and nipping mischievously.

Without asking, we knew the coyote mother was gone, and we begged Dad to let us keep the puppies. He didn't say much, maybe regretting he'd brought them home at all. He could have knocked them over the head back at their cave, before tossing their little bodies into the gunny sack.

We promised to feed and keep them clean, love and nurture them. We promised to tame their wild natures so they'd stay forever sweet and cuddly.

"But what about my chickens?" Mom said in a sympathetic tone. "Coyotes will eat kittens, too, and little pigs."

In the morning the pups were all gone from the tub. Jeanie and I searched till we found their six little sets of ears, each with a bit of scalp between, nailed up to dry on the wall behind the shop. Proof for the bounty then paid on coyotes of any age in Montana.

In fall and winter the coyotes seemed to be hungrier and bolder than ever; they moved in closer to the sheep corrals. Every night they howled hauntingly close by.

Dad would buy a 'killer horse,' take it out to a likely spot a mile or so from the ranch buildings, remote enough as to be no risk for pets. There he shot it and placed poison baits throughout the carcass.

As we got older, he let us watch as he shaped the poison baits. He took down the squat can of strychnine paste from where it was stored on a high rafter in the shop and pried off the lid. It always surprised us to see the violent pink of the substance and we wondered if it was natural strychnine or the color added to emphasize danger.

With gloves and a small wooden paddle he formed marble-size balls of beef fat and pressed a dose of poison into the middle of each. These he put into an old smelly cloth sack to take along to insert into the horse carcass or scatter close by.

> *Jeanie and I searched till we found their six little sets of ears, nailed up to dry.*

*We corral our sheep every night to protect them from coyote marauders.*

Western Ranch Life in a Forgotten Era

Dad was particular about odors, making sure to hide any human smells with his own creation of coyote scent.

He emphasized to us the risks of strychnine and to keep close watch on the baits until they were set into the muscle of the dead horse. It wouldn't do for a dog to eat from the little sack.

As the weeks went by, coyotes ate the poison and died. He rode out and carried the carcasses in behind the saddle. One winter I helped check the area around a dead horse laced with poison baits, earning money to purchase my first horse.

Dad threw the skinned carcasses, purple and grotesque, behind the shop to haul away during the first spring thaw. Their outstretched legs seemed to be in constant running position. Piled up inside the shop were their beautiful pelts, initially stretched wrong side out. Before spring he sold the coyote hides to the fur buyer in town. In those days prime winter coyote pelts were worth a lot of money.

Finally, the government came to the aid of sheep and cattlemen. In a major coyote eradication program beginning in the late 1940s, the federal government brought out its best hunters, trappers and a new, more effective poison named '10-80.' Amazingly, coyotes largely disappeared from range country.

Sheepmen had it easy for many years. No longer did they have to hire herders or even corral their sheep at night. No coyotes howled from the hills on a moonlit night.

Relentless foe of coyotes that he was, I guess Dad missed them when they were gone. Inevitably, the day came when he saw his first coyote in many years—a lone coyote running up the draw ahead of him. A lifelong hunter and trapper, ever dedicated to protecting sheep and young calves and ranchers from predators, Dad grabbed his rifle from the pickup rack behind his head and eased his sights onto the coyote.

An easy shot. Then he lowered the barrel and just sat there watching.

"It was so good to see a coyote again after all those years, I just couldn't kill him," he told us later, shaking his head.

Dad appreciated nature and the wildlife that surrounded us daily. Active in several wildlife and conservation groups, he was a strong promoter of the balance of nature. He taught us to see with long-distance eyes and understand what we saw, whether it was an eagle perched on a high point with a rabbit in its claws, a deer hidden in the brush and visible only by the telltale outline of an ear, or a coyote hunting mice with raised paw in long grass.

But the coyote Dad spared that day was not the last coyote, after all.

---

## Wolf killing packs

Our neighbor Grandpa Bill Hill, who as a young cowboy twice trailed up from Texas with big cattle herds, told us he once rode over a hill just north across the Yellowstone River from our ranch and saw a terrible scene he could never forget.

On the flat below, a pack of a dozen wolves were attacking a terrified bunch of horses. Several wolves cut off the lead to slow down the panicked horses, forcing them to mill in a circle.

At the rear, the old male wolf slashed at the back legs of a colt, cutting the ham strings. The colt half fell and was immediately set upon by three or four young wolves.

Other wolves tore savagely at another downed horse as it struggled to rise.

Bill grabbed his rifle from the scabbard and fired, scattering the wolves. But he was at too great a distance to kill more than one or two.

He arrived too late for most of the injured horses and, following rangeland code, quickly put them out of their misery.

When the federal predator programs ended, coyotes made a comeback and multiplied more than ever across their old ranges in the west and even extended into population centers up and down the east coast.

It seems the sheepman's lament still holds, perhaps more meaningful than ever.

"When the last man dies on this earth, a coyote will be there to pick his bones!"

## Purple Coyote Carcasses

In fall and winter, when the pelts were thick and rich, Dad skinned—he called it *peeling*—the coyotes he shot, trapped or poisoned, pulling and molding their skins to dry over wooden stretchers, leaving their naked carcasses to freeze in a pile behind the shop. In summer, coyote hides were no longer prime, so the dead were hauled away quickly with their hides intact before they began to smell.

*Nothing pretty about those grotesque skinned coyote carcasses, with their stringy purple muscles, small sleek blue heads and bulging eyes!* Long and thin, without hair and hide, they looked like surreal pets of aliens, put to sleep for extended travel through space. Really creepy.

And so we figured they'd be of great interest to our friends. Of course, most of our friends were country kids like us.

One Saturday morning a Miles City friend brought along his nine-year-old daughter to discuss some business with Dad. This man had grown up on a ranch but married into local town society. On reflection we suspected he might have brought Suzie along to get a little ranching education—and that day, she did.

She was a pretty girl, hair in ringlets, dressed in a clean, short frilly dress, not a smudge on it. In her hair perched a large pink bow, perfectly matching her dress. Her dainty black patent leather shoes held not even one smear of barnyard manure, of course.

As she told us about her dance lessons, she gracefully flung out her arms, arched her neck and back, and pointed her toes in a ballerina manner. Her speech fit the picture she played of a storybook princess—a pretty doll condescending to visit ranch kids.

While the men visited we knew it was our job to entertain Suzie. So we took her to see our latest alien find behind the shop.

*Coyote plans his attack.*

## Wolves and coyotes

Historically, in ranch country it was open season on predators of all kinds—the wolf, coyote and elusive mountain lion. The late 1800s and early 1900s brought large herds of cattle and sheep to replace the millions of buffalo that once grazed the hills and prairies.

All buffalo herds had 'their shepherds: the wolves'—as described by explorers. With buffalo gone, the deadly predators transferred their attentions to cattle, horses and sheep.

Wolves hunted in packs, usually a male and female pair with one or more years of offspring. Sometimes as many as fifteen hunted together in a pack, old-timers said.

*(continued)*

There they lay, three coyote carcasses at their most garish, thoroughly skinned and naked, in appalling shades of purple, red and blackened flesh. Their ropy muscles were strongly defined, no fat to spare. Thin, blackened legs seemed to reach out in great strides, as if still running. Each carcass ended in a long skinny tail, a sweep of segmented bones, each set onto the next, smaller and smaller out to the very tip, the muscles shriveled and dark. Most appalling, huge protruding eyeballs bulged out on each side of the elfin heads.

Susie rounded the corner of the shop with us.

"Eei...i...ii!" She screamed in shrill piercing cries.

We clamped our hands over our ears and leaped away.

*What was wrong?* We thought she must be in pain, or maybe dying. Then we noticed the horrified expression on her face, her gaze locked on the three naked coyotes, racing into eternity. Gone was her pretty little bow-like smile and calculated doll expression.

She screamed and jumped up and down and squealed. And she didn't stop screaming, even after we hastily led her back out of sight.

We were as shocked at Suzie's response as she was at the carcasses. In our family no one screamed unless there was a dire emergency. We hadn't intended to gross her out, just to entertain. We truly thought she'd be interested.

Suzie ran head-long into her father coming to investigate her screams. But on the way, to make matters worse, she didn't watch where she was running. Some cattle had just walked through the yard to the water tank, leaving a few splats of fresh manure. She slipped on one, almost fell and messed up her shiny black shoes.

She clung to her father, having lost interest in playing with us. Suzie never returned. Guess she preferred town life.

*Dad brings ewes to the corrals with feed bucket.*

Growing up in range country in the 1940s, we were accustomed to coyotes howling from the hills every night, usually at some distance.

FRANCIE

Still, it could be scary when you were alone, riding home near or after dark. Their lonesome howling sounded alarmingly near. They howled, yipped and barked, calling back and forth to each other from the ridges, seemingly closing in from all sides. If the wind lay still as it often did on crisp, cold nights, the coyote voices pierced the air even more clearly.

Though we kept telling ourselves

that coyotes never attack a person, certainly not a person on horseback, we felt uneasy. It was a scary sound, but melodious, too. Fear magnified when the haunting voices seemed to circle around ahead of us.

"Shut up! Get out of here!" Unfazed, Beverley yelled at them.

All sounds stopped abruptly when she shouted in that commanding tone of hers, followed by moments of startled silence in the darkness—then the night erupted into a cacophony of howls, barks and yipping that seemed closer and more menacing than ever.

Coyotes have no respect for people sleeping at midnight. Awakened suddenly one night, Beverley yanked up her upstairs bedroom window and scolded into the blackness of the night.

"Hey you! Git—Go on home out there!"

The silence lasted till morning, much to our delight. Even though our house was large and we all slept in different bedrooms we heard her. We teased Beverley forever after, for being so mean that even the coyotes tucked their tails between their legs and quietly slunk away.

Anyone who raised sheep had plenty of these predators lurking close by, keeping ranchers and herders on the alert. Coyotes were fast and ruthless.

With cattle, there was minimal risk, though sometimes during calving a hungry coyote might move in, snatching at the newborn while the cow was giving birth and unable to fight back.

Sheep are a different story: they're helpless. We had cattle—and sheep, too, from time to time—with plenty of coyotes hanging around and keeping us on edge.

Dad was ever vigilant in his efforts to keep the coyote population down in our end of the valley. Ours was the last place in the irrigated valley east of Miles City. Here the valley narrowed and then pinched in abruptly as the Yellowstone swung tight against our steep bluffs beyond the lower pasture. All the rough country above the ditch was rangeland with plenty of cover for lurking coyotes.

Dad was a crack shot and during his early years in Montana he worked as a government hunter and trapper. He grew up in the outdoor traditions of the day, so was more experienced at this even than most ranchers.

He regarded keeping the coyote explosion under control as, not only necessary for our own survival, but for our neighbors' as well. He took seriously this responsibility to the community.

His methods were gun, trap and poison. Sometimes all three at once, as when he placed poison baits in a dead horse carcass, set traps along nearby trails and carried a high-powered rifle for a quick shot whenever he came to check his baits.

> **Wolves** *(cont'd)*
> Because they moved in packs, wolves proved fairly easy targets and were soon gone from southeastern Montana ranges with settlement, except for a few hardy individuals.
>
> Local legends recognized such outlaw survivors as the Blue Blanket Island wolf and Old Three Toes.
>
> Old Three Toes richly deserved his reputation as a killer. His three-toed track was found in sheep corrals in a single night throughout a one hundred-mile circle across the corner of three states, this area of Montana and both Dakotas. He left scores of uneaten dead lambs in those corrals, and survived well past homestead days. Many ranchers tried to outwit Old Three Toes—a rancher on horseback ran him ninety-five miles during three days following his tracks in fresh snow, changing horses five times. But it was not until 1925 that a determined government hunter finally brought him down.
>
> With wolves gone, coyotes flourished. Their numbers mushroomed throughout the first half of the 20th century in the sparsely settled West. At last, a successful federal program to help ranchers, largely eliminated coyotes from eastern Montana ranges. It eradicated predators with the poison 10-80. After this poison was banned, inevitably, coyotes staged a comeback.

By spring Dad sold a harvest of prime pelts. Still, even more coyotes moved in to replenish the supply.

They seemed invincible.

## Bringing Home the Cows

Goosebumps ran down my arms as I listened to *The Bull* bellowing in his pen.

JEANIE

The bullpen in the south end of the horse barn was built of heavy two-by-eight timbers but, nevertheless, when he lost his cool, they shook and creaked and we feared they'd splinter into toothpicks.

Our mean, powerful black-and-white Holstein bull—luckily dehorned—paced back and forth, looming large in his pen. We called him only The Bull, though officially named something like *Aloysius-Americus St. Ferdinand III*.

What a terrifying animal he was!

When Francie and I were very young, around age six and eight, it was our job to bring in the cows for evening milking. This presented two problems.

First, the long walk to reach the cows and bring them home. Often our milk cows were more than a mile away and lost in the hills or down in the swamp. It might take us all of two hours to find and bring them in because it was so far—and because there were so many interesting things to do and see along the way.

At the end of this long journey we encountered the second problem, bringing the cows past the terrifying bull on our way to the milk barn. The Bull frightened us, because a few times he broke out of his pen into the larger horse barn when he heard the milk cows bawling for their calves as we brought them closer to the corrals.

*Our mean, powerful black-and-white Holstein bull paced back and forth, looming large in his pen.*

As small girls we entered the horse barn quietly, watching to see if he looked at us between the rails. If he saw us, he began pawing, rolling his head around in the dirt, bellowing and charging the rails—and we ran away quickly.

No one ever teased The Bull. Most of the time he was quiet, but when he got angry, for whatever reason, he attacked his pen and the splinters flew.

Sometimes he broke out his gate and ran loose in the horse barn. Sometimes he broke through his barn wall and ended up in the corral. Other times, let loose for breeding in the smaller corral, he broke out the corral gate.

When he escaped, The Bull claimed the large open territory between the house and ranch buildings as his own and charged every moving thing in that space. He ran the chickens back in their house. The cats fled. Our dog scooted to the house and up on the porch, whining and scratching on the door to be let inside. The birds in the cottonwoods stopped singing. Everything was silent.

Then, standing in the open yard alone, he began his bellowing challenge. That and the sound of his pawing up the hard packed dirt horrified us. He sent clods of what had been flat, hardened gumbo right over his head, defying anyone or anything to come out.

I had terrifying nightmares in which The Bull pawed, bellowed and charged my hiding place, while I stood frozen, too scared to run.

Mom told us that when The Bull broke out, if we girls were caught in the open near the house, we must get up quickly on the porch. But if he saw us running he might give chase, so if it was too far to the house, get into another building or behind a fence.

We did better than that. We ran into the house and didn't stop until we found a good view upstairs. There we peeked out the windows as the drama unfolded.

Dad, quick to respond, saddled our white horse, Eagle, snatched up a pitchfork, and ran the bull in. All the while we cheered from an upstairs window.

One day when The Bull saw him coming on horseback with a pitchfork and shouting, he learned the hard way to turn and run back into the corral, into the horse barn and on into his pen. As Dad and Eagle raced after him, he turned on them, ready to charge. Dad threw the pitchfork at him and it stuck in his rump. He ran into his pen with the long handle whipping back and forth. From then on, the impression of Dad with the pitchfork imprinted so strongly on The Bull, Dad could grab a pitchfork and run him in on foot.

Dad was the only one who could handle him. When he was gone one afternoon the hired man tried to chase The Bull back into his pen on foot. He bellowed, rolled his forehead around on the ground and pawed the dirt, kicking clods over his back, making low roaring sounds deep in his throat. This was his prelude to a coming charge.

The hired man got the corral gate shut and got out of there. Holstein bulls are dangerous. The Bull's father had killed a neighbor's hired man and this bull was just as mean.

Another day he got out when Dad was gone, and the hired man tried his luck on Eagle. In the wide space between the house and barns, The Bull pawed the ground and charged Eagle. He hit the

*Each evening at five, Jeanie and Francie bring in the cows for milking. Sometimes it's a long hike to find them in the lower pasture.*

**He scattered the chickens. The cats fled. Our dog scooted to the house and up on the porch, whining.**

*The nearest prairie dogs raised the alarm, yipping and stretching up tall on their hind legs to see better.*

horse squarely in the center of his chest, almost knocking the wind out of him, lifting both his front legs off the ground.

But Eagle didn't flinch. Mom was afraid The Bull would injure or maim both man and horse—or worse—so she called them off. We had to leave The Bull out until Dad got home.

We needed The Bull. His calves were genetically strong and large. His daughters gave lots of milk, important for Holsteins.

"Girls, time to go get the milk cows." Mom called to us every afternoon.

She allowed us plenty of time in case they were down in the lower pasture, or worse, the swamp. Also, she knew we might stop to watch the prairie dogs and investigate rocks we found along the way.

On our long trek Jeanie and I crossed a prairie dog town. The antics of the 'townspeople' there were irresistible. We probably wasted a lot of time, but it was such fun!

"Eeeka! Eeeka!"

FRANCIE

## Branding

Dad branded, castrated, dehorned, earmarked and kept the branding fire hot. Jeanie vaccinated. Beverley was good all-around help, while Anne and Francie kept the chutes full of calves. For an all-day job, neighbors came or Dad brought out State School boys to help.

*State School boys hold calf's kicking legs as Jeanie vaccinates, below. The calf enters a vertical calf table, which then clamps the calf and tips into a raised surface.*

*Our neighbors rope their calves, wrestle and hold them on the ground in traditional style, while Dad brands.*

*At right, Dad's just-right touch on the branding iron briefly sears the animal's hide but doesn't burn deeply.*

The nearest prairie dogs raised the alarm when they saw us coming, yipping and stretching up tall on their hind legs to see better. Sometimes they flung themselves into the air in their great excitement.

"Eeeka! Run for your holes!" was the message flashed through town.

They scurried from everywhere. Whether socializing with friends or enjoying a stem of new grass in a nearby draw, they dropped what they were doing, heeded the call and ran for the safety of home.

As we crept closer the earliest sentinels dropped down their holes out of sight, while others took up the cry. After awhile they popped up again, scolding from a safe distance, eyeing us warily as they went back to nibbling and socializing. When greeting another, they stood up together and hugged and kissed. Apparently that's how they connect socially, checking on mutual acceptance. We delighted in these friendly creatures and kept up a chittering conversation with them.

Along the way to find the cows, we also played with interesting rocks we found. There were sparkly granites, softly-colored sandstone rocks slid down from the buttes above, and round flat river stones washed smooth in an ancient channel of the Yellowstone. We inscribed them with our names and messages, using little pink and yellow 'writing chips' of colorful rock from the cow trail.

*Fossils of bone fragments from Jurassic formations under our feet gave glimpses of dinosaurs that once lived here.*

We speculated on the under-an-inland-sea-for-millions-of-years pieces of petrified wood we found, formed from giant redwoods, revealing their tree rings, wood grain and knots. Fossils of bone fragments and seashell imprints, kicked up from Jurassic formations under our feet, gave us glimpses of dinosaurs and other ancient creatures that once lived here and connected us to an earlier time of inland seas and lush jungle vegetation.

Our all-time favorite rocks were Montana moss agates, famously found in the Yellowstone River gravels and tributaries. When we picked up an agate we spit on it to clean off the dirt and reveal intricate designs and mossy scenes hidden within. We treasured them, but didn't necessarily treat them well.

"Ah, an agate!" Francie thrust a dull-coated fist-sized rock at me.

"Let's break it open and see what's inside!" I'd say.

We cracked it open between two rocks, moistened the broken edges with saliva, and admired a miniature landscape of dark trees, shrubs, lakes and brilliant red-orange sunsets, set in a clear background. Cut and polished, it could be fashioned into beautiful one-of-a-kind agate jewelry—rings, pendants, belt buckles—that is, if we hadn't already ruined the perfect cut.

We took our best rocks home to Mom who admired and explained them and maybe added them for awhile to our treasures on the kitchen window sill. Agates ended up in buckets in the shop, for 'someday' when they might be cut and polished. We threw discarded ones at errant cows.

Maybe we dawdled longer when we saw that the milk cows awaited us in the swamp. We hated when they grazed their way down into the swamp that followed the railroad tracks to the Yellowstone River. They seemed to like cooling their feet there in hot weather, feeding on the smelly, sub-irrigated plants. Or slogging on through the muck and cattails and grazing on the small island of grass beyond.

*We were happy with the change to Hereford bulls, center foreground. Peaked roof behind is the bull barn, home of the fearsome Holstein bull.*

This meant we had to wade into the dank water and black mud with its slippery, decomposing vegetation. Worse, the swarming mosquitoes were vicious there, especially in late summer.

We left the house about three o'clock, slathered with mosquito repellent because, in the swamp, mosquitoes swarmed so thickly it was difficult to see through clouds of them. The cows constantly lashed their tails and rubbed heads against each other.

Usually our gentle milk cows came willingly. The rocks we threw got quite a reaction from an offending cow. She sprang into a run, with her udder swinging heavily back and forth. Not good for a loaded cow.

Most days the cows wanted to go home for milking and they knew a scoop of oats was waiting. Besides, new baby calves waited to call to their moms from their pen in the milk barn. We liked petting those sweet new calves, with their soft and silky hair as we taught them to drink fresh milk from a bucket while the moms looked on.

> **There was the Bull, awaiting our arrival, bellowing moodily.**

Milk cows walk slowly. As little kids on those long treks so did Francie and I.

But on the last leg of our trip there was The Bull, awaiting our arrival. In building the bullpen, Dad left a few large spaces in his wall between the thick studs so The Bull didn't feel so shut in. As the cows passed in front of his pen, he watched us through those openings, bellowing moodily, making his plans. An uneasy feeling, to be sure. We all felt sorry for the lonely guy in solitary confinement, but no way did we want him out.

Holstein bulls are dairy animals and often dangerous, as opposed to the mild Hereford range bulls. To our relief, The Bull was sold with most of our dairy herd after the war. Luckily he didn't hurt anyone. At that time our range cows increased to the point where we needed Hereford bulls.

The long journeys to get the cows ended, too. Conveniently, in the smaller pasture near the ranch buildings we kept a few milk cows for family use. The range cattle took over the larger pastures and the dread swamp with its muck and mosquitoes.

## Milking Antics

One afternoon, a cow I was milking kicked so hard with her kickers on, she fell over, flat on her side—the side where I was milking. One-legged milking stools made for a quick get-away.

A contented cow is supposed to give milk without fussing, although some cows are never contented. For that kind we used 'kickers,' metal clamps attached to each leg at the back of her knee joined by a short chain. They held her back legs together with only a bit of space between—and prevented some unpleasant surprises. An ornery cow couldn't kick, but tried to and caused plenty of problems.

What an effort it was to milk cows during those early years and through World War II when we had so many!

One of our most tedious tasks—milking cows by hand. Twice a day. Early in the morning and at five in the afternoon. After evening milking, the cows stayed in the corral eating hay until the whole process began again early the next morning.

Only dairies milked with machines then but we certainly were not a dairy and never intended to be.

However, milking cows provided ready cash in hard times and for our family, a way to get back into raising range cattle. Also, we did it for the war effort during the early 1940s. Our creamery in Miles City pasteurized skim milk that was processed into powdered milk for our troops overseas.

During the early part of the war we milked as many as fifteen cows by hand—most of them high-producing Holsteins. Dad, Mom, Beverley and a hired man or two milked them, with Francie and I pitching in and getting better at it as we got older.

So who did the milking after that? Most of the time, if we had six or fewer cows, it was us girls—with maybe a new hired man thrown in—sometimes a new one who said he could, but couldn't, milk a cow.

*How hard could it be to milk a cow? Surely he could if we did it.*

*Lady Macbeth was just plain mean. She kicked hard— when least expected.*

When milking, Jeanie favors the cats by squirting them an occasional stream of warm milk.

Once in awhile a man raised on a farm came along and he was a blessing. However, since he was usually good at everything, he didn't stay long in the milk barn, but left us for more important work in the fields. If Mom had time, she might help.

Whenever we got a new hired man, Dad sent him to help us girls with the milking. When Dad asked if the man knew how to milk, he might say no, but was willing to learn.

ANNE

Or he might say yes, even though he'd never seen a cow except from a car window. After all, how hard could it be to milk a cow? Surely he could do what these kids could do.

Each time with a new hired hand, we girls exchanged glances, wary of what might come next. Here was fair game: a man who said he knew how to milk, but didn't.

A know-it-all? We'd see.

We developed a plan for that kind.

We brought the first five cows into the barn. Because our cow barn had only five stanchions we milked five at a time, then turned them out into the corral and brought in five more. They stepped over the gutter, a long, low trench to catch the manure. Then they stepped onto a low platform of clean, straw-covered floor, walked forward a few feet into the stanchion, stuck their heads through and began eating their oats.

To keep an ornery cow from backing out we closed a bar above her neck so she couldn't get her head free until we opened it. Sometimes it took time for a young cow, her first time milked, to accept this restraint.

She soon learned she liked putting up with us because the oats in her feed box was delicious. Most high-producing Holstein cows with heavy, full udders appreciated being milked, and peacefully ate their grain.

Each cow preferred having the same stanchion at each milking. If you switched her to a different stanchion she might be cranky and no one likes to milk a cranky cow, especially a hired man-in-training. A contented cow is the only way to go.

We owned some friendly ones with such mild and lovely personalities we could get away with anything. One was Morgiana, our best cow ever—I could milk her anywhere outside around the buildings, wherever I found her, and she stood quietly. The first of several we named for characters in Tales of the Arabian Nights, Morgiana was mom to sweet little white-faced Ali Baba.

Other cows were unpredictable or downright ornery.

One of these, Lady Macbeth, was just plain mean. When milked she kicked hard, especially when least expected. Jeanie had ways to avoid getting kicked, such as pressing her head tightly into a cow's flank, down low, just left of the udder, so she couldn't kick. If you did it just right it wasn't too painful.

But then, Jeanie excelled at milking cows. She wrapped her long fingers around the cow's teats producing a fast rhythm as the milk hit the bucket. You could tell when she was nearing a full bucket by the sound of the milk stream hitting the foam inside. She finished and started another cow long before the rest of us. Jeanie milked a lot from the age of eight right up until she left home to teach.

Jeanie also had a few tricks up her sleeve to make the tedious chore of milking more interesting.

One evening we brought the first flight of cows into the barn and hooked their stanchions.

The cats came running. We gave the cats a pan of milk after we finished the first cow. They meowed plaintively and we couldn't resist sending a stream of milk their way. Jeanie, who was an excellent shot, squirted them directly in the mouth. The cats loved this, and sat up on their haunches lapping up the long stream of milk. They were so cute, sitting there begging, catching the squirt of milk and then delicately licking the extra off their fur.

"You can milk this cow." Jeanie said one evening to the somewhat overbearing, self-important new hired hand.

Francie and I gasped.

Lady Macbeth! *Would she kick the hired man and send him sprawling across the floor? Would she put her foot in the bucket and shower him with milk? Or would she simply knock him off his one-legged milk stool?*

His test was about to begin.

Jeanie put the kickers on Lady Macbeth, none too happy at seeing an unfamiliar milker. She explained to the new hired man what she was doing.

This man had never milked and though he tried hard to get milk out of the teats, none came. It looked so easy the way Jeanie did it. Lady Macbeth squirmed and shifted her feet warningly.

Finally, he deigned to ask for help, and we showed him how to gently squeeze the teat from the top down. While he worked on that skill, we returned to our own cows.

Jeanie looked over at him and sent a stream of milk toward a cat in his direction, 'accidentally' hitting his leg.

"Oh, sorry."

Nothing is stickier than being coated in a drying stream of warm milk. What could he say? She just slipped of course.

He grunted and brushed at it disgustedly.

This alerted Lady Macbeth and she let him have it, kicking violently, sending the kickers flying and him sprawling in the gutter

*Jeanie's tricks made the tedious chore of milking more interesting.*

manure.

He got up and swore heatedly.

"I'm not milking that cow. No way!"

"Okay, I'll finish her."

Lady Macbeth knew she couldn't outsmart Jeanie.

Francie and I held our breath while we waited to see which cow Jeanie would give this arrogant hired man next.

"You can try this gentle cow," she said.

Once more we looked over in surprise. *White Face! Jeanie's choosing White Face! She has no mercy, but seems to think he deserves it. Of course she's not warning him.*

True, White Face was gentle and even easy to milk—but she had one bad habit. It concerned her personal hygiene. Half-Holstein and half-Hereford, White Face had the distinctive markings of that cross—solid black with white face, white legs and long white tassel at the end of her tail.

When White Face put her head in the stanchion, she usually let loose a stream of sloppy manure, not bothering to lift her tail. It ran down, coating her entire tail and into the gutter below. The long hairs on her once-white tassel, now brownish green, became a mop of icky goo. She used it to swat flies and unwary humans.

White Face observed her victim. If it were one of us ready to milk her, she might relax her tail as we carefully approached her with our milk stool and bucket. But if it was a stranger, or if one of us let down our guard, yuck!

The new hired man approached her with caution, wary of being kicked across the floor one more time. He didn't realize he was in for an entirely different experience.

White Face waited while the man struggled to milk. Finally she had enough. She twitched her tail impatiently.

Francie, Jeanie, and I paused our milking to watch the spectacle about to unfold. White Face drew back her tail for a mighty swing.

*Whap!*

It was a direct hit with her soppy green tail, right across the man's face and shoulders. Stunned, he fell backwards off his milk stool again.

We couldn't help it—we choked up.

The hired man swore violently, ran out of the barn to the water trough, where he pushed his face under the surface, splashing mightily.

We sprayed the cows for insects so they didn't need to swat. But for some who swatted anyway, like White Face, we tied up their tails.

For a bit of excitement we girls sometimes rode one of the last cows out of the barn. Released from the stanchion, she'd burst out

*It was a direct hit with her soppy green tail, right across his face.*

the barn door—you had to duck when you went under that low doorway—then sashay a few steps, stop and let her rider off.

One evening Francie climbed on one of these 'swatters' and forgot to untie her tail. Jeanie opened the stanchion.

"Wait!" she called. But too late.

The cow charged out the door, leaving the end of her tail still tied to a low rafter. From then on, sadly, she carried only a stump for swatting flies, her proud tassel gone.

> *The cow charged out the door, leaving the end of her tail still tied to a low rafter.*

During the war few hired men were available to help. The war took nearly all able-bodied men, except for older men and those deferred for war-effort needs, such as farming and ranching, who already had too much work of their own.

JEANIE

Later, injured or shell-shocked men came home and some of them worked for us. Partially rehabilitated, they were hitchhiking home, broke. Dad picked them up on the highway and offered them jobs. We had quite an assortment of these men who worked just long enough to buy a bus ticket home.

Once, toward the end of the war an especially fine young man, Tommy, helped us milk. He was eighteen and waiting to be drafted. With him milking, we could get all the cows finished quickly. That gave us unaccustomed free time after the milk was separated, so we worked in the house on our 4-H projects, helped Mom cook, made cookies and other fun things. It was nice to have that extra time.

*Lady Macbeth had a mind of her own when it came to milking.*

However, as time passed, we couldn't resist giving Tommy a bit of a bad time.

It was probably all my fault. I couldn't seem to go with the status quo. Ideas kept popping up for pranks to play while milking. It made the time go faster and required a bit of planning—more challenging, perhaps. Francie and Anne were willing accomplices. Our pranks were unrelenting. By now Beverley, being older, was usually off on a more difficult assignment.

One day Tommy came in and told Mom he was quitting. He hitchhiked to Miles City without even collecting his pay.

When Dad found out, he went after him and eventually found him in one of the twenty-three bars on Main Street. He was sitting up on a stool dejected, with his head in his hands and resting his elbows on the bar.

"What happened? Why are you quitting?" Dad asked. "I need you. Come on back to the ranch with me!"

Tommy just shook his head.

*Western Ranch Life in a Forgotten Era*

> *Amazingly, when we swung the pail over our heads, the milk held tight against its bottom.*

"Come back to the ranch with me, Tommy. You're the best hired man I've had in years!"

"Elmer, it's your daughters," Tommy finally admitted. "They're so mean to me!"

When he came home without Tommy, Dad had no mercy. He lectured us sternly.

"And furthermore—from now on, you girls will do all the milking by yourselves!"

And we did.

That week someone told Dad that Tommy joined the Army. Later we heard he distinguished himself in the Battle of the Bulge. In that fiercest of battles it was pure carnage and Tommy returned with a chest full of medals.

We felt deeply ashamed of ourselves and hoped never to have to meet up with him again. Though somehow we couldn't help but wonder.

*Tommy learned his survival skills from Army training and natural ability. And in some small way, did he also acquire a certain wariness born that summer on the Brink ranch?*

*Beverley, home from college, pitched in to help and charmed us all with her charisma and generosity.*

## Cream Separator Do's and Don'ts

Carrying two full, foaming three-gallon buckets of milk each, from the barn to the house after milking, led us to the next task—separating the milk.

JEANIE

But on the way we girls sometimes indulged an interesting trick we discovered. In fact, it operated on the same principle as the cream separator.

If one of us had a milk bucket only one-third full or less, we'd swing it over our heads and down in a great smooth circle, around and around, as fast as we could go.

Amazingly, the milk held tight against the pail's bottom and didn't spill out. Centrifugal force kept it up there, Mom told us, and—even as it whirled upside down at the top of the cycle—none fell.

Of course, we didn't dare slow down or bump the bucket.

This was the same principle that ran the cream separator, a centrifuge that separated skim milk from cream, whirling rapidly to sort out the different densities.

This marvelous machine stood in our separator room, a large room with an outside entrance toward the lawn and the open space between house and corrals.

It was the only machine that had a whole room named after it. About four feet tall, the separator looked like a giant grasshopper standing on its hind legs. The large round tank at the top where we poured in fresh milk was, we imagined, the insect's head. To operate it you gripped the cast iron handle, around twelve inches long, and turned it in a circle.

Running the cream separator became a twice-daily adventure. The worst part was getting the handle to turn in the first place. For a young kid it was almost impossible. But by starting slowly, with as much force as you could muster, you eventually got the handle going and gradually picked up speed until it was swinging around at seemingly a hundred miles an hour. The operator's arms spun 'round and 'round for dear life. At that speed, with momentum built up, we soon became exhausted.

We took turns. When one tired, the next one took over. But we couldn't just hand it over to another and collapse. No way.

If we let the handle stop turning or even hiccup a pause, the machine lost momentum and the person who slid onto the handle couldn't catch it up again. Cream mixed back into the freshly skimmed milk in the guts of the separator and whole milk came out both spouts. Milk spurted out the center of the machine, too, and splashed all over the floor. We had to start again—pour both mixtures, from both cans, back into the tank, mop up the floor and begin the whole operation over again.

So the second person slid her hands onto the handle next to the pair of hands already there, both spinning rapidly, and flew around at full speed together till the other gradually slid her hands off. The cream separator was a live monster.

One pause, one hiccup and it went into neutral, lost its momentum and all was lost once more.

When we were spinning around at high speed there was a problem with two of us on the handle. To make it work, we planted our four feet tight on a nine-inch spot of floor. Not easy. Sometimes the extreme confusion of it all hit one of us and she began to snicker. Soon we both dissolved in giggles, while flying around the circle together. Inevitably, an uneven pause followed and the machine slid into neutral, all momentum lost.

*Whoops!*

*Western Ranch Life in a Forgotten Era*

## The separator

The cream separator used centrifugal force to separate the skim milk from the cream by weight. As the handle turned, whole milk went spinning through the series of more than thirty lightweight aluminum disks stacked tightly together, each similar to a six-inch funnel with the spout cut off. These spinning disks moved the lighter cream to the top of the disks and out the cream spout into a five-gallon cream can while the heavier skim milk poured from the lower spout and into another five-gallon can.

Surprisingly, skim milk without the butterfat is heavier than cream, just as water is heavier than oil. Like oil, cream eventually rises to the top all by itself—the separator just makes it happen faster. It was a marvelous magical machine that couldn't be used for any other job.

To keep the process going, the handle must turn at an unbelievable high rate of speed by human power. Two of us girls took turns spinning the handle until all the milk was separated. We didn't dare pause during hand-off, when we spelled each other off, or the cream failed to separate and mixed back with the milk.

After we filled one milk or cream can successfully, we exchanged it for an empty one. This, too, had to be done smoothly while the milk and cream were coming out both spouts. We also took care not to spill milk on Mom's hardwood floors. If we did, we mopped it up right away.

Each time the separator was used, twice a day, everything had to be thoroughly cleaned and scalded with boiling water because milk sours fast.

A long five-foot by thirty-inch wide sink, about eight inches deep, and a charcoal gray was built under the window of the separator room, running hot and cold water. Made of zinc, it didn't rust and was convenient for sterilizing equipment.

Cleaning separator disks was difficult because within minutes, the cooling cream congealed on many of the top disks and had to be scrubbed off each one individually, before being scalded in boiling water. We needed lots of room to work when we were selling milk and cream and were grateful for that unique zinc sink.

*It was hard to catch up to speed when we rolled around the floor laughing so hard we cried, milk squirting everywhere.*

If Mom was near she grabbed the handle and took over till we got control.

If not, we tried frantically to hang on. But it was hard to catch up to speed when we were rolling around the floor in fetal position laughing so hard we cried. No one looks more ridiculous than when turning a cream separator that's squirting milk in all directions.

Sometimes a hired man helped separate, if he was handy.

Though it was hard to get help, all during the war there was Hermie when Dad needed a hired man for a few days or weeks, because the armed forces deferred him for mental ability. He lived with his parents in town and worked for us off and on when we needed someone for a simple job with great strength—like stacking hay. When he came, Dad always worked with him or put him under the tutorage of a competent hired man. They respected his calm nature and Superman strength. Hermie had heavy brow bones and black hair that sloped back and then witched around in all directions.

When he had a little spare time, Hermie liked to hang around where we girls were working. As soon as he finished work, both at noon and before supper, he came straight to the house and hiked his rear up on our clean pantry countertop. He sat up there, swinging his legs out above our flour and sugar bins and leering at us.

Our pantry was a hallway about ten feet long between kitchen and separator rooms. One entire side was our workbench, a countertop that extended the length of the pantry. On the other side, cabinets reached to the ceiling. It was an important part of food preparation.

Between the countertop and cabinets, down the middle of the pantry where Hermie swung his legs, was the narrow walkway that, because he was there, we couldn't use. While we tried to work, there sat Hermie, enjoying himself. Right on our countertop.

He acted like he'd never seen girls before. It isn't easy to help get a meal with someone sitting on the middle of your workspace, grinning at you.

I wanted to grab a broom and beat him about his head until he fled the house. Good manners were required in our home so I didn't; I just left the kitchen. Mom always knew why. When she needed my help badly, she cleared him out.

He liked the Brink girls and made no bones about it. He told the hired men he liked Beverley and Francie best because they worked hard—but not me, because he said I didn't. I was grateful for that.

Anne was too young for his notice. He admired Francie most, but one summer when Beverley was home from the University of Montana with all her glamour and charisma, he was charmed. She always added a sparkle wherever she went.

Ever generous, Beverley included Hermie and the other hired men in her conversations, no matter their smarts or personalities, an attribute I admired, but couldn't muster. When she was home, Hermie dragged his eyes off Francie. Francie was grateful and Beverley was not.

That summer our bachelor neighbors, the Winninghams, set him up to ask Beverley for a date. Since Hermie was such a nondesirable swain, they knew it would make Beverley furious. Maybe they intended to get even for all the pranks she and the rest of us played on them through the years.

Somehow she learned of their plan, possibly from Hermie himself, so made sure she wasn't close enough for him to talk to her. If he came into a room, she left by another door. Hermie tried for more than a month but couldn't seem to get close enough to ask Beverley for that date.

Then one day while we were separating milk, here came Hermie through the door preceded by his gleeful hee-hawing laugh. Beverley was right in the middle of a batch of milk, turning the separator handle as fast as she could.

Francie and I fled through the pantry, leaving her all alone with the cream separator and Hermie.

"Hey, wait! Come back!" she called.

Anne, being the smallest, was the last one through the door. As she ran out, she heard Beverley yelling at the top of her lungs for us to come back.

Then it was just a bunch of gibberish, but loud enough and continuous enough so Hermie couldn't talk to her. I peeked for an instant and there he stood beside her, wide mouth split into a goofy grin, looking dreamily down at Beverley spinning away on the cream separator.

Beverley was so conscientious, so responsible, that she couldn't just turn loose the handle and ruin the batch of milk. And she sure wouldn't let him take over and shift hands with him. She just stood there yelling and continuing to separate the milk, making sure Hermie couldn't talk to her.

Afterward we wanted to know exactly what happened. When we asked Beverley, she just grumbled.

But for some reason Hermie gave up trying to date her. And we were left wondering why he never came in and perched on our countertop again. Did Beverley finally lose her cool?

> *Beverley was so conscientious, she couldn't turn loose the handle— just stood there yelling.*

## Patrolman Thomas, Always Vigilant

"Your cattle are on the highway."

Our milk cows were often out on the highway in those days, tempted by new green grass growing in the damp barrow pits in the right-of-way where water collected.

JEANIE

Dropping everything, we rushed to get them off, running on foot as hard as we could go. And having so much land poorly fenced was dangerous for our cattle as well as for vehicles.

"Your cattle are on the highway," was a phrase that rang especially hard in our ears when we first moved to our ranch.

Previous owners failed to keep up the fences. Consequently they sagged in many places, with rotten posts and broken and tangled wires. Dad replaced and repaired fences when he had time, but since the highway cut around one side of our ranch and then swept up right through the middle of our fields into the hills, it was lots of hard work. The State Highway Department did not fence their highways so we had to do it. Eventually we had good tight protection for our cattle but it took several years.

We often moved our cattle across the highway from one side of our land to the other and we did it with care, crossing quickly when no cars were in sight. Occasionally we'd be caught by Mr. Thomas, Montana Highway Patrol Officer whose beat was that stretch of US Highway 10. He'd jump out of his patrol car and help us cross quicker.

"Hurry up, girls! Let's get those this bunch across! Turn that calf!"

His other concern was us girls driving vehicles on the highway. Highway laws in Montana were weaker at that time, but it was quite obvious we were too young to drive. It was up to the patrolman to keep the road safe.

"Your cows are on the highway!" Those dread words sent us hurrying to get them back in the pasture.

Shortly after Pearl Harbor was bombed, every available man who could, joined the military. Dad ran the ranch with Mom and four daughters, the oldest one not yet twelve.

When Patrolman Thomas drove down in the field where Dad was irrigating, Dad knew what was coming.

"Elmer, you've just got to keep your girls from driving your ranch vehicles on the highway! When I see a truck approaching and the driver so small she can scarcely see through the windshield, it's usually your truck and one of your daughters driving."

State officials and local ranchers liked Mr. Thomas and so did our folks, but he constantly worried us girls. We had to drive short distances, less than a mile on the asphalt between gates to get to our

Even underage, we girls need to drive our truck at times—crossing the highway between fields, from home to pasture, and on errands to neighbors.

fields and pastures on the other side or run an errand to a neighbor.

Most kids who grew up on ranches were responsible, but their parents often gave them greater tasks than the patrolman considered safe for their age.

During the war, the highway speed limit dropped to thirty-five miles per hour. Most drivers carefully kept to the speed limit to save gas for the war effort. Speeding was wasteful and unpatriotic; besides, gas was rationed.

Relatively few vehicles were on the road, even though this was a major U.S. highway and carried all traffic going east and west from Chicago and the west coast. Occasionally, a convoy of U.S. Army trucks and jeeps came through, but those vehicles, too, kept to the speed limit. Somber in their dark khaki, they travelled in measured pace across our ranch, unlike the military plane formations that sometimes swept overhead, seemingly coming in fast and low and perfectly synchronized over our rimrocks to the east.

While rationed, we could purchase all the gas we needed because farm and ranch production was important. We kept it stored in two big barrels built high up on an over-head rack. The hose and nozzles were like those in filling stations and were easy to use—simply lift the nozzle off the hook and let gravity flow.

Mr. Thomas knew our production was necessary for the war effort, but he still insisted our parents keep us off the highway. But was Dad listening? Yes, but we were too desperate for help.

He explained why his daughters had to run errands or drive across the highway to get to our fields. Dad did the work of several men and needed our help. Mr. Thomas, of course, was aware of that.

"Mr. Thomas is just doing his job," Mom said. "If he stops you, be nice to him and do what he says. Turn right around and come back."

*"Your cows are on the highway!" Dropping everything, we ran as hard as we could.*

One time, when I was trucking down the highway to the field with my neck stretched up so I could see through the steering wheel, we saw the dreaded black and white patrol car approaching us.

"Look! Here comes Mr. Thomas!" Francie cried.

"Duck!" Anne shouted, and disappeared under the dashboard. Francie did, too.

He came upon us so quickly, I didn't have time to think, and on impulse, my brain filed a quick camera shot of the highway ahead and I ducked, too. Mr. Thomas didn't stop and we thought he hadn't seen us. Relieved, we laughed nervously at our good luck.

But when we got home, Dad confronted us.

"Mr. Thomas was here this afternoon," he said. "This is exactly how he put it, 'Your daughters all ducked when I met them. Your

> "Your daughters all ducked. Your truck was driving itself down the highway!"

truck was driving itself down the highway.' Don't you girls try to fool Mr. Thomas!"

We knew Dad didn't want any monkey business, but our patrolman had become our nemesis.

Many times when we stopped on our driveway approach, ready to pull out onto the highway, here came Mr. Thomas. His black and white patrol car with its prominent Montana State seal on the door, bold and shiny, slowed right in front of us. We weren't yet on the highway but there he was, shooting his arm out his window signaling us to go back.

We dutifully went back home—for a while.

With no place to turn around, going back wasn't easy. Our driveway was long and curvy for about a quarter-mile, with a short cutbank and dry creek bed on one side and the irrigation ditch on the other. We were mortified that he caught us. But, because of him, we excelled at driving that big truck in reverse.

I suspect it tickled Dad to see our truck entering the yard backwards. He'd walk over to the cab, look at us with his eyes twinkling and ask what happened.

After a few minutes, giving due respect to the State of Montana,

## Feeding in Winter

*Feeding cattle in winter was a daily priority on our ranch. Dad believed in feeding well to keep cows in good shape for spring calving and milk production. Also, animals with plenty to eat and wind protection survive better through blizzards and hard winters. We fed prime alfalfa hay, supplemented at times with 'cake'—a pellet form of high protein grains mixed with molasses. We fed lambs dried beet pulp, a by-product shipped back from the Holly sugar factory. In sub-zero weather, starting the feeding vehicle, tractor or truck each morning for feeding hay, was an urgent concern. Kerosene heaters in our livestock water tanks kept the water from freezing.*

*Cattle feed on hay tossed off truck in our fields.*

*Hay scattered widely across field on snow allows cows to eat in small groups; thus, each gets a better chance for her share of hay.*

*An open winter with little snow permits feeding hay in a pasture also retains winter grazing from our high-protein grass that cure the stem.*

we went back and finished our errands. We wanted the highway safe, too, and we drove carefully or Dad wouldn't have trusted us.

The summer when she was twelve, Francie drove up the highway into our hills on a job assignment in our beautiful new GMC truck. She returned, driving around the last turn where the deep cuts on either side of the highway shallowed out. Dad, doing tractor work nearby, saw the truck fail to make the curve, fly off the highway and bounce with a series of bangs across the barrow pit and come to a jolting stop just before hitting the fence.

Dad ran over, reached up and jerked open the cab door. There sat Francie, in the driver's seat, furious.

"Here, take this!" she choked out, dropping the steering wheel down into his arms.

When she rounded the last curve, the steering wheel fell off into her lap. Apparently, some assembly-line worker forgot to add the bolt that secured the wheel to the shaft. Francie was indeed, lucky. It soon occurred to the rest of us that we were, too. We'd driven that truck for six months.

Driving was in my blood and I wanted to become a trucker when I grew up. Of course I knew it was a hopeless dream—our

> *When she rounded the last curve, the steering wheel fell off in her lap.*

Brockle-face cow and calf wait beside a corral feeder for grain and molasses pellet mix.

ad feeds cattle loose hay from a hay wagon using the raised farmhand attached to the front of John Deere tractor.

venty degrees below zero, and Dad wears his heavy fur cap he pitches hay to the cattle, below.

Cows brought into the corral in fall await inspection to separate any that need extra care.

parents expected us to graduate from college. I could see myself behind the wheel of one of those double-trailer Consolidated Freightway giants that traveled the highway from coast to coast.

Just before the war ended, factories making Army trucks changed production to farm vehicles and the first new truck delivered to Miles City was our GMC. I drove it proudly for hauling sugar beets and livestock. Breaking-in was considered important for a new truck so Beverley and I drove it to high school for several weeks. By then, old enough to drive legally, we took driver's training as soon as it was offered.

Thereafter, Patrolman Thomas gave us only a brief wave when we met him on the highway, unsmiling as always. But he told Mom he was proud of our driving ability.

We were in our element. Only now, after all those years, we were legal!

## Bloated Livestock

"Anne, there's a cow in the alfalfa!" Mom called.

I ran as fast as I could to reach the cow before she ate any more green alfalfa. If she had been in the field very long, she was sure to be bloating. Green alfalfa produces a gas in the cow's stomach causing it to swell—without help the cow could die. The more she ate, the worse the problem.

*Green alfalfa produces a gas in the cow's stomach causing it to swell.*

The many tons of hay we raised was all alfalfa and from time to time a cow or two broke through a fence and got into it, which they deemed delicious.

As I slowly chased the cow into the yard, Dad came to help get her into the corral. "I'll have to stick her," he told me as he appraised the swelling.

He stuck his jackknife straight down into her left side. The stab was precise—in the triangle between the left hip bone and where the stomach bulge normally started.

"Eeoow!" I couldn't help exclaiming as the nauseating odor from escaping gases hit me.

I stood away from the cow just in case a stream of green slime followed the foul smelling gas.

It didn't. I had chased this cow out of the field just in time, before she ate too much, so a simple knife stick was all it took. She should recover rapidly and live a long and productive life.

If a cow wasn't too badly bloated, it was important to walk her slowly as I did in moving this cow out of the field. She also must be kept from water. A bloating cow was thirsty and we found if she drank, it caused more gas and swelling. So we hurried to get them across the irrigation ditch and out of the pasture quickly. Often a cow burped up the excess gas and recovered without sticking.

However, if a cow had a bigger bloat, it required a larger incision and then both the stomach and hide were stitched separately by needle and thread. If she was already down, she lay on her right side with the left side bulging high. Then Dad cut a slash and dug in with a hand to pull out a squishy, stinking green mass of stomach contents, half-digested alfalfa. Sometimes the heart stopped beating and he got it started again—artificial respiration—by raising and pumping her front leg.

A cow may not be as strong after that and, when she recovered, was usually sold at the sales ring.

Later we learned if we cooled the cow down quickly—by pouring on buckets of water or spraying her with a hose—it helped reduce the swelling. Finally we became proficient at treating bloating cattle.

But that was long after we lost Morgiana.

Morgiana was our beloved milk cow. She was a black and white spotted Holstein, had a huge bag of milk, gentle disposition, and milked all year, almost until her next calf was due. Her calves were big and rangy and these half-Holstein half-Hereford steers grew up to be our next winter's meat. My sisters and I drank her milk throughout our childhood. Often she was our only milk cow.

Gentle Morgiana deserved our affection, but like all milk cow pets, she had her distinctive personality and a real stubborn streak. One of her failings was to stretch her neck as far as possible through barbwire fences and with her long tongue grab succulent plants from the other side of the fence. Her favorite was alfalfa.

Since we milked both morning and evening, our milk cows stayed close to home and didn't run with the range cattle. A long fence line separated their pasture land from our fields. Morgiana kept weakening the fence by pushing her large bulk against it as she stretched her neck through for those tasty morsels. Eventually the fence gave way and she climbed through into an alfalfa field.

Many times we chased her out just in time.

Then one morning it happened. Morgiana ate too much alfalfa.

"Anne come quick and help me!" I heard Francie calling from the yard.

When Francie found her, Morgiana's left side was blown up so high she could hardly walk.

Francie and I got her into the corral where we could work on her, but Dad was in Miles City, and Mom and I were the only ones to help. We had watched Dad stick bloated cows, but not closely enough to do it ourselves.

Our beloved milk cow Morgiana broke into the field one night and ate too much alfalfa. The bloat was too much for her and she died.

*Then one morning it happened. Morgiana ate too much alfalfa.*

Mom called the veterinarian, hoping he'd come in time to save our cow. He had a fast little foreign sports car and considered himself a dashing figure in this convertible, boasting on the speed of its powerful engine. His clinic was only twelve miles from the ranch, and with that car we expected speed.

"She's swelling higher and higher," Francie said, more frantic by the minute.

Morgiana's side rose as the gas from the lethal alfalfa mix expanded in her stomach. But the vet didn't come. She panted hard—the pressure crowded her lungs—she couldn't breathe, and still he didn't come.

Francie pulled out her jackknife and prepared to stick the tip into Morgiana. By this time the skin was so taut her knife point just slid off the hide.

Still no veterinarian.

Montana had no highway speed limit—where was he? His fast little sports car could make the easy highway trip from Miles City in a few minutes.

Francie desperately tried many times to get her jackknife through Morgiana's hide, but the knife continued to slip off. Mom came with a longer, sharper hunting knife, but neither she nor Francie could pierce the hide. I had never seen Francie so frustrated and upset.

Finally, Morgiana could no longer stand on her feet. She toppled over, her left side up. Francie bent over her and thrust the knife once more with all her strength into the cow's side. But the knife slid off.

Morgiana died.

Francie was rather shy except when talking with family and never spoke an unpleasant word or discourtesy to anyone. Twenty minutes later the veterinarian finally showed up. The top of his Jaguar was down; obviously he'd been enjoying the sun on his leisurely trip to our ranch.

By this time Francie was on her horse and met him head on. Looking down at him in his little car, she blistered him with scorn over his lack of professionalism and the sluggishness of his overrated car.

A few years later we adopted a new scientific method of feeding cattle on irrigated pastures, promoted as a way to increase livestock production and income by the U.S. Agricultural Experiment Station at Fort Keogh near Miles City.

Our folks believed strongly in the science of agriculture and trying out new methods. As a result, through the years we supplemented our cattle herd with pigs and sheep and fed a variety of

Sometimes it helped to walk bloating livestock.

**Francie desperately tried to get her jackknife into Morgiana's hide, but the knife kept slipping off.**

rations and food combinations. Scientists at the Ft. Keogh Experiment Station and our County Agents with their link to Montana State University research at Bozeman found Dad an eager listener and participant in their innovative ideas, even sometimes against his better judgment.

Unfortunately, this was the case with our irrigated pastures.

Their Huntley Pasture Mix for irrigated fields combined a blend of grasses and leafy browse, including alfalfa, which tested high in protein and produced animals with strong bones, health and vitality. All in much less space than turning cattle loose on the range. Usually thirty acres of good rangeland supported one cow-calf unit, while this new pasture mix was expected to support all our yearling heifers on thirty acres for six months.

We grew prime stands of alfalfa and alfalfa seed. But as a field grew old and no longer produced well, Dad planted other crops. The experts at the experiment station encouraged him to try their newly developed irrigated pasture mix in one of these fields.

Dad, knowing there was alfalfa in this new mix and that a small amount of old alfalfa would come up each spring, even though plowed under, was dubious. The technology for eradicating all alfalfa from a field before planting another crop was not yet developed.

His family, sadly, experienced bloated cattle years before when he ranched with his parents and brothers on productive Missouri River bottom land, taken for the Fort Peck Reservoir. His dad bought two expensive, highly-bred Angus bulls from Minnesota to improve their beef herd. A day or two later, both bulls were found dead in an alfalfa field, bloated and swelled so tight their legs stuck straight out.

Would this be a problem now?

It was. The irrigated pasture grew well—and so did the mixed-in alfalfa.

Usually it was not just one animal that bloated, but several. Dad watched cattle in that pasture like a hawk.

He puzzled over why sometimes the cows bloated and other times in what seemed like the same conditions, they didn't. He found it had a lot to do with moisture in the field and the age of the alfalfa plants. Also he noticed some cows were selective in their eating and searched the field, stopping to eat only alfalfa. They were more likely to bloat than the rest of the herd.

The scientists appreciated Dad's on-the-ground observations and analysis. Sometimes they brought Montana State College students on tours to study their irrigated pasture mixes. On one trip a friend's son was among the students. After they visited a number of ranches, it was time to stop at the Brinks. During lunch, their

*Despite Dad's misgivings, he tried the new scientific method of feeding cattle and sheep on irrigated pastures.*

*When an animal bloats the left side swells tight.*

instructor told them to be prepared to answer some difficult questions.

"Think of every question he might ask and have answers ready," he told them.

With everyone primed, they pulled into our place.

"No matter how prepared we were," our friend said, "we never anticipated the difficult questions Elmer asked. 'We'll study that and get back to you,' our instructors told him."

By this time I was in high school and experienced enough to be good help. I learned from Dad how to knife-stick a cow. The place to push the knife through the hide had to be just right to get into that stomach compartment where gases formed. At first it was hard to tell just where to insert the knife point.

That summer we put in yearling heifers, and my job was to ride around them every hour and check for bloat.

Over a month went by with no bloating, so Dad decided checking them once in the morning and once in the afternoon was enough. That translated: He had other chores for me between riding around the heifers.

One day Mom and I were in the yard when a car drove up. Two men shouted at us without getting out.

*The irrigated pasture grew well—and so did the mixed-in alfalfa.*

## Feeding Sheep

*Ewes and fall lambs leave the corral water tanks for their feeding grounds.*

*Purchased in the fall, ewes winter well with their heavy wool coats and shelter from our cottonwood trees, and produce a good lamb crop in spring.*

"Your cattle are bloating. One is already down and others are staggering. They're near the highway."

I grabbed the hunting knife and Mom ran for the truck.

Many heifers showed signs of bloat, so we had to get them out of the pasture fast. Mom said she'd get them out, if I'd work on the three in worst shape.

I stabbed the down heifer first and a mixture of gas and half chewed alfalfa shot out of her stomach. She didn't move and I glanced at her eyes just in time to see them glaze over.

She was dead.

By now the other two were down and I ran and stuck each one. Huge amounts of gas and sludgy alfalfa shot out of them and boiled out on me. Amid the slimy mass I couldn't find my incisions to enlarge them and as I looked down at their still faces I saw it was too late. Their eyes grew dull and they both died.

I could see Mom getting the rest of the herd successfully past the irrigation ditch and out of the pasture. She hurried to the house and reached Dad in town.

The previous summer I watched Dad quickly butcher a bloated steer and, being prime beef, it became our winter's supply of meat. So I decided to do that now.

I'd hunted and butchered my own deer for several years. This looked like the same thing, except much bigger—how much bigger I hardly realized until I tried to move those two. The hide on the first heifer was so thick I could scarcely get the knife in far enough to make an opening to pull out the internal organs.

It was a hot day—if I didn't get her bled and cleaned out in time, the meat would be ruined. She was so heavy I couldn't tip her up like a deer. I just wasn't strong enough.

As soon as I got her cleaned out, I turned to the next and repeated the process. The first one that died probably lay there too long, with the meat ruined, so I didn't try to butcher her.

I was totally exhausted. Besides, it was very embarrassing to be right by the highway with cars slowing down to watch me butchering cattle, all covered with blood, squirming intestines and green goop. I kept my head down so people might think it was one of my older sisters, not me.

Just as I was finishing the second heifer, here came Dad on the John Deere tractor. He lifted the semi-cleaned yearlings with the front hoist farmhand and took them into the barn. There we hosed water over the carcasses to cool them down. Dad finished cleaning, then immediately took them in to the butchering plant in Miles City where they were skinned, refrigerated whole, then cut and wrapped for the freezer.

*Dad feeds and gentles the registered Columbia ewes he buys when girls leave home.*

*First I stabbed the down heifer. A mix of gas and alfalfa shot from her stomach.*

"Mom, I've been thinking," Dad said that night at supper. "With Beverley, Jeanie and Francie gone, we've lost most of our dependable ranch help. Maybe we should cut down on our cattle and buy registered sheep."

Sheep are much smaller and easier to wrangle than cattle. Dad and Mom raised a lot of sheep in the Missouri River Breaks and knew how to handle them. I learned, but never liked them much. Unfortunately, sheep bloat just like cattle. Occasionally one bloated and Dad stuck it.

Then one day the unimaginable happened.

"Anne, come quick," Dad called, as he burst through the back door.

I was just getting ready to go on a date, but Dad needed help. Two dozen sheep were bloating and we'd have to stick most of them. That also meant sewing them up. They had come up to the barns for water and in their misery were spread out in the yard between the house and barns.

It was my job to hold them down and thread needles for Dad to sew them up, first the inner stomach, and then the outer hide.

Miles City girls in high school thought living on a ranch was romantic—like the movies—so I just smiled when they mentioned the latest B-grade western movie heroine. I wasn't about to tell them of my life holding down bloated sheep oozing partially digested alfalfa or butchering a yearling heifer in the blazing sun beside the highway.

Manure, and a foul reeking fluid of gassy half-chewed alfalfa, totally covered me from wrestling so many sheep when my date drove into the yard.

Dad and Mom often talked about giving up on irrigated pastures and using them for hay, moving the cattle and sheep to rangeland.

*Could this really happen? No more bloated livestock and no more getting plastered with green slime?*

That night we didn't lose even one sheep, but talk about scaring off a boyfriend!

Dad with his prize Columbia buck that won grand champion fleece at the Montana Wool show.

**I was ready to go on a date, but two dozen sheep were bloating.**

## Sally Was No Lady

"Sally! Get back here!" I yelled as Sally evaded me, dashing for the hills.

*'Here we go again.'* I said to myself with deep resignation.

ANNE

This time I didn't even try to catch her on foot. Queenie was in the corral and I quickly saddled her and took after Sally.

I trotted Queenie around the base of the hill since I expected

Sally to stop in the big coulee where the greenest grass grew this time of year. Sure enough, there she was, head down, munching contentedly.

She looked up when she heard us. Seeing me on horseback, she innocently flopped her ears forward, and slowly and sedately turned back to the corrals, entering regally into the milk barn.

But just before entering the barn she turned her head and looked at me.

*"See this is where I want to be. Why did you go to all that trouble to saddle a horse to bring me in?"*

Hah!

Sally, the only Guernsey milk cow we ever had, delighted in annoying me.

She had the most ornery personality a milk cow could ever have.

Still, she was beautiful. A golden fawn color highlighted deep bronze patches around her face, sides and ears. Her most remarkable feature was her ears. Twice the size of a respectable cow's ears, she used them to express her innermost feelings.

It was easy to know she didn't like people, especially me.

My older sisters had left the ranch by this time, either working or in college. It fell to me alone to take care of the milk cow.

Sally was easy to milk. She never kicked, but went right into the stanchion for her nightly grain. Her calf was only ten feet away in his pen, but once he got older, she no longer cared enough to come into the barn on her own.

Our milk cows, up to Sally's time, were docile animals that wanted only to lead quiet lives getting the best food and protection from winter's storms. In other words, they were pampered. They enjoyed their routine and seldom protested. This was not enough for our Guernsey. Sally definitely had a mind of her own.

Like some milk cows, Sally could hold her milk back so her udder didn't seem full. Once full, it's heavy, swings when the cow walks, and no doubt starts to hurt. Sally never let her milk down until she was in the stanchion and eating her grain.

Some milk cows, if in the home pasture, came to the barn at milking time, but not Sally. We milked at regular times, morning and night, and Sally knew it. But she made sure she was in a far corner of the pasture when that hour came.

The pasture nearest the house had many hills and coulees to hide in, also sandstone rocks and trees. Finding Sally was a challenge; getting her to the barn—a real event.

A milk cow should never run. Our parents cautioned us that running a milk cow could cause blood in the milk..

*Yuck.* Sally liked to run.

> *She had the most ornery personality a milk cow could ever have.*

Usually if a milk cow didn't come in to be milked, I walked out to get her. She'd see me coming and start toward the barnyard. Not Sally. She threw up her head, swung those big ears back against her neck, and started evasion tactics. I'd run to head her off, whereupon she turned around and went the other direction. It might be necessary to get a horse to get her in. Sometimes I brought her right to the corral gate and she dodged back and headed for the hills. All this time I was trying to keep her from running. If Dad was working nearby, he stopped what he was doing and helped me corral her.

Then there was her calf.

With the exception of our best cow Morgiana, who milked longer, our cows gave an abundance of milk for about ten months, then stopped producing and were turned out to pasture with the beef herd.

When they calved and the calves were put in a pen inside the milk barn, we could milk the mother. We left the corral and milk barn door open, and several times a day the cow came into the barn to check on her calf.

Teaching the milk cow's calf to drink milk from a bucket took skill. I first backed the calf into a corner, then holding the bucket in one hand, I put two fingers of the other hand into the calf's mouth. The calf eagerly sucked on my fingers while I jammed them, calf muzzle and all down into milk. In theory, the calf then started sucking the milk while I withdrew my fingers. In reality, it didn't always work that way.

What really happened was something like this.

"Okay, Calfie," I said. "Come get your supper."

Calfie responded with a little prance, bounced to the far side of his fresh straw bedding, dropping bits of manure. I eyeballed where these landed, knowing eventually I would need to get inside his pen and corner him, without sliding on this manure.

"Come calf," I coaxed, reluctant to crawl over the boards into his pen. No luck. Carefully I stepped into the pen, avoiding those yellow splats.

One hand held the calf bucket, the other reaching out to the calf trying to entice him nearer.

Little calves are so innocent-looking. Soft big brown eyes follow every movement. Coming close, they love having their baby fur caressed.

Making ridiculous mooing sounds to the calf, I approached him in his corner.

Suddenly he became overly friendly and ran at me, nuzzling my leg, wet muzzle slobbering against my pants while I frantically tried to push him away. I held out my fingers to him and he grabbed them

*Finding Sally was a challenge. Getting her to the barn— a real event.*

and started sucking, a natural feeding instinct. Guiding his mouth into the bucket, I helped him get his first taste of milk.

How lucky I was if he continued sucking milk after I removed my fingers! Usually, I removed my fingers and he knocked the bucket out of my hand, spilling milk all over me.

One day, Calfie was feeling especially exuberant. He jammed his head down into the milk, getting it in his nostrils. No calf ever liked getting milk up his nostrils. I understood—I didn't either.

His response was revolting. He raised his head suddenly from the bucket and turning toward me he snorted violently. This snort was filled with milk and nose goo. He was so fast I had no time to jump away. And his perfect aim blew all this snot onto my face. Calfie's milk slime made a direct hit.

*The calf ran at me, nuzzling my leg, wet muzzle slobbering over my pants.*

Blinded, I jumped out of the way and slipping on the fresh manure fell on my face in a calf pie. Calfie came over to me, covered in milk, and started sucking on my wet braids. Not only would I need a complete change of clothes and a bath, I also needed to wash my hair.

Sally seemed to know her calf was giving me a bad time. She came into the barn, looked at the mess I was, flipped her ears back and forth, and belched with satisfaction.

I went away to college, and the first time home, I asked Dad how he was getting along with Sally, expecting to hear of some new ornery behavior on her part.

"No problem at all," he told me chuckling. "Mom and I don't use as much milk with you gone, so each day I get what we need, then turn the calf on her to finish off. When he's done, I turn Sally out in the pasture. Now, late every afternoon, she comes into the barn to see her calf."

How simple! *How come I never got to do that?*

## Mangled by the Greyhound Bus

It was just a few minutes after five o'clock when I looked back and saw the Greyhound bus approaching in the distance. Dad noticed the bus too, and we hurried to clear the highway.

ANNE

Mom drove the truck, closing gates ahead and stopping traffic by waving her large red flag on the end of a broom handle, though only occasionally was there a vehicle on this lightly traveled road. Dad drove the sheep on foot with our part-collie dog, Jimmy, keeping them off the highway and down on grass as much as possible.

I met them in the afternoon as soon as high school was out and helped herd the sheep the final few miles to our ranch. My older sisters were in college or working, so Dad and Mom were otherwise alone, trailing them to the ranch ten miles away.

That fall sheep were selling low at the Miles City livestock auction ring, even though the Agricultural Experiment Station at Fort Keogh predicted a high price for wool in the next year. Since Dad was a big believer in their research, he thought feeding out a hundred ewes and their lambs, born in spring, could be profitable. We had enough feed left for these bred ewes in addition to wintering our cattle.

Darkness comes early in October, and we hurried the sheep as fast as we could to get them off the highway and into our ranch before dusk limited the drivers' vision.

What a relief to finally look ahead and see the end of the neighbor's field—our gate was at the end of their property and we could get the sheep off the highway at last. We had less than a mile to go and still an hour before sundown.

The right-of-way was very narrow at that place and part of the bunch spilled over onto the highway. Dad and I were trying to push them off the road and string them out farther down in the ditch when we saw the bus coming.

*We buy a hundred head of ewes that fall and trail them home from Miles City along the highway right-of-way.*

If we had a bell sheep, a lead sheep, it would have been easy. One of us could lead her ahead down off the road in the right-of-way and they'd all follow. But this new flock had no idea where we were taking them so it was especially hard to drive them. Sheep like to follow a leader.

How we missed Tootsie Mama!

Tootsie Mama was a lead bell sheep who helped us some years before when we fattened lambs. As a bum lamb, bottle fed, she was a pet and her greatest desire in life was to be close to us. We kept a collar and bell on her neck and only had to lead her to get the band of sheep to follow—an easy way to get them through a gate or into the corral. If we still had Tootsie Mama, it would be quick work to get the herd off the highway.

The Greyhound bus kept coming, bearing down fast. Surely it was slowing down—wasn't it?

*No, it seemed to be coming just as fast—racing on toward Mom and our high stock truck.*

Mom was a distance behind us with the truck and a flag, stopping traffic.

Suddenly I saw her jump from the truck and into the oncoming

*We hurried the sheep as fast as we could to get them off the highway before dusk.*

traffic lane frantically waving her big red flag over her head and running right down the highway toward the bus, directly in its path.

*Didn't the bus driver see her?*

Dad noticed her at the same time and also ran toward the huge bus yelling and waving his arms to get the driver's attention. The bus didn't slow down. It continued roaring toward us at top speed, its tires whining.

I frantically pushed at the sheep, trying to move them down into the ditch out of the Greyhound's path.

"It's not stopping—it's not stopping," Dad yelled.

"Jump clear, Marie! Jump clear!"

Terrified for Mom, I glanced back just as she leaped out of the path of the oncoming bus, her flag dragging.

I gasped. Somehow I managed to jump clear, too. The Greyhound roared past me and slammed into our little band of sheep at top speed without ever slowing down.

*No! No! No!*

Horrified, I watched. The front wheels bounced over the sheep and they balled up ahead of the back tires, slewing the bus off the road and right into the main herd down in the ditch. Sheep were packed under the back tires so they could no longer turn. The bus jolted to a stop.

*The air shrieks with terrible squeals and screams and piteous bleating of sheep while the bus lurches back and forth.*

The impact peeled the hides off some sheep and they lay there blatting and bleeding horribly, skinned alive.

Others it killed outright. Blood splattered all across the highway and the grass at the side of the road. The noise and stench of fresh blood and guts from the sheep sickened me. High-pitched squeals and deep, low moans filled the air with terror.

Dad ran to the bus and—as the driver appeared dazed in the doorway—he asked about the passengers.

"No, no one hurt on the bus. We're okay," he mumbled.

The accident happened in front of a neighbor's house, who called the highway patrol.

Dad turned his attention toward the sheep, pulling injured bodies from under the bus.

Mom came running holding her hands up to her face.

"We have to put them out of their misery. Oh please! Can't someone help them?" she cried.

Where was our beloved dog Jimmy?

"Jimmy! Jimmy!" I shouted, searching through the confusion.

He crept to my side, trembling, tail between his legs, bewildered by what was happening. I knelt down and hugged him—thankful he

*Jimmy crept to my side, trembling and bewildered, tail between his legs.*

Jimmy, our livestock dog, helps us bring the sheep home.

# Sheepherder holdup

After herding our feeder lambs in a field near the highway, our Bulgarian neighbor Teddy Wasso got a jolt that he'd dreaded for a lifetime. He was grazing the lambs on left-over crops after harvest—a one-week job before they went into the corrals for winter.

When the hired man came to take his place and corral the sheep for the night, Teddy got in his car that was parked by the highway and started off for his home.

As he told the story later: "Suddenly I hears this loud breathing behind me and a hand seizes my shoulder.

"*This is it!* I think. *All my life I know this going to happen. Someday I am going to get mugged, robbed, maybe stabbed or shot. Now here it is!*"

His heart racing, Teddy pulled over to the side of the road, stopped and sat there.

"*Why didn't I look in the back seat before getting in the car?* I tried to think how to defend myself, but I was frozen. I couldn't move.

"But nothing happened. The hand stays there on my shoulder. I was so scared I couldn't turn my head to look back.

"'Okay,' I says. 'What you want?'"

The hand pressed his shoulder, the silence broken only by heavy breathing in his ears.

"'I got no money. I take you where you want to go...'

"But the guy behind me—he don't say nothin,' just keeps breathing loud with his hand on my shoulder. I sits there, my heart beating like crazy.

"Then I feels something wet slap agin' my ear. Finally I gets up my nerve and turns around.

"There's your dog Jimmy! He's standing up on his back legs panting. His paw heavy on my shoulder, licking my face and smiling at me! I love that dog."

hadn't been killed—burying my face in his fur to keep from vomiting at the awful smell from the ripped-open and dead sheep, and to block out the dreadful sights and sounds.

The highway patrolman came quickly and took it all in, shaking his head sadly. Mom with her red flag, the mangled moaning sheep, the bus tipped partly sideways down in the ditch. Dead sheep both on the highway and down in the grass.

"Your family okay?"

He looked sharply at each of us, then boarded the bus to check on the passengers.

The patrolman understood the ranchers' view. As he said later, we had adequate help, three people and a dog working the sheep. A red flag. It was broad daylight, with the sun behind the bus, traveling east, no reason why the driver couldn't see us on the flat stretch of highway.

"The bus almost hit my wife," Dad told the officer when he stepped out of the bus. "She ran right there in front of him waving the red flag and he didn't even slow down! She could have been hit! She jumped clear just in time."

"He wasn't looking—I wonder why," the patrolman said. "You did all the right things."

Under Dad's direction the officer started shooting sheep. He had to kill eleven badly injured ewes. Twelve more were killed outright by the bus.

He and Dad pulled the horribly mangled sheep out from under the bus, directing the driver to back up a few feet so they could pry the dead ones from the back tires.

"No," the bus driver kept saying. "I didn't see anyone flagging. I didn't see any sheep."

That seemed impossible.

The patrolman quizzed the passengers again to see if anything unusual could have diverted the driver's attention from the road. The riders in the front-most seats claimed to be sleeping.

*Just nine miles out of the stop in Miles City and they were sleeping?*

Other riders said they saw or heard nothing.

The patrolman told us there must have been some disturbance on the bus to distract the driver and silence the passengers.

A short time later a check for the twenty-three sheep came from the Greyhound Bus Company, but we never knew the results of their investigation. Was the driver reprimanded for his lack of attention? Greyhound with its wonderfully safe record and motto of *'Leave the Driving to Us,'* apparently concluded the fault was theirs and we had done everything possible to avoid the accident.

*The devastating sight and smell and desperate cries of those wounded sheep stayed with me for years.*

*Montana Stirrups, Sage and Shenanigans*

# 9  Rural School

## Long Walk Home

"I'm going home." Beverley announced in her strong five-year-old voice.

"You can't go home—it's ten miles," countered the teacher. "And we're not going to take you!"

Thus began Beverley's first weeks of country school.

When Beverley turned five years old in April 1934 she was so precocious it was nearly impossible for Mom to keep her busy. As a teacher Mom realized that, yes, Beverley was ready for school.

This was a tough decision for Mom and Dad. The nearest school was a one-room rural school ten miles from our ranch through uninhabited rangeland and badlands. Home school was unheard of at that time, so somehow they must find a close neighbor who would board Beverley during the school week as was the custom in ranch country.

During those depression times, our folks lived in the Missouri Breaks south of Fort Peck Dam and sixty miles from Jordan with three small children, and times were hard. Buffalo and other native grasses grew during wet years, but this was the worst of the drought. Nothing much grew on this land to feed our sheep except sagebrush and a few scrubby cedar trees in the draws.

Ranches were far apart, and homes available did not provide the kind of life Mom and Dad desired for their oldest child. The teacher and her husband, who lived at the schoolhouse, agreed that Bev could board with them during the week and go home after school every Friday, or so, depending on road conditions. After a rainstorm the muddy gumbo road was impassable.

This was wild rough country, with the few ranches between the school and our ranch tucked out of sight in draws and coulees, down along creek bottoms, and not visible from the main rutted, dirt road. Dirt roads branched off here and there from the main

*"You can't go home—
it's ten miles. And we're not
going to take you!"
the teacher warned.*

one.

So Beverley started school that September when she was only five. The oldest boy, a bully, made life miserable for her.

"Buster is so mean. He makes fun of me and he's always twisting my arms and hitting me!" she told the teacher.

"You gotta stick up for yourself!" the husband scolded.

"He hurts me worse when I fight back! Make him stop," she implored her teacher.

But the teacher did nothing, allowing the bully free rein to taunt her as he pleased.

"I'm going home." Beverley finally told the teacher after being in school a little over a week.

"You can't go home, it's too far," the teacher told her, "and we're not going to drive way over there."

But Buster kept bothering Beverley, so she made her decision one especially difficult day after the other children had gone home.

"*I am going home!* You won't stop that boy from being mean to me. You don't help me," Beverley shot back as she ran out the door.

"You can't go home till Friday when your parents come," they called after her.

Bev ran down the dusty car track in the direction she came from home.

There were no other ranches in sight, just a great void of dry grass, hills, sagebrush, endless vacant miles, rattlesnakes, coyotes and maybe even a few mountain lions.

"Come back! Come back!" the teacher's husband came out of the schoolhouse and yelled at her to return.

Beverley ran over the hill and down the road until she could no longer hear him calling. When she could run no longer, she walked—and walked—on and on. She remembered the fine dust settling over her shoes and legs. But Beverley was determined to get home and she didn't care how dusty or lonely the road was.

Only five years old, she'd travelled on that road only a few times before.

Without meeting anyone, she just kept going and going, sure that home was just over the next hill. When roads branched off she tried to stay on the heaviest traveled road, although it was hard to tell which track to take. Sometimes she saw tire treads or horse tracks in the dust and followed them.

The sun, low in the west, made long shadows grow across the dirt in front of her and sometimes a horny toad skittered out of the trail ahead. She thought she'd get home before dark.

Beverley kept looking back to see if the teacher and her husband were chasing her. She expected them to run her down and catch her.

*Beverley ran down
the dusty car track.
No ranches in sight, just
a great void of dry grass, hills,
sagebrush, endless vacant
miles, rattlesnakes, coyotes
and maybe
a few mountain lions.*

She made plans to hide behind a sagebrush or run into a gully—and then keep running. They'd never catch her. But no one came looking for her.

Suddenly it was dark. Then thousands of stars came out and the sky was beautiful.

"The stars are our friends," she remembered Mom saying. Crickets chirped in the dry grass near the road.

She told us later she wasn't afraid, even when the coyotes yipped and howled across the badlands.

"Coyotes are afraid of people," she told herself staunchly.

She could see better when the half-moon grew brighter. And still she walked resolutely down the dim road. Home wasn't very far away, she thought, maybe over the next rise, where familiar buttes silhouetted against the starry sky.

Her exhaustion grew but she kept going. If only she could curl up in the rutted road and go to sleep. She was so tired—but she didn't want to stop. She wished someone was with her.

Finally, through the darkness, a pinpoint of light shown down a draw and across a dry creek under a hill. Soon a rutted road turned off that way. She followed it. She made out the dim light shining through the window of a ranch house and went down and knocked on the door.

"Why, it's a little girl!" a teenage girl opened the door. "Where did you come from? Where are your folks?"

A streak of feeble light from the kerosene lamp on the table backlit her. She stepped aside and Beverley went in.

"Mamma!" called the girl. "Come here! A little girl!"

At first Beverley only told them her first name, but considering the sparsely inhabited Missouri River breaks, the family quickly identified her. A five-year-old child doesn't often show up on someone's doorstep out in that wild land without parents close behind. When Bev told them she ran away from school and wasn't going back, they decided not to return her to school.

The strange thing, though, was that the family kept her for three days, even though they knew who she was and where she lived. By then they knew the teacher and her husband were not looking for her, and our parents believed she was at school, so no one returned this wandering child to her home.

The family and especially the teenage girl, gave her so much attention that Beverley wanted to stay. She liked being spoiled, a feel-good time for her.

Finally guilt, knowing our parents picked up Beverley on Fridays, and the desire to do the right thing pressured the family to tell Bev she should go home. It wasn't right for her to stay any longer.

*"Coyotes are afraid of people," she told herself staunchly.*

*More than ready for school, Beverley began first grade while still five years old.*

Western Ranch Life in a Forgotten Era

"I'll take you in our car," the older girl said.

She drove Beverley on for miles until they came nearly to the top of the big hill where she saw our house below on the flat. She and Bev walked nearly to the top of the hill.

The teenager didn't want to face our parents. She knew it wasn't right of them to keep Beverley so long. She hid behind a sagebrush.

"I'll watch from up here," she said. "And you run home. I'll be sure you get inside your house safely."

Mom and Dad were working outside in the yard near the house with the hired man.

I was three and little Francie was only a baby. I looked across the dried grassy slope and way up a hill. The vast spaces of earth and sky intrigued me. Where did it end? Our part of the world was huge.

Then I saw something—far, far away on that dusty road a little dot coming down the hill, the only moving object in sight under a blue sky that went on and on forever above the rough Missouri Breaks.

I watched it for a while. I wondered what that spot was and why it moved.

*Maybe a dog? A coyote?*

Suddenly, I saw something familiar about that dot.

"There's Beverley!" I cried.

Everyone looked up, astounded—she was supposed to be in school, ten miles away. Then we all ran as hard as we could to meet her.

Beverley was crying in great sobs. We asked her what the matter was, and she said she saw a big bug, but I knew big bugs didn't scare Beverley.

I was so glad to see her. I missed her a lot and now here she was. Later, she told me there was no bug. She was crying because she was scared Mom would be mad at her for running away from school and for staying to play with the big girl instead of coming right home.

Our parents could hardly believe that the teacher simply let Beverley go without looking for her.

*Why did they let her run out alone into that wild country where there were coyotes, rattlesnakes and miles of rugged badlands? Surely a death sentence for so small a child!*

Incredibly, neither the teacher nor her husband ever went after her.

The teacher simply told the other pupils, "Beverley went home."

When one dad inquired, the teacher told him our parents came and got her. If he knew she ran off, a community search would have found her. And they would have fired the teacher.

Without a doubt, our parents went to the school and had a talk

*Boarding with an unsympathetic teacher, Beverley has her fill of bullying and sets off for home, ten miles away.*

*"Why, it's a little girl! Where did you come from? Where are your folks?"*

with that irresponsible couple and likely with the school board, too. I remember how seriously they discussed it.

After that, Mom taught Bev through the first and the second grade, too, in one year at home.

When we moved to Fairview the next fall to winter our sheep, Beverley and I attended town school for a time. At age seven she was more than ready for third grade and I started first.

But after Bev's experience I didn't think much of teachers. Still, it was fun walking those few blocks to school, gathering friends along the way.

*Whack! The first kid hit the posts broadside.*

## Playground Creativity

Edwin stood by the swings, a six-foot plank in his hand, with the other students gathered around, as Jeanie, Anne and I came into the schoolyard. [FRANCIE]

He laid the plank lengthwise across the swing and jumped up on one end.

"See, you stand here, with most of the board out in front of you. Like this," he explained. "Pump it up good and high."

We held our breath. We knew there was more.

"Pump it up good and high! Then, when you get high enough to clear the posts, quick flip the board around. Aim it back between the posts, and when you get high on the other side, flip it back and go through again."

He swept through the posts a couple of times, gaining momentum. Then he was ready.

"Watch this!"

We looked on in amazement as he swung high, twisted his body, flipped the board around and shot it back between the posts. Up and away he went, high on the far side, then twisting and flipping again and back through the posts. He did this several times, pumping hard and fast to keep it all under control.

We each took a turn, beginning with the older ones.

Edwin shouted, "Higher, higher! Get up higher. Now! Flip around. Quick!"

"I can't."

"Sure, you can. Do it now! Okay—well go on through again and catch it next time. Now!"

"Help!"

*Whack! The first kid hit the posts broadside.*
*Rrrup! The second kid let the front end of the plank dip too low. It dug*

Edwin Hill turns his inspiration and sense of fun to the playground—from a flying trapeze and long planks on the swings to a water geyser spurting from the pump. Here with sister Dorothy.

*Western Ranch Life in a Forgotten Era*

*into the ground and flipped its rider.*

But after a few misses, bumps and bruises, we had it. And for the rest of the day, someone stood on the plank at every chance, joyfully swinging high and wide, flipping it around, and firing it back between the posts.

Edwin Hill was wonderfully creative about our playground equipment as he got into the upper grades. Living in the school yard he had plenty of free time, even after he'd chopped and carried in the day's wood and hauled water from the pump for both his mother and the teacher.

He began to invent new ways to enhance our play.

Another day the board was much longer. A heavy two-by-ten plank some ten feet long that Ed begged and borrowed from a neighbor's building project.

It became a two-person missile to shoot between the posts.

"You both gotta stand out a little ways from the center, like this. With about three feet behind you. Hang on tight to the chain. Pump together to get it up high, then you flip around and go back through."

> "Hang on to the chain. Pump it up high, then flip around fast and go back through."

This was more difficult. We had to get it going so high it felt dangerous on our high swings. Keeping eye contact, we worked together, half-scared, giggling nervously.

"Ready? Okay. Let's do it—now!"

Again came misses and whacks. Bumps and bruises and a kid or two flying through the air. But once we got it, the ride was great fun.

Surprisingly, injuries were slight, and the teacher, with the windows all on the other side of school, didn't seem to notice. (By this time our teacher was no longer the conscientious highly-professional Miss Wyss who kept an unobtrusive vigilance indoors and out.)

On other days new inventions appeared with nailed-on stumps: the single sit-down and pump style; two stumps for double sit-down flights; a stump nailed to one end for a smaller child to sit while a bigger kid pumped the swing high.

At the top of the merry-go-round Ed tied a rope with a narrow board seat. He called it the *flying trapeze*. To fly through the air, the rider stood on the bench straddling the rope, holding tight, then kicked off while others pushed at high speed 'round and 'round.

As the merry-go-round picked up speed, the rope rocketed up and out with the rider aboard. The longer the rope, the farther out you flew.

This made me dizzy even watching. I tried to stay away. But the lure was hard to resist when everyone else soared high.

*How can I give up my turn at a wild ride on the flying trapeze? Or pushing friends around when they deserve it too?* I couldn't.

It was great fun for a short time. Even jerking the whole merry-go-round apparatus back and forth in a 'bucking' motion to whole-hearted laughter was fun. But invariably I regretted it. Suddenly I'd feel so dizzy and nauseous I could hardly stand. And it lasted for hours.

Edwin invented the water pump trick, too.

One morning after pumping the day's water, he brought out a wooden plug he'd carved to fit and hammered it snug into the pump spout.

When we got to school he told us, "Okay, now. We take turns. Each of you pump as hard and long as you can and don't stop in between—keep the pump going."

With our school pump, we had to pump the handle up and down a long time to bring up the deep groundwater to the surface. Finally we heard the water gurgling up and pumped furiously, several of us working the handle at once.

"Whoosh!"

Suddenly a big force of water hit the plug and erupted out the top, shooting high in the air like a geyser. We burst into shrieks and laugher, running into each other as we scrambled to escape the cold shower.

Ever close by, eagerly watching our antics and gazing into our faces, was our beloved and beautiful Kelly, Art Hill's dog, a red Irish Setter. He came trotting down their lane every morning to greet each kid and hang out with us in the schoolyard.

Kelly didn't mind the water shower in the least, seemed to join in the laughter with a grin and leaned in close for his strokes.

Teacher called us in by ringing the hand school bell she kept on her desk. We took pride in that bell and, even when focused on a game, reacted instantly and ran for the door when it rang.

Even Kelly knew what the bell meant and retired to a shady spot.

We felt affronted and insulted when—instead of ringing the bell—one of our teachers took to rapping on the window with the scissors when she wanted us to come in. Not always did we respond to that, so she'd rap more sharply until we feared the window might break—and came in.

During school hours the rule was simple: never, ever go outside the school yard. The only exception—if you were assigned to go out on the highway to check our school mailbox. Our community's five or six mailboxes were mounted on a wagon wheel, like a lazy Susan, at the corner, just outside the fence. Neighbors could drive up, spin the wheel and open their mailbox from the car window.

We were lucky at Tusler to have more play equipment than many

## Mobile home

During the school year the Hill kids and their mother Esther moved down from the hills to live at one side of the schoolyard in their long trailer house, built like a Conestoga covered wagon used by pioneers. Like that, it was on wheels with a double layer of canvas arching over the top. Wood skirting enclosed the wheels and base to keep out the cold wind.

The Hills' grandparents ranched and farmed five or six miles farther up in the hills, on the open range plateau above our irrigated valley. Years before they had bought a dry-land homestead, now surrounded by another rancher's section of open range.

Fortunately, a large flowing spring was on their land, providing 'all kinds of good water,' as Grandpa Hill marveled. He often told us with a chuckle, "That cattle rancher tried every way to get my spring!"

Their covered wagon home in the schoolyard was efficiently arranged, all in one room, with stove in front by the door and beds in the back. Usually it was plenty warm, too warm on a hot day. But in the coldest part of winter the small cook stove couldn't keep up. Frost built up on the canvas walls, then thawed and dripped into pans set in strategic places. Sometimes it dripped onto the beds.

Edwin recalls one bitterly cold January when the temperature kept dropping, farther and farther below zero.

"When word came that a major cold snap was on its way one evening we accepted the Brinks' offer to come over and stay with them a few days," he said. "What a relief that was, and we enjoyed several days of fun with the Brink kids too."

> The dads dug a jumping pit where we practiced for the county track meet each spring.

country schools: two high-hung swings, a simple trapeze climbing bar and amazingly, that full merry-go-round with circus top. Nearby the dads had dug a jumping pit and filled it with sawdust where we practiced for the county track meet each spring.

On the open side of the school was a wide area of beaten dirt where generations of children had played baseball and running games. This was a great place for Pom-Pom Pull-Away, Spider Web and in winter snow, Fox and Geese.

We played Anti-I-Over by throwing the ball over the lower roof of the teacherage from one team to the other.

"Anti-Anti-Over," we yelled as our team threw, and then prepared to dodge and run if they caught the ball and came for us.

As a little kid my heart beat fast. I watched fearfully for them to come charging around one corner. *Or would they come from the other side?*

I was all too easily caught.

We played Fox and Geese on a double wagon wheel pattern trampled out in the snow, and sometimes pulled the little ones around on a sled, although sledding was better in the hills at our homes. When the snow was just right, warm enough to stick, we built snowmen and snow forts and fired forbidden snowballs at each other.

When we tired of familiar games, we opened the glass doors to our library shelves and took down the big brown Game Book. There we discovered new games that looked interesting and tried them out. Some proved to be keepers and others were quickly forgotten.

Then came a new diversion, another of Edwin's bright ideas. Granted it was not his best idea—and it went on much too long. But undeniably, he found willing conspirators.

Ed was in seventh grade when he started going down to the Yellowstone River and exploring after school. There he discovered punkwood.

Mary and I were eighth graders and only five of us left in school when we started smoking punkwood. Even Anne and Dorothy, much younger, were game to try this new sport.

Punkwood is fun stuff. Light, dry and porous, it breaks cleanly with a pop.

Formed of small floating tree branches that beach along the banks for months and probably years, it soaks up water and sand till nothing is left inside except thin cellulose channels.

I wondered if Ed's mother knew how much time he spent down

*Our school picnic and hike, when nine students attend Tusler school*

there, collecting punkwood. Our parents taught us the Yellowstone, like the Missouri, was a dangerous river, even in winter with deceptively thick ice.

Ed showed us how to break off a punk, about four or five inches long, light it with a wooden match—usually it took several matches—while trying to suck air through the punk and start it smoldering.

We tapped out what sand we could before lighting up. It was hard work sucking air through a sand-filled punk.

We tried all sizes, from the diameter of a pencil to an inch or more and soon found the bigger pieces were harder to draw smoke through. They also filled our mouths with more smoke and left a stronger bite on the tongue. However, they did last longer.

I much preferred the slimmer smokes. They bit the tongue, too, but not so powerfully.

The five of us smoked punkwood every day through that fall and winter—at noon, recess and before school. We stood in a sheltered, hazy corner of the schoolhouse puffing away. Sometimes we pretended sophistication—in a silly and hilarious kind of way—as our interpretation of movie stars.

No one scolded us for smoking or told us to stop. Not our new young teacher. Not the neighbors, who must have seen us huddled in corners, never at play. Not our parents, who may not have known right away. All our dads smoked, of course.

At first, we kept out of sight of the teacher. But since she paid little attention, we grew bolder and, during windy or stormy weather, even went into the schoolroom to light up our punks in the red pot-bellied heating stove.

Our pretty, red-haired teacher, fresh from a session of summer school, continually amazed us. She faced no discipline problems—our parents wouldn't allow it—so her job was relatively easy.

She let us do whatever we wanted. Our earlier teachers ran a tight ship. They wrote out daily assignments and held recitations in each subject for each grade. Under their instruction, we researched and wrote reports and took tests. At the year's end, we knew most everything in all our textbooks, plus a whole lot more.

With this teacher things were different. It was her first year and we made it easy for her. The five of us were good students, enjoyed learning and knew we had to finish all our books before summer vacation. So we did, working more or less independently, helping the younger ones when needed. We didn't want to bother our good-natured teacher who seemed to spend most of her time drawing illustrations for children's books. When we had time we enjoyed watching her art take shape.

*Ed was in seventh grade when he started exploring the river. There he discovered punkwood.*

> *Though few in number, we Tusler students worked hard to keep up our school honor.*

Maybe when our parents found out we were smoking punkwood they considered it harmless—this was in the 1940's before smoking was a known hazard. Or maybe they expected we'd learn a lesson from it.

We did learn a lesson all right—but it took all fall and winter.

When spring came, with the need to practice for trackmeet, we stopped smoking for good. I was ready to quit long before, but felt I had to stick with it when the others did. A matter of pride. Probably they felt the same way.

We didn't smoke again, but had our fill for a lifetime. For years I could feel the bitter bite of punkwood smoke burning my tongue.

In rural schools a big event came in early May—trackmeet. The trackmeet brought together all the country school kids from Custer County to the fairgrounds horseracing track in Miles City.

We began practicing as soon as the snow melted. Even Miss Wyss came out to help us, keeping time, moving the bamboo pole up its notches for high jumping, explaining the rules and getting us started. One of the dads renewed our sawdust pit, where we practiced both high and long jumps.

Though few in number, we Tusler students worked hard to keep up the honor of our school. We could all run and won awards for the 50, 100 and 200 yard dash. Jeanie's and Mary's specialty was high jump, mine was long jump.

Two grades got thrown together in the events, so competing from the older grade, the even numbers, was a big advantage. I felt devastated the spring of my eighth grade year, when my hopes for winning both a race and the long jump were dashed by my horse accident.

Sidelined, I could only stand by and cheer on my schoolmates. That was the year our school, competing with much larger country schools, our rivals, brought home the second place plaque—without me.

After the last relays were run and ribbons awarded, we ate lunch.

We all guzzled that once-a-year treat—a bottle of fizzy orange or grape pop—along with eating a steamed hot dog and Dixie cup of vanilla ice cream, with a flat wooden spoon.

A rare treat of store-bought picnic food for us unsophisticated country kids!

We looked forward all year to our single bottle of soda pop. *Cold and delicious orange crush! Mmmm.*

# The School Trek

In our first years at Tusler we walked to school, cutting straight down through our fields, climbing the fence and on through Art Hill's fields. Beverley, Jeanie and I, in our clean print dresses, each carrying our lunch in a tin Karo syrup pail.

This was our shortest and most direct route, a distance of about a mile and a half.

"Come on!" called Beverley leading the way each day.

We were never, ever late when we walked. Bev loved school, kept a good pace and looked forward to getting there.

Mom gave strict orders to come straight home after school, but sometimes Jeanie and I dawdled—especially when a sunny day melted snow into rivulets running down a dry creek bed of painted rocks. Or we stopped to pick the dandelions that grew extra long in an alfalfa field and braided them into golden flower leis.

Walking to and from school we learned a great deal—in the most natural way. We followed the example of our parents in paying attention to our environment and appreciating the rhythms of this land that they loved.

"A water snake!" Jeanie cried as she rushed to pick him up. "Look how his bright stripes run out to his nose."

We studied the worms, spiders and insects along the way. We respected the tiny black ants with their finely-built homes and the stinging bites of big red-and-black ants when diverted from their hard work of carrying little rocks and sticks for house repair.

Fuzzy brown and red caterpillars—*fuzzy bears,* we called them—became pets as they crawled up our arms. What beautiful kind of butterfly or moth would they become? My all-wise older sisters were open to discovery. They speculated, talked it over and even valued my opinion. So we carried a Fuzzy Bear to school where Miss Wyss put it in a jar with a twig and helped us look it up in the encyclopedia.

In deepest winter the great snowy owl flew south out of Canada and regarded us solemnly from a bare cottonwood branch. Spring found us answering back at friendly, bright-eyed, squawking magpies.

We delighted in the first meadowlark's song and paused to locate that much-loved minstrel of the plains, perched on a weathered fence post, head cocked, watching us.

"It's time," he sang out in his lovely liquid melody, "time to plant your *sweet* potatoes."

Our dryland neighbor Grandpa Hill (no relation to Art Hill) taught us those words to the meadowlark's song, though we hardly

*"Look—a water snake!" Jeanie cried as she ran to pick him up.*

*We walk to school in our dresses and winter stockings, carrying our lunch in Karo syrup pails. From left: Francie, Beverley, Anne, Jeanie.*

knew sweet potatoes. That lyric came north from Texas where he punched cows, we thought, or maybe Missouri, his boyhood home. However inappropriate to eastern Montana—did a sweet potato ever take root here?—the words stuck with us. They so perfectly matched the meadowlark's song with his lingering trill on the word *sweet*.

After Jeanie and I bought our 'Victory' bikes, toward the end of the war, we rode the highway route to school during fall and spring seasons, a distance of about two miles.

Mr. Parke Krumpe welded a seat behind my bike so Anne could ride there. He was very concerned it be safe for Anne; he feared she might catch her feet in the spokes. Feet could be permanently injured this way, he said, as he bolted on a ledge for each foot and cautioned Anne to keep her feet there.

Riding our bikes to school was great fun, coasting downhill much of the way. Going home went slower. Those old one-speed Victory bikes could be hard work pumping uphill. Built during the war, they had narrow tires that skimped on the rubber. The only bikes we ever had, bought with money we earned hoeing sugar beets, they were tough and lasted many years.

It was so hard to pump up that long slope of old Highway 10 and on up our gravel driveway that sometimes Anne had to get off and walk. She never objected though. She loved riding on the back of my bike and knew I was trying my best to pump up the hill. It was slow going and Francie and I could pedal only at the speed of walking. Keeping our balance wasn't always possible, going so slow. Many times we all got off and walked.

One fall day the Hill kids walked part way home with us, as they often did. We were walking our bikes, visiting and joking, when we came to the big bend in the highway—the point where they usually left us to return home to the schoolyard.

A broad swamp crossed under the road at the curve, filled with cattails right up to the highway on both sides. As we lingered by the swamp, chatting and laughing, we noticed the cattails were ripe to bursting. Running our fingers up under the fuzz at the top, we could strip off the seed heads and send them exploding into billowing mini-clouds. It was delightful to watch and we had great fun.

Then someone noticed that whenever one of the few cars passed over it, the fuzz flew up and burst into a small cloud around the car. More fuzz on the road could be that much more interesting, we thought.

Mary, Edwin and Dorothy Hill normally didn't do questionable things like this, but they didn't try to stop us, either. Soon they

*Jeanie and Francie ride our Victory bikes to school, with Anne on a seat welded to back of Jeanie's bike.*

**We noticed the cattails were ripe to bursting. The fuzz flew up when cars hit it.**

joined in and we all had a great time spreading a layer of cattail fuzz all across the highway.

Finally, when we had the road layered about five or six inches deep, here came a car.

As the driver hit the cattail fuzz, it rose up in a gigantic cloud all around her and she couldn't see. I thought she was going to hit the ditch. Suddenly we realized this could be dangerous, a blanket of fluff on a right-angle curve.

Luckily nothing bad happened—except she stopped and gave us a severe scolding.

As Edwin remembers it, "She didn't see our booby trap until too late and plowed into the cattails, sending a fog of fuzz into the air. The car screeched to a stop and backed up to where we were standing. Before we could move a woman stepped out of the car, her face red as a beet. Using rough street language she really gave us a dressing down, pointing out how dangerous this trick was.

*"What if another car had been coming from the other direction? Her visibility was all but eliminated by the fuzz.* She demanded we clean up this mess so future traffic would not suffer the danger she had."

As she drove off in a huff, we cleared the highway in near silence. Much embarrassed and chagrined, we agreed we'd acted dangerously, indeed. Fortunately the woman wasn't travelling fast with the wartime speed limit thirty-five miles per hour.

The Hill kids headed home and so did we, so shell-shocked we didn't even tell our folks. We knew Mom would feel bad and probably say, "Oh my, she could have had a serious accident."

We never tried that trick again.

*She drove off in a huff and we cleared the highway in near silence, embarrassed.*

We took three routes to school at different times, in different weather: through the fields, across Art Hill's pasture or around by the highway. The quickest way was to cut straight through the fields—but only after harvest. When crops or alfalfa grew in the fields, spring and fall, we skirted them and hiked ditch banks along the fence line, then walked the highway shoulder for the last half-mile.

For variety and when fields were muddy we cut through both our pasture and Art Hill's, above the ditch, then down their long tree-lined lane to school. This route led us through silver sagebrush flats and hidden coulees promising green grass and peeping flowers in spring, past weathered rocks and buttes painted and shaded in endless variety with each passing cloud, along soft gray fence posts, dusty cattle trails and the winding irrigation ditch brim-full of water in season, but otherwise dry.

It was scary for a first grader when Art Hill's big Herefords came to hover around and check us out. Beverley picked up a stick and shouted at them, but not too loudly because after all we were the intruders. We didn't want to get banned from these pastures by the Hills or Ben Thal, their kindly hired man.

I was horrified too of the big cattle-dipping vat, which naturally Beverley and Jeanie had to investigate from time to time. Down at the bottom, a big rectangle of turgid red water reflected our faces when we leaned over. The dipping vat was built with a moveable corral that trapped five or six cows, then dropped them down with pulleys. The animals sank entirely below the surface of the mysterious, dark and deadly solution that killed their grubs and insects.

I imagined the terror the poor cattle experienced at being half-drowned this way.

The dipping vat figured in my worst nightmares that year, vivid dreams in which I hid from evil pursuers, fled, then fell toward the dark waters.

Going home through the pasture we picked the earliest spring flowers, yellow sweet peas, ground-hugging white mayflowers, purple loco weed. Mom's advice on preserving the fragile plains flowers rang ever in our ears.

*I imagined their terror, when the dipping vat dropped cows below the surface into that dark and deadly solution.*

"Never pick all the flowers," she told us. "Leave at least two for every one you pick—and don't take all the prettiest ones. Then we'll find more beautiful flowers again next year."

With the hint of spring came time to bare our legs. We rolled down our long brown stockings and rolled up our long underwear. And when we could, we took off shoes and socks and went barefoot.

The change of seasons fascinated: our sudden extremes of heat and cold, the melting warmth of Indian Summer, winters of heavy snow or none, the joy of a chinook in January, and crisp clean smells of springtime.

In September, a month of golden sunshine, brisk mornings and chill night air gave way to a week of drizzle, then rain, maybe even whisks of snow. Sunny days served to heighten the pleasure of these weeks of golden Indian Summer that extended through October and even into November. The mellow harvest moon rose early, sometimes on our way home, reminding us to keep going and stop dawdling. A blustery fall cold spell came and went and we savored summer warmth as days grew short and darkness came early.

Soon enough winter's crisp chill brought back that long underwear and our long brown stockings. We wore dresses to school, but pulled our snow pants up under them walking to school and in the playground.

In the coldest part of winter, when the thermometer hovered

around zero or lower, Mom or Dad usually gave us a ride to school or came to take us home. Problem was, the car was a hard start on those frigid mornings, and we sometimes came to school late.

The colder the morning, the harder the start. Sometimes Dad set an ash pan filled with red-hot ashes under the motor to warm it up in the garage. When cold and bitter wind coupled with short days made the walk dangerous our folks picked us up.

During the worst blizzards, we stayed home.

One morning when it was thirty below zero and the car didn't start, our parents said it was too cold for school. But, as a seventh grader, I determined to go. After all, the Hill kids and our teacher would be there—they lived on the school grounds.

The sun was shining and the air clear and still, so I set off through the fields and never looked back, the air so cold it pinched in my nostrils.

Everyone expressed surprise to see me, as if an Arctic explorer had showed up in their midst. By the time school was out, the car started and Dad came and picked me up.

On Monday mornings Grandpa Hill sometimes stopped and gave us a ride to school on the way to taking his grandkids and their mom to their trailer house in the schoolyard. We girls rode in the back of the big truck with Mary, Edwin and Dorothy. We all arrived at school in style, standing and waving at friends. On crisp mornings we snuggled down in the truck box under their big buffalo robe.

*Monday mornings Grandpa Hill sometimes stops to pick us up when he brings Edwin, Mary and Dorothy to school for the week. We stand in the back of the truck or, in winter, sit huddled in quilts.*

For high school, Beverley stayed in town two or three years, as did Jeanie her freshman year. After that, they drove. Because of the war they kept to the speed limit of thirty-five miles an hour to save gas and wear and tear on the vehicles. Worn tires could not be replaced, so icy highways became hard to navigate with slick treads and sometimes they slid in the ditch.

"We didn't much like going so slow, but we kept to that speed, not going a mile past thirty-five so as not to anger Mr. Thomas, the patrolman. He kept watch on us," said Jeanie.

When Mary and I finished eighth grade, the Hill kids moved to town with their mother. Jeanie and I drove the ten miles to Miles City and back each day.

None of us rode horseback to Tusler until Anne remained the only one left. Then she rode Spotty for a year.

Travelling to and from school each day was a major challenge of our growing up years—whether by foot, bike, car, truck or on horseback. In some ways it seemed to us that our successful daily

Western Ranch Life in a Forgotten Era

treks depended solely on our own efforts.

But in truth, of course, our parents were totally involved. Like other country people, our folk's first priority through the years was getting their kids to school. They overcame many problems and made it clear to us that education was critically important in our lives and we must get there on time.

Whatever the difficulty, Dad and Mom arranged for us to get to school—and no doubt watched and waited anxiously for our return.

## Horseback to School

"A new family is moving to the CBC ranch and Tusler School will open again," Mom told me one evening in a pleased voice.

"You'll be coming back home for school this fall."

"Yipee!" I cried.

So Tusler School reopened during my sixth grade, after one year being closed. With two new kids we had five students again.

For the first time I went to Tusler without my older sisters and I missed them.

Mom and Dad decided I should ride our kid-sized pinto back and forth to school instead of walking. It was a fun ride, a mile and a half through the pastures, although Spotty and I didn't always see eye to eye. He had his tricks.

Spotty was a quick little pony with a lively disposition and subject to quick turns. He seemed pleased if he caught me napping. He seldom bucked, just flipped around quick so I lost my stirrups. With my feet out of the stirrups I was entirely at his mercy. Another flip and I hit the ground.

This was 1948, a peak year for the coyote population and they were hungry. They killed and ate whatever they could—deer, rabbits, mice and sheep. First they singled a deer out of the herd, hunting together in family packs of three or four, and took turns chasing it or other wildlife in wide circles until it was exhausted, then closing in for the kill. Killing sheep was easy: they simply jumped one, grabbed its throat and all leaped in to rip their victim to shreds.

Our cows chased coyotes away from their calves. When several cows were bellowing and tossing their heads, protecting their young, the coyotes slunk away quickly. Still, they picked off a newborn or sick calf now and then, or even a downed cow struggling with a difficult birth.

Some horses instinctively hate coyotes and try to run them off,

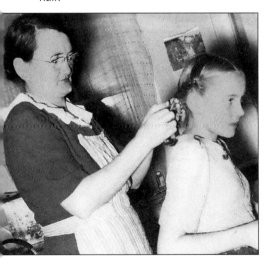

*Getting ready for school, Mom braids Anne's hair.*

turning and kicking at them when they get close. Spotty was one of these. In a pasture several horses will take after a coyote, while it runs flat-out for the fence. Maybe it's their instinct to protect themselves from wolves, which were vicious to horses. Still, some horses seem to make a game of it and take delight in kicking up their heels at coyotes.

Spotty hated coyotes. But at the same time, he held this gleeful thought that chasing a coyote made a good excuse to dump his rider—me. He was quicker than I at spotting a coyote. Before I knew it was there he'd throw up his head, spin around to better see the coyote and send me flying.

So there I'd be—early in the morning and half asleep as I trotted along on my way to school, one hand holding the reins and the other my lunch sack—while Spotty looked out for coyotes and trouble.

That fall morning he had it easy. A couple of coyotes ran out of the coulee just ahead. Spotty snorted, threw up his head and crow-hopped sideways to get rid of me, the better to pursue them.

I lost my stirrups.

My lunch flew in the air and I grabbed for the saddle shoulder. Too late—there was only air beneath me. Slam! I hit ground—smash on top of my lunch. Darn! Not for the first time I came to school with flattened sandwich and cookie crumbles.

Each morning Spotty and I arrived at the neighbor's barn near the school, where I shut him into a stall to wait for me until school was out. With my skirt still stuffed into pants, I walked down the lane to the schoolyard. There I stopped off at the girls' outhouse to pull off my pants, ready for another day at school.

A week later I counted five coyotes eating a deer carcass at the base of a gumbo butte some distance away. Spotty saw them first, threw his head up with ears forward, and swished his rear end, like he intended to take off.

All five coyotes looked up and saw us. One ran off, zigzagging up the draw as if dodging bullets. Another followed in the same zigzag pattern, looking back over his shoulder. They must have thought I carried a rifle.

The other three trotted a short distance, then returned to their kill, still watching me. I wondered if these were last year's pups, nearly full-grown. The old pair knew the risks of staying at their kill in broad daylight, but couldn't get their hungry young to leave. One by one they dashed in, grabbed a bite, then stood back to watch me warily as they chewed.

I jumped in the saddle, stretching up as tall as I could in the stirrups, waving my arms.

Anne rides Spotty to school each day in sixth grade.

*Spotty hated coyotes. Also, chasing a coyote made a good excuse to start bucking.*

Western Ranch Life in a Forgotten Era

> When I got home I reported the day's coyote sightings to Dad.

"Get out of here. Scram!" I yelled.

Reluctantly the coyotes trotted off. Spotty wanted to follow and tried to buck me off, but could only manage a few crow-hops. I was ready for him this time and jerked the reins back, forcing his head into his chest.

When I got home I reported the day's coyote sightings to Dad. Next morning he went to what was left of the deer kill with rifle and strychnine baits. The coyotes ran off when they saw him, zigzagging, but Dad anticipated their turns and killed two. Then he placed strychnine baits in the deer carcass. Later he came back to check nearby draws and picked up two dead coyotes.

There were so many coyotes that year that neighbors encouraged Dad to kill coyotes in their pastures too. If they had a dead animal, they called for him to bait it with strychnine.

Finally, in response to the desperation of ranchers, the government got serious about stopping the devastation on ranching caused by starving coyotes in our area. This was the year federal trappers brought out a new highly-effective poison, called 10-80. Within a short time coyotes grew scarce.

Students at Tusler one-room school. At left, Edwin is third, then Mary, standing, and next Jeanie. In row of desks at right, Francie is second and Dorothy fourth. Beverley had finished eighth grade and Anne yet to begin.

Meanwhile, Spotty and I continued our ride through the back pasture to school, where he sometimes pretended to glimpse a coyote off in the distance. With no coyotes to chase, Spotty was somewhat subdued, but nevertheless kept this thought foremost in his mind: dump Anne.

But I learned to be more vigilant, no sleeping on the trail.

Next year Tusler closed again, this time for good, and I returned to town school for junior high. I was ready for another school this time. Country school was not as much fun without my creative and caring sisters and our friends, the Hill kids.

Despite ending my morning and afternoon adventures with Spotty, I looked forward to town school and the peaceful ten-mile ride in the car with my sisters.

## Intensive Learning at Tusler

"Jacob! Don't you *ever, ever* do that again!"

Miss Wyss seemed to fluff up like an angry mother hen as she faced the seventh grade boy.

All we students in the one-room country school held our breath. Her unprecedented angry outburst shocked us.

Never, in all the years she taught us, had Miss Wyss scolded anyone before. She didn't need to. Daily she maintained a serene classroom and firm discipline without raising her voice.

But abruptly that day, she launched a tirade at poor Jacob.

Then, just as suddenly, she stopped scolding and burst out laughing—she of the formal teaching habits and impeccable standards— and she couldn't stop laughing, though her face flushed red and tears seeped down her cheeks.

In our surprise and relief, we laughed, too. But uncertainly, because we didn't know what might happen next.

Nothing happened, as it turned out, and she calmly returned to our recitations. It was a memorable moment, never repeated. We speculated on a possible memory from childhood. Or maybe seeing our shocked faces—and the shock on Jacob's face—chilled her anger and set off laughter instead, restoring her usual cheerful demeanor.

We three older sisters, Beverley, Jeanie and I, attended Tusler, the one-room country school eight miles east of Miles City, from March 1939, when we transferred from the Burns school near Savage, until we each finished eighth grade. Anne, four years behind me, attended there from first grade until the school finally closed for lack of students.

Twelve students attended Tusler after we came, including Beverley in fifth grade, Jeanie in third and me in first. We four Brink girls and the three Hill kids became the long-timers at Tusler School. Other students came to live for a few years at the two railroad section homes and the leased CBC ranch, then left as others took their places, with ever dwindling numbers.

Mary Hill and I held down the first grade, and for most of our eight years were the only two in our class and best friends. Others weren't so lucky. Beverley's friends moved away. Jeanie had only boys in her grade and that intermittently, coming and going, and not always the greatest of students. Mary's brother Edwin, a year younger than us, spent most of his years alone in his class. Anne was alone, too, at first. The next year Mrs. Schlosser moved her up so she could join Dorothy Hill in the third. Anne enjoyed that and so did Dorothy.

*Country schools, where neighbors gathered, were the hub of social life.*

## Dwindling students at Tusler

When we came to Tusler School, named for Henry Tusler an early cattleman, in March 1939, twelve students attended there, eight grades in one room with one teacher, Miss Clara Wyss. It was the most students during the years we were there, though likely even more attended in homestead days—with more young couples, on smaller farms and with more children per family.

From that point on, student numbers dwindled until the spring of 1948 when, with only three students, Tusler closed for good.

We four Brink girls and the three Hill kids were the long-timers at Tusler during those ten years.

Most other students came for only a year or two and then moved on. They included: Donald and Kay Lackner (living at the Milwaukee Railroad section house), Vernon Schluter (at the Northern Pacific Railroad section house), the Eisenhower girls, Dick and Duane Lenning (purchased Art Hill's ranch), the Frank Hill kids (Janet, David and Jean—grandchildren of Art Hills' and unrelated to Mary, Edwin and Dorothy Hill), and the Hays kids and Fredricksons (their families leased the CBC ranch; the Hays kids rode horseback three miles to school).

During this time four teachers taught at Tusler:
- Clara Wyss (fall 1938, or earlier, through spring 1943)
- Mrs. Schlosser (fall 1943 through spring 1945)
- Betty Grace Nichols (fall 1945 through spring 1946, when Tusler School closed temporarily for a year)
- Mrs. Audrey Herigstad (fall 1947

*(continued)*

Miss Wyss taught us five years until the spring of 1943. Engaged to a soldier, she left us to marry her David when he came home early from the war. A true professional, she was a wonderful, dedicated teacher who started us right and kept us all on track, just where we needed to be.

Country schools were the hub of social life in rural communities. Ours was the center where neighbors gathered as they brought kids to school and for the occasional card parties, Christmas programs, and most of all—end-of-school picnics.

For the Christmas program we practiced weeks, memorizing all the four, five and six verses of the Christmas songs in the Golden songbook. We put on short plays, recited poems and sang solos, duets and choral numbers, with parts for everyone down to the smallest child.

One spring the formidable Mrs. Schlosser, our next teacher for two years after Miss Wyss left—in a daring departure from the small plays we acted out at Christmas—challenged us in a real drama with several acts and elaborate props including a two-story tower.

The play was *Bluebeard*, murderer of many wives. We acted our roles with relish, enjoying his sinister, secret room that dripped with blood, skulls and bones. The new bride's screaming discovery of Bluebeard's crimes proved the high point of our drama. Imprisoned in the tower by her furious husband, she hoped her brothers were racing to save her life.

"Sister Anne, Sister Anne, do you see them coming?" she called from a high barred window.

With a 'Sister Anne' of our own, Jeanie and I kept alive that memorable line for use in critical moments, such as when we got the car stuck in a pasture ten miles from home.

"Sister Anne, Sister Anne," one of us would call out. "Do you see anyone coming?"

Little Anne giggled when she heard this and shook her head. We loved the grizzly drama and so did the audience.

However, some must have wondered aloud if all the violence was good for the little ones, because our mother, a former high school English teacher, rose in Mrs. Schlosser's defense.

"Literature is good for children," she said firmly.

For the end-of-school picnic every year, neighbors took a break before haying and came together to visit, enjoy the day and exchange news and crop information. The schoolyard blossomed as men set up plank tables on sawhorses in the shade of the big leafy green cottonwoods that framed our schoolyard and women piled on food.

We brought a dishpan full of Mom's favorite sandwiches nestled in our best dish towel. To prepare these, we first clamped the hand meat grinder onto the kitchen table and ground a quart jar of canned beef, venison or chicken with a dill pickle and bit of onion. Mixed with Mom's mayonnaise, made that morning, and spread on freshly baked bread, the sandwiches were delicious and filling. In a festive mood, Mom boiled and beat seven-minute frosting into a white froth, then lavished it onto a layer cake with fancy swirls.

Men took charge of metal freezers filled with homemade ice cream packed in salty ice in wooden tubs. You could eat your fill. But I had Mom's warning that my first year I'd eaten way too much.

"Remember, you had a terrible stomach ache from all the ice cream you ate."

After we feasted, the mothers talked quietly under shady cottonwoods with a watchful eye on small children. Men started up a ball game, joined by some of the younger women.

The older teenagers, our golden youth, hung around together, flirting and laughing—enormously romantic to our shyly admiring eyes. But when they reached eighteen, these young men went off to war. One of our own, Jimmy Stengel, an only child, was killed. Another came home badly wounded.

Again, on the Fourth of July, our community celebrated with a picnic in the schoolyard. At other times our families threw lively card parties, dances and box socials at Tusler to raise money for various causes.

At one memorable country dance Mary Hill's father—temporarily home from his shipbuilding war work on the west coast—pounded the piano keys in a spirited beat, half-turned on the bench, laughing, singing and calling out to dancers. He had removed the front of the upright piano and the whole instrument leapt and danced, its white-padded levers flying.

As part of our war effort we students sold Red Cross stamps to the neighbors every Christmas. At noon or after school we walked to the nearest homes—Art Hills and the Winninghams—with our sheets of stamps and an envelope for Red Cross money, carefully counted. We sold to our parents, too, and the more distant neighbors as we could reach them.

Each sheet held 100 stamps at a penny each, ten stamps to a strip. Most people bought one or two strips at ten cents apiece and sealed the backs of their Christmas cards with the pretty stamps. Rarely did people buy a full sheet of 100 stamps. *After all, it cost a whole dollar and how could anyone use so many?*

Our teacher kept careful track of these dimes, nickels and pennies, and turned them in to the Red Cross when she went back to town each weekend. Not much money for all our work, but no

> **Tusler** *(cont'd)*
> through spring 1948, when the school closed for good)
>
> Beverley, Jeanie, Francie and Mary Hill finished the eighth grade at Tusler country school. Our younger siblings completed the eighth grade in town schools.
>
> Anne attended Washington School in Miles City for fifth grade, driving the ten miles into town each day with Jeanie and Francie at Custer County High School. The next year Anne went back to Tusler country school when a new family moved in, then returned to town school for seventh and eighth grades.
>
> At sixteen, Beverley was already in her sophomore year at the University of Montana; she skipped first grade and a second year by combining her final high school year with a full year at Custer County Junior College.

*We relished Bluebeard's sinister, secret room that dripped with blood, skulls and bones.*

doubt it made an impressive total when combined with collections from school children across the entire country. Most of all it gave us a visible way to help bring comfort to our soldiers.

Our school day began each morning when teacher rang the bell and we stood on the front steps raising the flag and pledging allegiance. We then sang by our desks without accompaniment, *"The bombs bursting in air, gave proof through the night that our flag was still there."*

We knew all the patriotic songs by heart—all the verses—from the Golden Book. Next, Miss Wyss called us first graders up to the recitation bench in front of her desk. Mary and I spelled out our words and haltingly read out loud. Older grades took our places on the bench, separately, one subject at a time through the day, as the rest of us worked at our desks.

As we studied, we heard their recitations. Even in the lowest grades, we learned some of what each class in school was learning. It was a marvelous system, in its way, if perhaps distracting at times. When our turn came to move on to the next grade, we were more than ready. We already knew much of the material—and if we had trouble, knew just the right students to ask. We'd heard it all before.

We brought lunch from home packed in tin Karo syrup pails—sandwich, oatmeal cookie, and maybe an apple or tomato in season. More often than I liked, Mom made our sandwiches of sliced head cheese that she canned in pint jars at pig butchering time. Nutritious food that filled our stomachs, nonetheless.

The best time of day came after lunch when our teacher read to us—even though, tantalizingly, only one chapter at a time. How can we ever forget our anticipation of the wonderful adventures of *Treasure Island*, *Huckleberry Finn* and *Call of the Wild*, experienced in that magical half-hour?

George Washington hung above the blackboard with his tight-lipped smile. Facing him, a brooding Lincoln challenged us to ponder how it feels when your country is torn and bloody and it's your job to save it. A pull-down globe and roll-up maps set us dreaming that someday we'd travel to faraway places.

Country school inspired us with stirring poetry, too. *Old Ironsides* and *O Captain, My Captain*. Even the interminable poem *Lady of the Lake* turned out to be rather fun, after all, because for several afternoons Mrs. Schlosser sent Mary and I to read it aloud in the teacherage, her private living quarters attached to the school.

On Friday afternoons, under Miss Wyss's direction, we tackled handicrafts—Edwin in the cloakroom with hammer and saw and we girls embroidering dishtowels. It was a good break from hot September afternoons when lazy flies buzzed at the windows and always one gigantic fly, caught in a sudden bumbling tizzy against the glass,

*It was a marvelous system, if perhaps distracting at times.*

drew the attention of every straying eye in school.

Our next teacher, Mrs. Schlosser, discerned Jeanie's art talent and as an eighth grader assigned her to paint a mural above one of the blackboards. Immersed in western history, Jeanie chose real historical figures from our locality—plains Indians, a missionary, a trapper and Captain Clark of the Lewis and Clark expedition. All of them had traversed the trails or floated the Yellowstone right outside our schoolhouse door.

She painted her figures more than two feet tall, working intently for weeks in her spare time. With a background of buttes, leafy green cottonwoods and winding blue river, her large mural filled the twelve-foot stretch above the blackboard.

Mary and I immediately aspired to fill the matching blank space above the other blackboard. When our turn came, we chose our class topic: Egyptian history. Certainly, covered wagons and pioneer families would have been more appropriate to eastern Montana, but Mrs. Schlosser let us choose. We painted the stilted Egyptians with relish. And we had a marvelous time with the stiff, colorful, two-dimensional figures of early Egyptian art, their faces in rigid profile.

Edwin's opportunity came the next year. He painted the Great Seal of the State of Montana—four feet in diameter—on a large space above the wainscoting on another wall. The three murals added a stunning focus to our schoolroom and inspired daydreaming students.

Unlike most local rural schoolhouses—invariably painted

*Jeanie painted her mural with historic figures from our valley.*

*Winners of the Jaycees' scholarship contest. Francie, a first grader, is seated third from left, Beverley second in back row and Jeanie, at center right, in striped dress. All three wear braids. Mary Hill, seated far right, was also recognized for scholarship.*

white—Tusler was a lovely shade of slate blue-gray with white trim. On the northeast, set side-by-side, where they'd catch the best light, but not the sun, were seven long double-hung windows. Wired for electricity, our school was lit on dark days by two naked light bulbs hanging from the high ceiling. Our water bucket sat on the counter in the cloak room, freshly filled by Edwin every day.

In the utility corner of the schoolyard, firewood was piled, ashes dumped and water drawn at the red pump. Two outhouses separated the distance across the back fence—girls to the left, boys to the right. Such were the essentials of a 1940s country school without running water or furnace.

During the week, our teacher lived in a two-room teacherage, attached at the end of the big schoolroom. Her separate entrance opened to the pump and woodpile.

A merry-go-round, two high swings and a metal trapeze bar livened up the large playground. Packed hard by the feet of three or four generations of running children, it was shaded by big overhanging cottonwoods growing just outside the fence.

Tusler School, named for the early cattleman Henry Tusler, was not built 'square with the world.' Instead it paralleled the highway—the original Yellowstone Trail—which in turn paralleled the Yellowstone River as it flowed northeast at that point. Maybe that's why, forever after, I puzzled over directions. We knew the 'big north country' was across the river and we lived on the south side. However, we actually lived southeast of the river.

Just outside the schoolyard ran Kelly Creek—sometimes a trickle, sometimes bank-full in flood stage. It drained a large section of the badlands above and flowed into the Yellowstone River a half mile or so below the school.

One spring day Mary saved my life in a flooded Kelly Creek. Melting snow and rain had caused the creek to burst over its banks, sending raging flood waters across the highway a half-foot deep. It plunged with a roaring waterfall into the deep canal on the other side.

During school hours we weren't allowed to leave the schoolyard to investigate this fascinating scene. But when the final bell rang we all dashed out the gate to watch the rushing waters bursting over the road. Without hesitation we pulled off our shoes and socks and waded into the current flowing deep across the highway.

Small waterfalls made their way through twisted tree roots, and dark sluggish piles of debris surged across the highway. Fascinated by the murky, swirling waters, we poked at piles of sticks and branches that collected at the edges and caught in low-hanging cot-

> *Such were the essentials of a 1940s country school without running water or furnace.*

tonwood branches.

On the upper side of the highway, a swirling vortex of brown water sucked down with a roar into the culvert under the highway. Large broken branches stood briefly on end, then disappeared into the void.

We threw sticks toward the culvert watching as some shot out the other side into the deep canal on the other side. Others apparently got stuck in culvert debris.

The water was so muddy in flood stage we couldn't see where it was deep or shallow.

Unaware of the danger of crumbling banks, I stepped out on what should have been solid ground.

Suddenly I dropped straight down into deep, dark swirling waters. The shock of icy water was so abrupt, I hardly knew what happened until, horrified, I realized a powerful undertow was sucking me down toward the culvert. I lashed out with arms and legs, but could get no footing. In that swift current there was nothing solid to brace my feet against. Nothing to grab onto but soggy sticks swirling into the void.

Jeanie said that I disappeared entirely from sight.

"One minute we were walking along together—Mary and you in front, Anne and I behind. The next instant you were gone!"

I struggled underwater, panicked, kicking my feet, unable to gain traction. In that awful moment I realized the danger of being sucked into the culvert. No way could I save myself from that. I'd be trapped somewhere in that long tunnel by the debris of swirling branches, tumbleweeds and snags. Or, even if fate hurled me out the other end alive, I'd be helpless in the deep, dark frothing waters of the dredged-out, fast-moving canal.

Though only a fifth grader, Mary acted swiftly. She leaned over, swept an arm through the dark brown floodwaters and came up with my braids clutched in one hand.

I felt that sudden tug. Then I was being pulled back toward the bank.

Tall, smart and calm, Mary was the conscience of our playground when mischief was afoot, always insisting on doing right. Luckily for me, she did the right thing this time. Cool and fast, she held tight to my braids till Jeanie came to the rescue. Together they dragged me toward a shrub near the bank where I grabbed on, pressed my feet into the mud and, with their help, scrambled out.

Drenched and bedraggled, we ran into the schoolhouse. Mrs. Schlosser was not at all sympathetic with her dripping students. She handed us towels and sent us home still wet.

*One spring day Mary saved my life in the flooded Kelly Creek.*

Officially Tusler was a *Superior School*. It said so on a metal plaque over our front door. We didn't know just what this meant. *That we had windows, swings, blackboards and enough books for everyone?*

Anyway, we felt special and tried to live up to it. Miss Wyss made us feel that we had important future leadership responsibilities for the betterment of our communities and nation.

*What else is required of good schools?*

Then, as now, it seemed the important criteria was bringing together dedicated teachers with faith in their students, parents who stood with the teacher and valued education even though many had little themselves, and young people eager to learn and meet the challenges that came their way in changing times.

Fortunate that these came together at Tusler country school, we learned—not just our lessons—but how to live good lives, accept responsibility and make a difference in our world.

> "All are good students. Any daughter of yours is certainly capable of teaching them."

## Three Students on Deadman Road

"Marie, I've got a delightful little place with three students out in the Crow Rock country. I need a teacher there."

JEANIE

As I came on the porch, I stopped suddenly with my hand on the screen door. Mrs. Whittenhove, the County Superintendent, sat at the kitchen table with Mom.

"All are good students. Any daughter of yours is certainly capable of teaching them."

They both looked up as I opened the door.

I'd just finished my second summer session at Western Montana College of Education in Dillon and attended a year of junior college. Francie, in her final year of high school, would soon start college, Beverley was a junior at the University in Missoula, a major in journalism, and Anne was in the eighth grade.

Already August, and with a critical shortage of teachers everywhere that year, rural schools were especially desperate.

"Let me think about this," Mom said as she walked her visitor to her car.

And then Mom asked me. "Jeanie, would you like to skip college this year and teach school instead?"

My mom, the inveterate educator, asking me to stop college for a year? That made me pause. But, of course, I wanted to do it. Right then. But...

"What about Beverley?" I asked.

"No, you know how she attracts every man around. She might

marry the first one who asked her." Mom chuckled.

I thought how wrong—but mothers worry. Beverley would never do that, but with her charisma, she'd be a magnet for eligible bachelors in the whole of Crow Rock country and maybe also Jordan and Circle. Women would gather, too, with all sorts of interesting projects. She'd be swamped.

"I *really* want to!" I said, thinking how much better than when I taught Sunday School.

*Just think, me—a real teacher way out there in the wide open spaces. Living by myself! Teaching students! Just like in a book! Hmm. I'll take my horse, Dusty, and maybe a cat and dog. And some canned vegetables and…go!*

Mrs. Whittenhove drove me out there from Miles City. At eleven miles, we turned off the highway onto Deadman Road. A straight twenty-eight miles north from there on a three-track washboardy dirt and gravel road was tooth-rattling.

The car traveling north took two tracks on the right side of the road, and the car going south took the two tracks on the other side. That's four tracks. Not so fast…being a *three-track road*, both cars traveled the middle track *at the same time*. Somebody had to move over. On blind curves and hills, both needed to pull out of the center track, but it often didn't happen, I discovered. Some drivers were gamblers, some distracted, some just plain poor drivers. And everyone drove at high speed to gather up the great distances.

*Will we ever get there?* I wondered, gravel pinging on the under carriage and leaving a roiling cloud of dust reaching forever behind us.

We went straight for miles before coming to a right-angle turn. Those were section-line corners and you slammed on the brakes because the rough washboards might skitter the car sidewise right off the road. We crossed hills and coulees and avoided some bad washouts where the road caved off.

Jeanie practices teaching with Francie and others on a day that desks sit outside for refinishing Tusler school floor.

Then we went up and over a high divide and suddenly way down there a tiny schoolhouse— a white dot—dusted in sunshine on the prairie.

We pulled up in front. It was only about twelve by fifteen feet. The sun beat down hot and the south-facing schoolhouse door had blown open revealing hooks on the wall of the tiny coatroom—and a long, fat rattlesnake lolling comfortably on the sunny floor.

*Here I am,* I said to myself. *I'm supposed to be a lady-like teacher. I'll stay back and see what she does with that rattlesnake.*

"Oh my," Mrs. Whittenhove said. "We've got to do something about that rattlesnake."

An old shed stood alone behind the school. She ran back there

**A long, fat rattlesnake lolled comfortably on the sunny schoolroom floor.**

Western Ranch Life in a Forgotten Era

and returned with a scoop shovel in her hands and beat that snake to smithereens.

*She's ruining the skin, she's ruining the skin! Don't do that! He was dead after the first whap.*

She scooped it up, ran over to the lip of the coulee and flung it off the shovel. It disappeared, turning over and over until it vanished down there among the gray sagebrush.

I tried not to act too interested, because I didn't want her to know I could tweak off those rattles in a second and save the snake's skin for Francie to tan. She might not like my casual attitude toward snakes.

Then she showed me the school and said it was designated by Montana as an 'Isolated Rural School.' It seemed to me that overstated the obvious. This Fairview School was miles from nowhere. As far as the eye could see across the yellow plains—no sign of human habitation.

She told me I could keep a horse on one side of the shed and coal on the other.

There was a little building about thirty feet away called the teacherage, where I'd live. Inside, the bed was probably six feet long, with a two-foot closet at the end, against the other wall at the foot of the bed. My little house was eight feet wide at the most and ten long. It was long enough for the bed's width, a table under the window and two chairs. On the other side was another small window. Three shelves held my canned goods next to the big black Monarch kitchen range, with its silver decorations and impressive lion claw legs. And there was a coal scuttle, a long narrow pail with a sloping spout. That's all the furniture; it wouldn't hold anything else.

The huge stove was the elephant in the room and often gave off too much heat. Even though my house wasn't insulated, on some of the coldest days when the blizzard wind blew through the cracks in the walls, I opened the door to cool off the little place. Instant air conditioning.

Since I didn't have a car, Francie and Anne—with sometimes the Hill kids along—took me out on Sunday evenings and came back to get me on Friday afternoon.

Deadman Road was either muddy or slippery as only eastern Montana gumbo can be. Sometimes it was icy and snow packed. We never knew when we drove that road what it would be like. The name originated when, in earlier days, someone found a man, said to be a sheepherder, frozen to death there.

One Sunday at a bitterly cold twenty-five degrees below zero, we left the ranch, drove through Miles City and picked up two friends. Nothing like company to warm the chill.

> **She returned with a scoop shovel and beat that snake to smithereens.**

An unexpected blizzard blew up so suddenly we could hardly see. We were within a half mile of the Billings ranch—the old N Bar N—and stuck in a snowdrift lying across the road. The wind howled and packed the snow so hard the car couldn't plow through it, or move either forward or backward.

Usually in those circumstances we stayed with the car, bundled in all the wool blankets we carried for just such emergencies until someone found us.

This time I knew the N Bar N ranch was only a half-mile away, even though we couldn't see it through the blowing snow. We fought into the howling wind to Billings' house with that blizzard hitting us squarely in our faces. Even a half-mile was too far to walk in that terrible storm, but we struggled on.

They saw us coming and opened the door. We practically fell inside. When the heat hit us, we suddenly stung all over from frostbite on parts of our faces and legs.

After we warmed up, Mr. Billings took me in one of his big trucks to my school. Then he pulled the car out and sent the others on their way back to Miles City.

Quickly I started a fire in my teacherage and the school building so it could warm during the night. It was twenty-five degrees below zero in both buildings, but the Monarch in the teacherage warmed the little place in a hurry. The school took longer. I sat on that big potbellied stove in the schoolroom for forty-five minutes before it got too hot. All the while my Coleman lantern fizzed along in the otherwise silent, icy room. The darkness outside was vast and windy with stinging pellets of ice.

My sisters, not surprisingly in that storm, had run into more trouble and were stuck in a drift again. They expected to spend the night all bundled up in the car, but eventually a cattle truck came along and they headed to Miles City, leaving the car empty.

At home Mom and Dad worried. They knew the blizzard was too severe for the family car.

"I hope they have the good sense to stay in the car," Dad said.

He headed out to meet them in the truck, or to find them stuck—as he suspected.

Anne and Francie watched for Dad as they rode in with the trucker, since they knew he'd come searching. Luckily, they met on the road.

The next day, the sun shining with all its might, Dad went back and dug the car out.

A number of harsh blizzards swept through that winter and often the roads into Miles City were blocked. With their big tractors, parents of my school children kept the road passable between their

*When the heat hit, we suddenly stung all over from frostbite on our faces and legs.*

> Then one Saturday,
> I heard a strange sound
> and ran out. A little yellow
> Piper Cub circled overhead.

homes and my school, so school was held as usual even with the sub-zero cold and blowing winds. We didn't miss a day.

My radio predicted fifty degrees below zero one night, and I was delighted to experience that. My thermometer didn't go down that far, but a famous author, writing about the Yukon, said the test for fifty degrees below zero, was when spittle cracked when it hit the ice of a lake. I went out and stood under the northern lights and spit and spit. Either it didn't quite get to fifty below, or my spittle was hotter than the old miners, but mine never froze before it hit the ground—but I almost did.

Nevertheless, even if I suggested they play indoors, my students ran out at recess—they defied the weather and returned, noisy, joyful and invigorated—my dear little ones.

Because of stormy weather, when I'd been out there several weeks without getting to town, I ran low on food. My children's parents occasionally sent goodies with the kids, but I was too embarrassed to tell them about my meager grocery supply.

The blizzard finally stopped and the temperature warmed. The radio said the roads would be blocked for some time and I was getting tired of eating canned beans and biscuits, but life had been interesting. What with the subzero weather and the flickering, twisting northern lights that often writhed in magenta, blue, gold and green streamers above my head—lighting the night—while I went back and forth feeding my stoves with coal from the shed.

*Jeanie visits one of her students on Deadman Road and their pet antelope.*

Then one Saturday, I heard a steady hum in the sky. I ran outside and looked up.

A little yellow Piper Cub circled overhead.

My folks had sent out a box filled with beef and other groceries with family friends Albert Lahn and Chub Askins, the pilot.

Though it was too windy to land the plane, it was easy to make a drop.

Anne, who was about twelve or thirteen at the time, was riding with them. It was her first flight and she said how small I looked down there waving up at them.

When I agreed to teach out there, little did I know the difficulties in getting to my isolated school on Deadman Road or the joys of teaching my three lovely children.

## Nothing as Fearful

Our mother loved a challenge, especially an intellectual one. Her passion was teaching. So when she could combine a mental challenge with her love of teaching, she was delighted.

One fall Mom was finishing up her canning when a rancher from the eastern part of our county came to the door. Would my mother come teach their rural school? Probably not. She had the usual busy activities of ranch life as well as harvesting crops and preparing for the fall cattle trail. She felt she was needed at home.

But Mr. Beardsley persisted. "We need you right away," he told her.

"But hasn't your school already started?"

"The teacher we hired didn't work out," he said brusquely. "Yes, it's unusual to let a teacher go once school starts, but a situation arose that made that step necessary."

Mr. Beardsley was a long time rancher in the area and a county commissioner. His tone indicated he wasn't going to go into the why she was terminated, but would Mom please come to Cottonwood School and finish the rest of the term?

There were only seven students and all from families who prized education. Then he mentioned the clincher. One third grade boy, while smart, never had anyone who could teach him to read. What a challenge! She was as excited as a puppy.

"I'd like to come," she said. "Just let me talk to Elmer."

My folks admired the Beardsleys and other ranchers whose children attended that school. Teachers were scarce and what a shame to have all those smart ranch children missing school or getting an inadequate teacher. Mom and Dad discussed it, then decided that Dad and I could manage while she was gone. The other girls were off at college or working and I was driving myself to high school each day.

Cottonwood School was about sixty miles from our ranch, so Mom would stay at the school teacherage, living quarters for the teacher, during the week. Then return to the ranch for the weekends.

Sunday night or Monday morning, Dad or I drove her out and then returned Friday after school to get her.

Most of the time this chore fell to me. Of course I didn't mind. It was fun driving by myself through the Pine Hills and Powder River

*The clincher— a third grade boy who, while smart, never had anyone who could teach him to read.*

Mom and five students at Cottonwood rural school.

Western Ranch Life in a Forgotten Era

> *Mom enjoyed those ranch kids who were smart and eager to learn.*

Mom helps her students practice high jump for the county trackmeet. A smooth bamboo pole is ideal for setting the mark, but if it breaks, a prickly cottonwood sapling does the job.

Breaks.

I soon found joy in driving our low-slung Studebaker around the sharpest corners at good speed. Although when Mom was with me, I took the curves slowly.

"Anne can drive safely," I remember hearing my dad say.

Driving that narrow, lonely highway out to Mom's school was a great learning experience. Tires were not made as well as today and neither were the roads.

Too many flats.

I hated to hear that dreaded bang—then fight the wheel to keep the car on the road. The Studebaker held the road well, and I learned it wasn't necessary to slow down on the curves, even if a tire blew. Front tires of course were the worst. But even then, that car handled well. A flat tire meant I had to change it myself since there were few travelers on that deserted highway to stop and help.

Our folks made sure we knew how to change a flat and kept a spare tire in the trunk with full air pressure. Problem was, I didn't want to be late for school if it happened on a Monday morning. I was still shy coming from a country school; the thought of walking in late with my classmates staring, spurred me to become an expert at quick tire changes. My dirty hands I hid behind my books as I entered the classroom.

High school let out a little early on Fridays, so I arrived at Mom's school while the kids were still there. She let me help some of her students with a project or two and I enjoyed helping them build a medieval village in the table sandbox, or construct a world globe out of papier-mâché on the rounded end of the propane tank. She energized her students with marvelous ideas.

Mom enjoyed those ranch kids who were smart and eager to learn. She soon discovered they were a little behind in studies, no doubt due to inexperienced teaching. Still, it did seem a harsh reason to replace that teacher so quickly.

Mom quickly diagnosed the problem of the third grader who couldn't read. She admired his active mind; he had learned to be the school clown with his lightening fast quips. Surprisingly, he was great in math. She noticed when he wrote his math numbers, he did them backwards. The riddle was solved! She sought help from a consulting firm that specialized in learning disabilities, arranging for him to be professionally tested as well as purchasing the specialized texts for him. She rose to the challenge of teaching her first student with mirror-reading dyslexia, as it was called then.

Mom slowly began to hear of curious incidents related to the teacher she was replacing.

Then one night there came a loud banging on the outside school door. Mom leaped out of bed and ran into the schoolroom, thinking there was an emergency. Her long black hair streamed down from the loosened coiled bun she wore and her teeth lay in a water glass on the dresser.

Mom had been in a terrible Model T accident while in college. Roadways were dirt back then. She and her brother Chet were returning to their parents homestead near Lewistown when their car crashed off the end of a washed-out bridge. Mom wasn't expected to live.

She was taken to a new hospital in the nearest town, run by Catholic nuns. The hospital board thought it would reflect badly on them if one of their first patients died, so they quickly discharged her. The Sisters took her into their convent and nursed her back to health.

Among other injuries, all Mom's teeth were broken beyond repair. False teeth replaced them.

The banging continued.

"What is it?" she called through lips that had no teeth to support them. "What's happened?"

"Let me in," a strange rough man's voice answered. "Why don't you let me in?"

And he banged even harder on the door.

"Who are you, and why are you here?" Mom asked.

"Let me in or I'll knock this door down!"

The schoolhouse door opened to the outside and a bent nail, hammered into the casing and turned against the door, held it shut. There was no way she could barricade it. No telephone. Nor was there any way for her to reach help. The nearest ranch was too far. She was alone.

Terrified, Mom retreated to her room, slammed the door and pulled the dresser against it.

The banging continued. Momentary quiet—then his face appeared at her teacherage window. This time he cursed loudly and shook his fist.

Mom was not one to cower

## A teaching tradition

We came from a tradition of teaching. Our grandmother, Louisa Catherine Mosier Barrett, born in 1866, taught many years near Madison, South Dakota, and in Oklahoma Territory before she married at the age of thirty-three. She gave birth to four children in quick succession within six years and didn't teach again, as few women then did after they married. But she remained a teacher at heart.

Our mother, Marie Barrett Brink, put herself through college majoring in mathematics, English and Spanish. She had a great passion for helping students and her teaching career spanned nearly forty years. She taught five or six years in high school, at Winnet and Harlem, Montana, before her marriage, and later in Jordan.

When Beverley was a year old and no teacher applied for the country school on the Missouri River bottomland that would become Ft. Peck Dam near their first homestead, she agreed to take it. Every day she rode

*(continued)*

*Mom heads for rural school on Monday morning in red-and-white Chevy that replaced our Studebaker, a family favorite.*

*Western Ranch Life in a Forgotten Era*

## Tradition (cont'd)

horseback to school, a mile and a half, holding her baby on the saddle before her. That spring the ice jammed in the Missouri River and marooned the little schoolhouse in rising waters. Dad rescued Mom and Beverley with a rowboat. She told us the story of her terror, as he maneuvered the small boat between huge, swift-moving icebergs that ground against the side of the boat and how she frantically tried to keep little Beverley safe and dry.

Later, after her three older girls left home, Mom resumed teaching.

Unfortunately, living near Miles City, she met sexual discrimination against married female teachers. The Miles City school board held that hiring married women unfairly took salaries away from men, who were expected to be the sole family bread winners, though of course unmarried men benefitted from this discrimination. Although Mom qualified in both elementary and high school education and had coached girls' basketball, she was refused a position in Miles City schools during the '50s and '60s.

Thus, she turned to teaching rural schools and loyally supported the ranching communities where she taught at Locate, Kircher, Cottonwood and Mildred, only moving to a new location if there were seventh or eighth-graders in another school who needed extra coaching to succeed in high school.

All four of us girls became involved in teaching, in one way or another, even when focused on other careers. Jeanie had a long career in elementary education and Beverley taught college journalism several years. Francie and Anne taught in a variety of venues, at times in high school and college and in adult education programs. None of us ever stopped learning and teaching.

The three Hill kids, Mary, Edwin and Dorothy, became teachers, too. We all gained a solid foundation in good education and how it should be taught from our growing up years at Tusler School.

in a corner and accept whatever came her way. What defense was available? If he broke through the window, could she hit him with the table lamp? Suppose she missed?

Suddenly she realized her greatest defense. Apparently he was expecting her to be a glamorous young woman with a smiling face and come-hither eyes. She could change that!

Drawing her long black hair around her face, she glowered, made her face as mean as possible, and pressed her nose flat against the window glass.

"Get out of here!" she shrieked at him.

The man stumbled back from the window, looking at her with horror. What happened to the beautiful young teacher? He whirled around and ran for the road. She watched as he went over the hill, then listened until she heard an eighteen-wheeler start up and speed away on the highway.

The next morning, when Mr. Beardsley brought his children to school, Mom told him of her night visitor. Very apologetic, Mr. Beardsley said he and the other dads would, that very day, string a simple phone line between school and the nearest ranch to give her access to help if she or any of the children ever needed it.

When Mom came home that weekend, she told us about her visitor.

"I scared him away!" she gleefully said.

Dad was not amused, though he had a good laugh at Mom's method of getting rid of the trucker. He brought out the .38 pistol and insisted Mom take it to her teacherage, even though she said she wouldn't shoot it. She was afraid of guns. Still, she approved of Dad teaching us gun safety and accuracy at an early age.

"Just shoot wildly, don't even try to aim," he told her. "There's nothing as fearful to a man as a panicked woman shooting a gun!"

Mom finished out the school year. She had no more night visitors, and the students at the end of the term were well ahead of their grade levels. It was understood this was due to Mom's exceptional teaching.

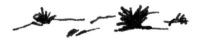

# 10  Field Work

### Wheatfield Fire

A woman came driving in from the highway one hot afternoon, careened around the corner honking her horn and skidded to a stop, laying a smokescreen of dust over me as I worked in the corral.  [JEANIE]

"Your wheat field's on fire! Do something! Do something quick!" she cried as she jumped out of her car.

Dad running from the shop, quickly grabbed up an armful of gunny sacks from the granary and dumped them in the water tank.

"Fill the water cans and bring as many wet sacks as you can," he said.

Then he took off on the tractor with lots of wet gunny sacks piled and dripping from the tines of the buck rake.

We scooped out five gallon cans of water and set them in the car trunk.

Mom and my sisters arrived and, working fast, we drenched more gunny sacks and piled them on the water cans.

"Anne, we need you to stay home." Mom turned to Anne, then a fifth grader. "Call all the neighbors. Tell them there's a fire in the wheat by the highway and we need help. Hurry!"

"Okay." I saw the disappointment in Anne's face. She wanted to come and help, too, but she didn't protest. "Then can I walk down there?"

"No, you'll need to stay by the phone in case someone calls. Don't leave the house until I come back."

The phone lines had just reached our house that year—I was glad for that.

We jumped in the car and followed Dad.

As we turned the corner we saw roiling black smoke. Orange flames snaked across the ground stubble and at each burning shock

> *As we turned the corner we saw roiling black smoke. Orange flames snaked across the stubble and flames leapt for the sky at every burning shock of wheat.*

flames leapt for the sky. Already nearly a third of the field was burning and blackened behind the racing fire. Wisps of smoke rose from smoldering piles of grain.

The wheat had been cut just a few days before with the binder kicking out twine-tied bundles. Then a neighbor helped Dad shock the grain, propping eight to ten wheat bundles upright, leaning against each other.

Those beautiful golden shocks now stood thickly across the field through the unburned section.

But along the front fire line, caught into tall pyres of flame, the dry shocks made perfect tinder. Flames leapt across the field and torched the wheat standing in shocks. Sparks flew from small explosions of ripe wheat kernels and rapidly spread to the wheat stubble ahead. Even though the stubble was short, it burned quickly and spread to the next row of shocks.

Dad drove down along the line of flames with the buck rake lowered. The wet gunny sacks dripped and dragged across the fire and mostly put it out as he drove. Some curled around the belly of the tractor.

It was intensely hot. When I first saw it I knew it was too much for just us. I thought it would burn the entire wheat field and then go on to the rangeland above—a disastrous prairie fire.

We couldn't waste time thinking.

We jumped out of the car and started beating flames with wet gunny sacks. Mom and we girls followed behind Dad, swatting out flames that passed under the tractor, and with our swinging wet burlap bags, beat fires that flared up on both sides.

*Neighbors or strangers, they did not hesitate. They grabbed things from their vehicles and started helping.*

"Elmer! Elmer! You'll catch the tractor on fire!" Mom shouted, terrified, imagining the gas tank exploding under him.

He didn't even hear her. The entire length of the field was on fire, and Dad drove down the advancing fire line all the way to the end, then turned back and did it again, concentrating on the flames.

Our thirsty burlap bags were heavy with water, taking all our strength to keep swinging at those flames.

Smoke swirled around us and stung our eyes as we gasped for breathable air.

Soon I saw through the smoke and fiercely burning flames—other cars and trucks arriving.

*Thank you! Oh thank you!*

Through the beating of my gunny sacks I could see those people did not even hesitate, but grabbed things from their vehicles and started helping. Our neighbors, of course, knew just what to do, but amazingly, so did strangers passing by on the highway who stopped to help. Some of them came from big cities and had never seen a fire like this. They instinctively pitched in wherever they were needed.

"Everyone I called said they'd come right away," Anne told us later. "The closest neighbors arrived before I finished calling. But it was really scary waiting and not being able to see what was happening."

Beverley and Francie were pulling smoldering shocks of wheat apart, separating unburned bundles from burning ones, beating out flames with heavy wet burlap bags where bundles could be salvaged and spreading others out on blackened ground to let them burn.

The tractor moved so fast and with us following behind, we eventually put out a wide swath across the entire field.

Wide enough for no wind-blown sparks to jump our fire line.

All of a sudden the fire went out. We paused and gulped deep breaths of clean air. Only mopping up and flare-ups remained.

I wiped the stinging sweat from my eyes and blackened my face even more with my sooty hands. Everyone's face and hands were soot-blackened, too.

Beverley, her eyes bloodshot holes in her blackened face, came running across the burned area to get us.

"I want you to see this!" she said. "Come look what's at the edge of the swamp."

She took us to a crisp, blackened rattlesnake. He was burnt into a coil, his head locked in striking position about ten inches above his coils. He must have been desperately fighting and striking at the fire. His mouth was open wide and we could see his fangs. His coils formed a stable fried-out platform and burning to a crisp made him so stiff that he couldn't fall over.

He faced the fire and fought to his death, even though he could have zipped off into the green swamp only a few feet away and saved his life.

Thankfully no one was hurt in the fire. We came away with a fresh awareness of the devastation that one burning stub of a cigarette—carelessly thrown out a car window—can cause.

Uncle Will and Dad pitch bundles into threshing machine. Grain goes into truck, while the straw blows into a strawstack. Our big Rumley tractor powers thresher from end of moving belt.

Beverley, at right, shocks silage cane with friend and hired man

Western Ranch Life in a Forgotten Era

# Prisoners of War on our Ranch

"Oh no! Here they come! Hurry up, Francie, run!"

"I can't! I can't run!" Francie shot back. She carried two nearly full three-gallon buckets of milk and so did I.

Heavy army trucks roared into the ranch yard, loaded with captured Nazi soldiers standing in the back. They stopped under the cottonwood trees near the shop.

Comments in a strange language flew at us like barbs.

Caught between the house and corrals, after finishing milking, we weren't expecting the first truck load yet, so it was a shock.

We were embarrassed by the German taunts—as we perceived them—and scared, too. War movies taught us something of the atrocities they committed across Europe. We knew about their ruthlessness.

These were Rommels' crack troops, Hitler's Master Race. As prisoners of war they came to harvest our beets. Even while prisoners, these tall, vigorous, young German men planned to win the war. England was battered by their bombs. The Allied countries were fighting back but Hitler continued conquering more countries in his quest to control the world. German U-boats lapped our shores.

Sugar beets, raised during the war years, a main ingredient in explosives, required a lot of hand work. Since all our available beet workers entered the military, the government brought us prisoners the Allies captured when we engaged Rommel's army in North Africa. Our sugar beets made the gunpowder for bullets that captured them.

Several U.S. guards jumped out of the trucks, long bayonets fixed on the ends of their rifles. The prisoners came to attention but stayed in the trucks while the American officer in charge conferred with Dad about the day's work.

*Our ranch is being invaded!*

Even the American armed guards with their khaki uniforms, rifles and bayonets seemed dangerous to us.

Then Dad led them down to the beet fields where they were briefed. Soon over a hundred POWs with six or eight armed guards swarmed over our fields.

We watched them warily from a distance.

From then on, in the morning when they came, Beverley and Mom made sandwiches for the prisoners to eat at noon. Dad insisted we send five-gallon cans of milk along for their lunch.

"Germans grow up with dairies. They like to drink milk," he said. And they did.

The German captain—seemingly the prisoner in charge—came

> *Soon over a hundred German Prisoners of War and six or eight armed guards swarmed over our beet fields.*

*During WWII this Milwaukee railroad bridge is guarded on the south side, at left, by men in our community. Kinsey volunteers from that area guard it on the north side of the river.*

in the house every morning to get the sandwiches and milk for their lunch while the guards and POWs waited in trucks. He was a college professor from Frankfurt and spoke perfect English, a good conversationalist. Because of his knowledge, interests and intelligence, Mom and Beverley enjoyed talking with him as they finished putting together the lunch.

Busy on the ranch, Mom seldom had the chance to talk with academics and she immensely enjoyed this change from normal days. It was so neat for me to listen to them—from another room—while Beverley and Mom finished the sandwiches and got everything ready for a hundred and ten prisoners and their guards.

That year, Hitler's Master Race prisoners looked around for ranches to own, now that they were going to win the war. That was really scary because it was obvious they liked our ranch and the war was not going well.

Bayonets prickling out at the end of their rifles, the guards looked ready to do battle at a moment's notice. There was a fear that a POW might escape and carry information back to Germany. All day long they guarded those prisoners standing at attention, gripping their rifles at each end of the fields. In late afternoons, they climbed back into the trucks and returned to their heavily guarded compound, the Custer County Fairgrounds just west of Miles City.

One night a prisoner escaped. The local radio warned people to lock their doors and take keys from their cars. Everyone reacted in alarm. Even in Miles City, most people left their houses and cars unlocked, with keys in the ignition. But not this night. The town was on lock-down. Few ventured outside.

FRANCIE

However, the escaped prisoner was soon found. He was strolling up Main Street, window shopping.

Efficient, hard workers, the German prisoners topped beets, tossed them in piles and loaded trucks and did everything speedily and well.

By the next year, 1944, how the tide of war had changed! The

*Hitler's Master Race prisoners looked around for ranches to own, since they intended to win the war. It was scary— the war was not going well.*

prisoners were old men and young boys; the people Hitler didn't want in his army at first, but later had to gather up when his dwindling military was bereft of soldiers. One man was eighty-four.

We girls weren't supposed to be anywhere near the fields or in sight of prisoners. While they worked in a field farther away, we usually worked in the garden or with the cattle.

One day we were picking up potatoes in a field near the house, loading them in sacks after Dad had plowed them out.

While we worked, we were horrified to look up and see a German prisoner coming over a small rise right toward us. We girls were thinking we'd have to run.

"Oh, he's just a boy!" Mom said.

He wore clothes far too big for him. His pants were rolled up and also his shirtsleeves. His shirttail hung out of his pants, the hem hanging below his knees. Mom was touched with sympathy toward him and we girls grouped cautiously around her.

As he came up to us he was trying to say, "Where is the guard?"

That much we could understand in his broken English. This little boy looked so worried because they couldn't find their guards and it was almost time to go back to their compound.

The guards had learned the prisoners didn't need them, so sometimes they borrowed our shotguns, our water spaniel, Rex, and hunted pheasants in our lower pasture. Or sometimes a guard simply curled up in the shade under a haystack and went to sleep.

Mom, Beverley and the little prisoner drove down to the lower pasture guard-hunting. It was then he told them he was twelve years old. A small twelve, at that. And he'd already fought a war and been captured. Mom could hardly believe it. They soon caught up with the guards, and brought them back to their responsibilities.

The guards were good to the prisoners and the prisoners were good to them. The POWs worried the guards would be severely disciplined for their lax ways, as would certainly have happened in Nazi Germany. For sleeping on the job these same men perhaps saw German guards shot on the spot.

The American Commander of the prison camp made rounds once a day to the farms and ranches where the POWs worked. Our prisoners saw him coming long before he arrived—driving the 35 mph speed limit imposed to save gas during the war, as the highway circled over a mile around and through our ranch before reaching the field gate. And it was a long way down through the fields.

"Wake up the guard! Hier Kommt der Kommendant!" You could hear the shout coming up from the fields a half mile away.

The warning passed from one field to the next, "Wake up the guard! Kommt der Kommendant!"

> *We were horrified to look up and see a German prisoner coming over a small rise right toward us. "Oh, he's just a boy!" Mom said.*

The nearest prisoner woke up the guard and got him standing up with his rifle. The guard brushed hay off his uniform and made sure he was clear eyed to greet his superior officer.

The routine always tickled Dad.

These last prisoners knew Germany was going to lose the war. Their country was bombed out. Its fields and factories decimated. Towns ruined.

They looked around at our green valley, the silver Yellowstone River, its sparkling curves in the distance, the grassy-green hillsides with grazing cattle and acres and acres of irrigated fields. They knew it was good land and many wanted to stay.

Several asked Dad to sponsor them so they could remain in the United States or come back to live.

"No. You've got to go back and rebuild your country. Germans are hard working. You're innovative and responsible. You've got to rebuild Germany," Dad said. "Do it the right way this time—cooperate with other countries and don't try to take over their lands. You can do it and I think you will."

Before the POWs, we had other workers, but none as efficient as the Germans.

The first year of the war, with other ranches raising beets for the first time, the labor shortage caused an emergency in the sugar beet fields.

Many of the American Indians who thinned and hoed our beets joined the service so none returned for harvest.

*Where could local beet growers get help for topping the sugar beets?*
*Late one night we had a government visitor.*

"We recruited teachers and older student volunteers to help top beets," he said. "They'll be here at your ranch in the morning."

Custer County High School closed for the duration and a group of students and teachers spread over our fields and topped beets every day until harvest was over.

The big football players were the best toppers, Dad said. They had large hands and could grab a heavy beet with one hand and lop off the top with the other, wielding that machete-like knife. The beet knife—sharp as a scalpel—had a three-inch hook in the front end for hooking the beet and a twelve-inch blade. A wicked and heavy tool.

You could hear them in the fields, "Hey Tom! Catch!"

Laughter and shouts filled the air as a trio of husky football players tossed beets back and forth.

Then the voice of a weary teacher, "Guys! We're here to get these beets loaded in that truck. Remember, sugar beets make

---

## How to top beets

Topping beets was dangerous because it required speed while using a large, sharp machete-like knife. The knife had a hook on the end for picking up beets, which could be more than a foot long and six to eight inches wide at the top.

The beet toppers came down the row after the large creamy-white beets were lifted from the ground by a plow-like machine. A tractor pulled it, turned the beets up and laid them on their sides on top of the furrow. With the heavy beet knife in one hand the toppers hooked the beet and lifted it up. Seizing the beet in the other hand, they sliced the leaves from the root in one quick motion of the knife. A fast slicing speed was necessary as the worker moved quickly up the row.

It was important to chop as closely as possible to the green top, not cutting off or wasting too much beet root, yet not keeping any of the green. Each time, the hand that held the beet was in danger. The workers dropped the beet on one side of the furrow and tossed the green tops on the other side to be loaded and hauled later for winter livestock feed.

The truck came along the furrow and workers scooped up the beets with short, blunt-tined beet forks and tossed them into the back of the truck. When the truck was full and almost overflowing it made the trip to the Holly Sugar Beet Company dock eight miles down the railroad.

We enjoyed riding along to the beet dump. Each truck waited in line to unload. When our turn came, we drove up a steep ramp and a big machine hooked onto our truck box. It was a dramatic moment when the truck box lifted, tipped sideways and the sugar beets tumbled down

*(continued)*

> **Beets** (cont'd)
> into the railroad car below, to be shipped to the Holly Sugar Refinery at Hardin.
>
> When our driver didn't come home after a night on the town, Jeanie hauled beets. What fun! It was the delight of her life.
>
> Our parents didn't let us top beets because they said the beets were too big and heavy. But we believed they were really afraid we'd take a swing with that sharp heavy beet knife machete and chop off a hand instead of a beet top.

*The long freights rumbled by on both railroads going east with their ominous shipments of khaki-camouflaged tanks, jeeps and anti-aircraft guns. Troop trains silhouetted soldiers in every window.*

bombs!"

"Those football boys sure did top fast. They had fun doing it, too," Dad chuckled.

He was amused by some of the teachers. "And then there was that sweet little gray-haired lady who stood hitting two beets against each other to get the dirt off. She almost beat them to death!"

Many of the teachers were unfamiliar with such physical work. Every day it became more painful as they bent over to pick up a beet. Years later those same teachers told us about the difficulties they had with topping beets during World War II on the Brink Ranch.

When that emergency ended, the field agents wisely decided it was far too dangerous for these inexperienced kids and teachers to top beets. Even though, luckily, no one was injured, they could easily miss and, in a second, sever a hand.

For most of the beet work we came to rely on Mexican labor. First, Mexican-Americans made their way up from New Mexico and Texas for hand field work. They soon went into the military. Eventually the government stepped in and transported Mexican Nationals, directly from Mexico to the farms and ranches with sugar beet allotments in our area.

When war was declared in December 1941, the farmers and ranchers in our irrigated valley began raising sugar beets. It was 'for the war effort,' a way we helped on the home front when our young men went off to war and powerful Nazi and Japanese forces overran Europe and Asia. As Americans we had a deep sense of our national responsibility to work together to win the war.

We saw the urgency of war every day as the long freights rumbled by on both railroads going east with their ominous shipments of khaki-camouflaged tanks, jeeps and anti-aircraft guns. We counted the cars as they rumbled east on the Northern Pacific and the Milwaukee Railroads that followed the Yellowstone River skirting of our ranch. Troop trains, too, roared by, silhouetting in every window a soldier wearing military cap and uniform.

A community responsibility was guarding our side of the Milwaukee bridge from sabotage all through the war. On the other side of the Yellowstone, Kinsey farmers and ranchers kept guard.

For us, even as kids, World War II was no distant drumbeat, it was here and it included us. Now.

Dad built a house of railroad ties for our Mexican beet workers. It had a kitchen stove at one end and built-in bunks at the other. A flowing artesian well ran continually just a short distance from their door.

[FRANCIE]

The Mexican workers, about six to eight men at a time, came in June for thinning beets. They went on to other farms when they finished ours and then came back for our first and second hoeing; then perhaps again for beet harvest. By the time harvest finished, it was late fall and they were bussed back to Mexico.

Dad paid them a small portion during the summer, but most of their pay was held until they returned to Mexico. This was a good arrangement for the workers: the rancher provided the food they picked up at the grocery, a place to live that was perhaps better than they had in Mexico, and they returned to their families with considerable money. A good deal for them and a good deal for the United States.

Every Saturday night these Mexican workers went to town—and introduced a new challenge to our law enforcement officials. Our local sheriff, deputies and Miles City policemen knew how to cope with rowdy cowboys and ranch workers who came into town on a Saturday night to drink and engage in their boisterous non-life-threatening fun. But they didn't know just how to handle these knife-wielding, foreign-speaking Mexicans.

Whoever was in charge of this imported labor was careful to send friendly groups out to work together on each farm. But when they all came to Miles City on Saturday night, many brought along their life-long grudges. After a few drinks, knife fights frequently erupted. Our police didn't know how to disarm a man with a knife because the Mexicans used them so quickly and efficiently. In a flash one could cut another's throat. It was a quandary for our law officers.

If our Mexican beet workers didn't come home the next day or two, Dad drove into town and bailed out those who were in jail—as he did from time to time for other hired men, as well.

We were surprised that first year when American Indians, from the Sioux or Cheyenne tribes, came to help since they usually didn't like working in fields—traditionally men were hunters. They didn't seem to like hoeing much, but they did work and brought their families along to help.

Mom said we should go out and see the babies with their round faces, black hair and beautiful big brown eyes. As a small child she lived near an Indian reservation in South Dakota and played with the children.

So one day she brought a pail of cookies and took us to visit in the field. We waited at the end for the workers to come up their rows. Two tightly-wrapped babies were lying on blankets there at the end, under the shade of a cottonwood tree.

*All through the war, men in our community guarded our side of the Milwaukee railroad bridge from sabotage.*

*Our first truckload of sacked grain. Truck was later cut down to build our Bug.*

Western Ranch Life in a Forgotten Era

*Above the tractor motor came another, more ominous sound, but faintly. "Ee-l-mer, hel-lp... Putt, putt, putt Ee-l-mer..."*

After the mothers worked down one row and back up the other they stopped to nurse and tend their babies. The rows might be a quarter mile long, but the babies were good and waited quietly for their mothers. We never heard them cry or even whimper. They just looked at us with their big round brown eyes.

When the war ended so did our sugar beet program and we planted our acres back to alfalfa and grain crops.

Mom told me, many years after the POWs topped our beets, after the war they got a letter from the German professor.

JEANIE

"If your oldest daughter isn't spoken for, I would like to speak for her hand in marriage," he wrote.

Dad refused to answer the letter. He and Mom thought the man was just looking for a way to come and live in the United States. They decided not to tell Beverley, knowing she would be horrified to think someone wanted to use her like chattel.

Later on I told Beverley about the letter. Sure enough, she was furious.

"It was just like that professor to act as if he were buying a cow!" She was as outraged then as she would have been at the time.

*Going fast, the back tires catch in a ditch and flip the tractor over, pinning hired man beneath.*

## Overturned John Deere

One summer, we had a hired man who was a bit reckless and loved to drive machinery at top speed, a dangerous combination on a ranch.

FRANCIE

No matter how many times Dad cautioned him about safety, Elliot would take chances, especially if he thought he was out of sight. Employing this man proved to be a potentially serious liability.

One afternoon, I was on my way to the granary to get a bucket of feed for the evening milking, when Dad arrived home from town and headed over to help.

Then suddenly he stopped in his tracks and lifted his head, listening intently.

"What's that?" He cocked his ear toward the field beyond the line of cottonwood trees.

At first I heard only the tractor and thought the hired man was still working in the field. Then I heard the strange sound too.

Over and above the putt-putting tractor came another, more ominous sound, but faintly.

*"Ee-l-mer, hel-lp ... Putt, putt, putt ... Ee-l-mer, hel-lp... Putt, putt, putt... Ee-l-mer, hel-lp..."*

Chilled, I dropped the feed bucket and we both ran.

On the other side of the cottonwoods we saw a horrific scene. The John Deere tractor had flipped over, its big back wheels stuck in a field ditch and the front wheels spinning in the air, the motor still running.

Elliot was pinned underneath, pressed into the dirt of the ditch bank, moaning and calling out hoarsely. *"Ee-l-mer, hel-lp!"*

"Okay, Elliot, just relax. We'll get you out," Dad called as we came closer.

"Turn off the tractor," the hired man gasped, "before the gas explodes!"

Dad shut off the ignition and surveyed the situation. There was no safe way to pull him out. "Just lie still. Try to relax. You'll be okay. We'll get the ambulance."

Finished in one field, Elliot had tried to take a shortcut across a ditch to another field. It was easy to see that he hit the ditch too fast with the front wheels, the extra speed spun them into the air as the back wheels powered forward into the ditch where they stuck, flipping the tractor over on top of Elliot.

I was working on a 4-H sewing project in the kitchen, with Mom looking over my shoulder, when we heard the alarming sound of Dad skidding the car to a stop by the door. He hollered to Mom as he ran up the step.

"Call the ambulance! Elliot's had an accident with the tractor. Bring the car down. Take the road by the haystacks. And Anne, bring your camera."

A seventh grader, I had recently documented a school Earth Science project using my cherished little Brownie, and was pleased that he thought I should use it. I snatched up my camera as Mom made the phone call. Dad picked up some tools and jumped in the truck. We all raced for the field.

I couldn't believe what I saw. The tractor was upside down with Elliot underneath, in and out of consciousness. No longer able to talk, he just made horrible groaning noises. The tractor steering wheel was across his pelvis, and half dug into the ground.

Mom assured him the ambulance was coming from town, then she drove back to the house to guide the ambulance to the site. Dad told me to take pictures of everything: the tractor, tracks, ditch, and where the wheels had thrown up soft dirt as he jerked the tractor into a right turn to cross the ditch.

Neighbors had heard Mom's phone call on our party line for the ambulance, and they appeared out of nowhere to help. They'd brought large metal bars to add to the one we had in our truck. They

*I dropped the feed bucket and we both ran.*

planned to use these to lift the tractor off of Elliot, but they needed to wait for medical help first.

It seemed like forever, but finally Mom came driving through the field with the ambulance behind. In those days ambulances were not very well equipped. This one, run by a Miles City funeral home, served two purposes: an emergency medical vehicle and a hearse. It was a long station wagon, sporting glossy wooden sides with a stretcher inside. The driver worked as the funeral home director. His assistant, Mr. Gourd, was the high school shop teacher, who taught first aid.

They pulled up beside the tractor and shook their heads. How could anyone live with a tractor on top of him? Fortunately Elliot had left the rake he was dragging behind the tractor in the field, otherwise that machinery could have impaled him as the tractor turned over. The men used their metal bars to lift and hold the tractor up a few inches while the ambulance crew wedged a wooden pallet underneath Elliot and finally got him out.

There was no blood, yet internal bleeding in his abdomen was massive. Elliot's pelvis was crushed by the steering wheel and his legs were injured. We didn't think he would live, but he was a strong man. He spent several weeks in the hospital and when his pelvic bones healed he returned to active life, though ranch work was no longer his desire.

Two or three years later, Dad was stopped in Miles City by a lawyer on the street.

"Elmer," he said, "We're going to own your ranch."

This lawyer had contacted Elliot and convinced him to sue for negligence against Mom and Dad. He seemed gleeful that he would "get the ranch."

Dad asked a few neighbors who'd been there if they'd testify that he used safe methods. They all said they'd never attempt to cross the deep ditch at that point and instead would cross where the ditch leveled out—and very slowly. They were willing to testify that Elliot was at fault in the way he drove. The pictures I snapped were important to the folk's lawyer, showing that the dirt Elliot threw up as he turned a right angle were done at high speed.

Not long after the lawyer filed Elliot's claim, Elliot drove out to our ranch.

"Elmer," he said, "I know I was driving too fast and should not have tried to cross where I did."

"But still you're suing us!"

"No, I changed my mind. I don't want you losing your ranch because of me. Besides, that shyster lawyer would probably wind up

*Between tire and flywheel, the tractor seat on ground shows where hired man was pinned.*

**Neighbors came from nowhere to help, bringing large metal bars to lift the tractor.**

with it one way or another."

The lawyer was not pleased with Elliot's honesty, but Elliot insisted the trial be dropped.

The dangers of ranching and farming have long been documented. However, as we were taught at an early age, "Safety first and you'll live longer."

Elliot was just plain lucky. His ride off the ranch that day could have been in the same vehicle—as a hearse.

## Child Truck Drivers

Jeanie was eight the first time Dad let her drive in the field—and I was six, her lesser half. On the pedals.

Dad and the hired man were loading beet tops in a field just east of Miles City to take home for our cattle. They were tired of jumping in and out of the truck, moving it along beside them as they worked up and down the rows, while throwing in the green beet tops with pitchforks. This was before we raised beets ourselves, before the war.

Mom took Jeanie and me out there after school.

Jeanie's legs couldn't reach the clutch and brake pedals, so my job was to sit on the cab floorboard and push them in for her.

Dad put the truck in the low, low gear called compound and aimed it between the beet rows. Then, standing on the running board, he gave us instructions, mostly meaningless to me.

"In compound, the truck'll go fast enough, Jeanie. If not, you'll need to shift to low. Tell Francie when to push in the clutch and hold it there while you shift. I'll be right outside if you need me."

There I crouched under her feet with the three pedals before me, puzzling over the differences between them. Both the clutch and brake had to be pushed in all the way, not easy for my small hands. But the gas pedal—just a touch was more than enough to make the engine roar!

Jeanie perched on the edge of her seat, intent on driving straight down the row, so short she had to peer out the windshield between the spokes of the steering wheel. Sometimes her dangling feet kicked me.

She steered and shifted and shouted to me above the motor noise.

"Brake! Clutch! Give it the gas!"

I sat on the floorboard pushing with all my might. Unable to see anything, I was scared we'd crash and had only the sketchiest idea

*Jeanie and Francie, ages eight and six, honed their half-driver skills in the beet field.*

*I never thought that on my first driving job I'd be crouched under the dash.*

what to do.

"Girls, if you're in low and the truck shudders, give it more gas. Jeanie'll tell you when to push on the gas, Francie. But don't push it in too fast." Dad said, one of the times he jumped on the running board.

The clutch was hardest. It took all my strength.

"Clutch, clutch, clutch! You gotta push it in harder," called Jeanie.

I heard the gears grinding and pushed until she could shift, then let the clutch snap back.

*Whoops! Too quick. Should have eased it back out.*

Reverse was hard if we got ahead and had to back up. That took all three pedals—and in the right order, please.

Jeanie loved it, of course.

She had sat in the truck often enough, parked down under the trees. So small we could barely see the top of her head above the dashboard, moving the steering wheel back and forth, reaching for the shift knob, stretching one foot way down to brake and keeping an eagle eye on the imaginary scene ahead.

Sometimes I sat beside her—both our feet dangling. Which was fine with me—a relaxed passenger on a pretend ride.

I never thought then that on my first driving job I'd be crouched under the dash.

"Jeanie, don't run into anything!" My sister heard that plaintive call every so often from the dark depths under the dashboard.

I was terrified we'd crash. But of course, that was impossible—no obstacle stood in the entire field—and we probably drove all of two miles an hour.

Even then at age six, I knew that in our family we weren't supposed to be scared to take on a new job, even if we had no idea how to do it.

*Just keep going, figure it out and you'll learn as you go.*

We stayed out in the beet field till dark—two or three hours, both of us honing our half-driver skills. On that day Jeanie and I—she above and me below—successfully maneuvered our way through the beet field on our first trucking job.

Jeanie and Francie drove our truck with sides dropped for loading sugar beet tops, shown here with load of sand.

## Hayfield Highs

One day, when I was ten, Dad said, "Jeanie, today I want you to drive the truck for haying.

I leaped at it. Driving around through all those hay fields—what fun! I was in the middle of a 4-H sewing project and put it down quickly.

"Not so fast," Dad said.

He knew what our interests and capabilities were. For years he watched me slip into any truck parked at our ranch. I sometimes got it moving even if it wasn't our truck. Of course, the owner came running.

Sometimes Francie was with me and we'd quickly turn off the ignition, climb out and run away skipping and jumping before the owner got to us. It helped that I was small, blonde and skinny and looked quite harmless.

"Listen." Dad said. "You'll back up the truck to raise the stacker with a load of hay and drop it on the stack. Then you'll drive forward to lower it so I can buck on more hay. You won't be driving around—only forward and backward."

Then he relented.

"But you can drive out to the field and back by yourself. I'll take the tractor and Beverley will run the side-delivery-rake. When you get to the stack, we'll see what kind of a truck-driver you are."

Beverley, a seasoned tractor hand at twelve, was already out in the field and liked her job. I was eager to join the crew.

When I got to the stack, Dad connected the loops on each end of the steel cable to the front of the truck and the stacker. I could see I was not going anywhere.

He bucked up a pile of hay to begin the stack and our stacker, Bob, a neighbor, got on it with a pitchfork, moved it around and tromped it down to make a solid stack base.

Dad rode with me the first time I put up a load.

## Afalfa seed

We put up lots of high quality alfalfa so it was normally a big crop during the summer. Usually we cut two crops of hay; sometimes three in an ideal season.

On good years an abundant crop of alfalfa seed was much more profitable than hay. If it looked like the seed would set well, Dad left the second cutting of special fields for seed. Or sometimes he even nurtured the first cutting for seed and cut no hay at all from that field.

Expensive alfalfa seed was an important cash crop, but could be risky. If it failed, we lost the hay crop too, because the plants dried up when left to mature.

The tiny, reddish seeds knocked off the stems easily as they ripened. We never rode or walked through a seed field; but we couldn't keep deer out at night and as they increased in later years, they trampled down more seed than they ate, ruining thousands of dollars worth.

Also, if it hailed before we harvested, the tiny green leaves got battered to the ground along with the tiny, precious seeds.

There went our cash crop and hay crop, too.

*Dad on tractor shoves hay from buckrake onto overshot stacker teeth.*

## Baling Hay

Dad usually drove the tractor for baling. Our family could bale hay alone until it came to stacking the heavy bales. Our baler made bales a heavy eighty to 100 pounds, about four feet long, a foot high and eighteen inches across, pretty heavy for us girls.

Our first baler took two people to operate behind the tractor, sitting on round iron seats, one on each side. Jeanie and Mom, when she had time, rode the baler, pushing wires through a slot, upon which the machine, at the opportune time, grabbed the wires, wound them around the bale and twisted them tight. Two strong parallel wires stretched around the long way. Francie mowed alfalfa and Beverley windrowed it with the side-delivery rake. Anne, too allergic to alfalfa to be in the field, did the necessary inside and livestock jobs.

The baler, hitched to the back of our heavier John Deere tractor, scooped in the hay, rammed it into an endless cube as the tractor driver straddled the windrow. A gigantic knife sliced the hay into rectangles and compressed it into a bale.

An inch-thick iron plate constantly jammed the hay together and released for more hay. It went "bang!" when it shoved the hay together and "ugh!" as it retreated.

"Bang!—ugh!—bang!—ugh!—bang!" it went, all day long.

Beverley worked opposite Mom one day when eight inches of angry rattlesnake mashed into the bale came toward her. Beverley yelled and just before the snake reached her, rolled backward off her baler seat.

Dad stopped the tractor. They killed the mangled snake and dug the rest of its body out of the compressed hay—not an easy task. That bit of excitement kept everyone wide awake all that season.

"Start out slow backwards and gradually speed up. You can see Bob signaling now—he wants the hay up front. Start backing but go slow so it'll drop straight down and slide off the tines. When the stacker reaches its peak height—like now, slowly press on the brake.

"See how the teeth drop the hay straight down? The faster you back up, the farther back on the stack the hay lands. Watch Bob. He'll signal where he wants the hay to land.

"But don't go back too far or you'll break the stacker."

Lesson over, Dad jumped out, got on the tractor and wheeled out across the field for another load.

*So there is a deadly catch to this truck-driving. The stacker might break.*

*I'm committed to a thirty-foot distance, go forward, go back. All day long. Forward and back. Forward and back as the haystack grows.*

The rack of wooden tines loaded with hay rose into the air like a carnival ride between tall masts. When a tall, brown weed came into my left peripheral vision, I made that my signal to brake. I slammed on the brakes and the rack flipped over and dumped the hay down onto the stack.

Meanwhile, Beverley, and later Francie, on the lighter John Deere tractor, drove around the field with the side-delivery rake scooping up dry, three-day-old, now pale green, sweet-smelling alfalfa and expertly rolled it up in lush long windrows that streamed, one after another along the length of the field. It was an art the way she wheeled that tractor around so fast and so precisely.

It looked like fun, but I liked my big, black truck.

The trick for me was to know where Bob wants the hay. He pointed to the front of the stack.

*Okay, he wants it at the front of the stack near the stacker poles. If that's where he wants it that's where he'll get it.*

*Possibly.*

After awhile I learned the procedure for front and middle. Bob was patient.

*I guess he remembers his own kids learning to drive—both boys, now in the military.* As soon as it looked like war, they had signed up.

By noon, I was better at flinging the hay in the right depressions in the stack. After we ate, we girls did the usual things, helped Mom clean up the kitchen and wash dishes while the men took a half-hour break. Then back to work.

By evening, I could second-guess Bob, and knew about where he wanted the next load. Sometimes I got it there and sometimes not—but I was improving.

My worn-out right leg shook steadily, but it still worked. I didn't even care.

*How exciting!*

*Montana Stirrups, Sage and Shenanigans*

From then on, stacking hay was my job.

"Rattlesnake! Rattlesnake around here someplace!"

This time, Dad's good friend, Fred Lahn was on top of the stack, yelling.

"I can smell it," he called down.

Of course he couldn't see the snake's dusty body in all that loose hay. Fred was struck by a rattler in his younger days, almost died and since then, smelled a rattlesnake before he saw it, he told us. The venom left its human victim with an acute sense of rattler-smell. It also caused him some permanent nerve damage.

No wonder we stopped until we found the rattler if someone saw, heard or smelled one.

Everyone searched for the snake at the base of the stack because sometimes they were thrown clear off. But Fred found the snake up there, dug it out and tossed it over the edge for Dad to kill.

It was tangled up with hay and awfully mad, working its fangs and trying to strike at everything—struggling to coil up in the loose hay.

After several years using the overshot stacker, we bought a baler and switched to baled hay.

Baled hay was more concentrated, higher quality and easier to feed to cattle. However we stacked our hay, loose or baled, it all needed to be unstacked again, hauled out and spread to feed our livestock every day throughout the winter.

A remarkable addition to loading bales on the truck was the invention of the bale elevator. As a teenager I drove down the line of bales with the elevator attached to the front of the truck. The bales rode up the elevator chain chute and neatly dropped onto the truck bed. It usually took two men on the truck to keep up with the elevator.

One time three teen boys from the State School in Miles City made up my crew. We kept them working through a scorching hot day when the sun beat down unmercifully on them.

I told them they deserved a break.

"Tomorrow, if you get up at dawn and we start early, you can lie around during the middle of the day in the shade under the cottonwood trees. Then we'll start work again in the afternoon when it cools down and work until dark."

However, it didn't work out that way.

Dawn came cool and cloudy. We were out there working at sunup, but the sun didn't come out at all that day. At noon it was still cool and nice working so we only stopped to

*Stacker arcs into the air as Jeanie backs the truck, out of photo at right, flipping hay onto stack. Hired man pitches the hay toward edges to square up the stack.*

*New baler does not require riders nor hand tying of twine, as did our first one.*

Western Ranch Life in a Forgotten Era

*Hay buyer Joe Erdelt of Terry loads bales with State School boys.*

eat. We worked all the way through until evening, enjoying the unusual cool day. The boys were hard workers and good sports about it.

Another time, two young ranchers' sons, cousins, deferred from the military for ranch work, made up my crew. They liked stacking hay and did it with great gusto because the alternative for them was jail. Caught driving drunk several times, they finally landed in our county jail for a six-week sentence. We were so desperate for help that Dad made a deal with the sheriff. During the week, Monday through Friday, these two cousins lived on our ranch and stacked hay bales. On Friday evening they returned to jail. They stayed behind bars until Sunday night when Dad picked them up and brought them home to work. What a good deal for them and especially for us! They sang and had a good time—the happiest hired men we ever had.

When we brought in a load, I stopped the truck against the bale stack where they wanted it and went to the house to do other things while they unloaded. They were expert at knitting together a stable stack, interlocking alternate layers—something that took practice.

After a time, I heard, "Hey, Boss Lady! Boss Lady! Ready to go, Boss Lady!"

All set to hit the fields again.

Later, when most of us girls left home, Dad contracted with ranchers, first Joe Erdelt, then Matt and Pete Theilen, to bring in their own balers and crews and bale our hay on shares. We raised prime alfalfa and, because these ranchers ran purebred cattle, they wanted the best hay for their cows during the winter.

Francie and I missed ranch work when we went away to college. We came home during summers when we could and helped.

There's nothing like the smell of honey-sweet alfalfa and the satisfaction of putting up our winter's supply of hay.

*Men unload and stack bales near corrals.*

*Hay buyers, the Thielen brothers of Plevna, load four trucks of our prime alfalfa for their cattle. Assistants, Jeanie stands on running board of first truck and Beverley sits on fender of second.*

## Bee City

Looking up at the sky, hearing distant buzzing, Francie, Anne and I watched a dark brown comet stream through the air from the direction of the bee hives toward a big cottonwood tree near the house. Its dark tail zig-zagged behind it—a call to action.

JEANIE

"Dad! Dad! The bees are swarming!"

The swarm landed on the most tip-top branch and formed into a gigantic brown teardrop. The top of the swarm clung to a small twig while the tail petered out to a point several feet below, each bee clinging onto the bodies of the ones above it and buzzing its wings. We ran to the machine shop to get Dad.

We had to act fast. In a few minutes, they'd be gone on the next lap of their flight.

"Quick!" he said. "Tell Beverley to get on her bee clothes! Tell Mom to get two buckets of water and stand under the swarm!"

He jumped into his bee clothes and grabbed an empty beehive he kept in the shop and headed for the tree.

"Go get my rifle and some shells!"

"Mom! Beverley!" We yelled.

Our family went into red alert.

By the time we came with the rifle, Dad had on his bee veil and was lighting the smoker.

The queen bee allows only one queen in her hive—herself.

Whenever a new female bee hatches, the queen bites off her head, if she can. But the young queen is fast and her vengeance simple—she takes half the worker bees with her. She launches into the sky, at the head of a cohesive dark cloud.

Down along the Yellowstone River lots of old cottonwoods with holes in them were just right for a swarm. And that's where they ultimately headed—if we didn't get them first.

While we swung into action, Mom protested.

"Oh, Elmer, we surely don't need another hive! Oh, no!"

But, oh, yes we did! He chuckled.

One more hive, then one more and, yes, another! Pollinating the alfalfa blossoms was what our bees did best. It took many hives to ensure pollination of our alfalfa fields. So we captured our bees when they swarmed, if we saw them go and could follow their flight, and added a new hive in the collection.

Mom's assignment was to throw water on the landing bees to stun them as they fell from the tree. She feared their stings but we didn't have a third bee veil or set of bee clothes for her.

"Stand here, Marie. Right here, under the swarm."

*Dad shows little Anne the wonders—and delightful taste—of a honey-filled frame.*

### Bee hives

Dad and his brothers captured escaped honey bees when they raised alfalfa and cattle on the rich Missouri River bottomlands of their first homestead.

In spring we stacked the hives two supers high. As those frames filled with honey, another super was added, comparable to a story on a house. We painted the outside of the rectangular wooden supers white to decrease the summer heat inside. Our eight or ten white hives grew through the summer, four or five supers high, some shorter, some taller, virtually a bee city. Each super held about nine or ten frames, slid in from the top, each with a golden wax base of slightly raised hex shapes, the beginning cells for the bees to build upon.

From that point thousands of bees flew out across our fields bringing home nectar. They changed it to honey and loaded the honeycombs full. Alfalfa seed was our best cash crop due to the pollination of the bees. Even though we worked with bees, it was rare that anyone got stung. Bee veils, bee garments and the smoker were effective, and the rest of us stood clear.

*Our bee colony of hives ensures that the beautiful field of purple alfalfa blossoms just beyond the ditch will get pollinated for seed.*

"Oh, No! Not there! The bees will land on top of me!"

"No they won't." He reassured her. "You need to be this close to them or you'll miss them with the water."

Mom gingerly edged closer with her water buckets.

We knew this scenario. Mom striving to be brave as she stood there, under the tall cottonwood, with her two buckets of water waiting; Dad holding his fire till she was ready.

Beverley, wearing the other set of bee clothes, stood resolutely nearby. She handled the smoker, which was similar to a small metal watering can with bellows, puffing the bellows gently to keep the wad of burlap smoldering inside.

*That branch is so thin, how can it possibly hold the whole swarm without breaking?*

Dad fired his high-powered rifle once, the branch broke, and the whole swarm of bees came plunging down to earth.

Mom shrieked and threw the first bucket of water on the mass of bees as they landed. She shrieked again and emptied the second bucket, then ran to the house as fast as she could. Mom was the front man, the sacrificial lamb, wearing her short-sleeved dress with no protection at all for her face, arms or legs.

As she threw the water, Beverley worked the smoker, wheezing out clouds of choking, acrid smoke, calming and numbing the bees before they recovered from the water shock.

Dad searched with gloved fingers through the pile of sleepy bees to find the young female, the future queen of all these wild renegades, and slipped her into the hive. She was easy to spot—giant-sized, compared to her workers.

The first bees to recover followed the queen inside and then the whole bunch of bees came to their senses and trouped in with their little guns stashed inside their tails like the dutiful little soldiers they were.

We set the new hive next to the others in Bee City, out behind the barns by the closest alfalfa field and a new working hive was born. Each bee in the new hive instinctively knew its duties and fit into their happy, cooperative community.

Our 'beet house,' down in the flowing-well pasture near the highway morphed into a 'bee house,' at the end of the war, the perfect place for processing honey. We built the small house of railroad ties for Mexican nationals who worked our sugar beets. It stood near a giant cottonwood tree.

The war over, beets were finished on our ranch, too.

This house had only one big room, but a nearby artesian well ran a good stream of soft water, day and night, from its open pipe.

By the hot days of August when the alfalfa and wild sweet

## Alfalfa clover honey

Honey was highly prized during the war and commanded good prices. With sugar rationed, the demand grew and many women learned to cook with it. Honey from alfalfa and the yellow sweet clover, which grows wild in the area, is of the highest quality, and is known as clover honey.

Ours was almost clear, light golden, and of a delicate flavor. We girls enjoyed eating honeycomb, too, the honey still contained in its perfectly formed hexagonal wax cells. We chewed the sweet honey out of the wax, till only a wad of chewing gum remained.

clover finished blooming, the bees had filled their supers with honey. Then it was time to process. We extracted the honey, always leaving plenty intact for winter food for the bees wintering in the hive.

In the middle of the bee house stood the large round extractor, forever going round and round, flinging from the comb the warm honey, the consistency of water, against the walls and us. Once in the extractor, the honey hurled onto the drum walls with such force some of it turned to a lighter-than-air mist that rose up out of the extractor and pervaded the room.

The bee house started out clean—spick and span from our scrubbing each night—tables, chairs, racks, floor and doorknobs scrubbed spotless.

Then, hour by hour, things got stickier and sticker and the honey haze grew heavier and heavier. Even the air tasted sickeningly sweet. The lighting in the room dimmed as the afternoon wore on and the honey haze thickened.

Inexplicably, our mood grew ever more hilarious as we got stickier and stickier and passed the point of misery from the sickly sweetness everywhere.

Our feet stuck to the floor and, when we pulled them up, made crackling sounds. Our clothes felt sticky all over and honey coated everything we touched. We breathed in the too-sweet honey fumes that coated our faces.

Then someone cracked a joke and everyone broke down laughing almost uncontrollably. The worse it got, the funnier it seemed. Even Mom smiled as she worked. Still, I hated going back each morning for the three or four days it took to extract. My sisters dragged me along. At first I tried to keep aloof from all that honey—but it was no use.

By late afternoon we were having great fun again, impossible as it was.

When the day was over, the little house was a disaster and so were we. We walked slower and slower on that sticky floor.

*Yuck!*

It was harder and harder to pick up our feet. Every time we moved, we had to peel our shoes off the floor. None of us dared lean against a wall—we might stick there. The state of the floor and the walls, was the worst part—or maybe the haze

At the end of the day, we cleaned everything spick and span for the following morning. Next afternoon, when we could hardly see through the honey haze and were all stuck to the floor, the cleaning process began all over again.

We distributed our honey among our neighbors and gave it away until one summer Beverley decided to go into the retail bee busi-

*It's time to bring in the honey-filled frames and supers to extract a golden harvest.*

*From a hive stacked five supers high, Beverley lifts out a honey-filled frame, holding a handy smoker to calm angry bees. Note her face veil on the ground.*

*We needed bees for pollinating alfalfa seed, but not all that work of processing honey.*

### Decapping and extracting

The extractor, gas-motor driven, was a large metal drum in the middle of the room.

Totally amazing to us was how the bees built up their perfect little hexagonal beeswax cells onto the base of both sides of each frame, filled them with the precious liquid, then sealed over each cell with a thin layer of wax. Geometrically perfect!

Mom and Beverley decapped by slicing off the thin wax from the top of the honeycombed cells on both sides of each frame with a hot, hollow steam-filled knife connected by a rubber hose to a pot of boiling water on the stove.

We stood the decapped honey-filled frames upright around the inner wire basket, dripping honey. Centrifugal force spun the honey from the swiftly rotating basket and down into the drum bottom, where we removed it by opening the spigot drain. When the bucket filled, we poured it into a five-gallon cream can.

The honeycombed wax base on each frame, swept free of its sweet harvest, remained intact and ready for the next year.

---

ness. She ordered beautifully colored labels and sold enough honey to help finance her senior year in journalism at the university. Of course we all had to pitch in. It definitely wasn't a one-man job. We needed each other for courage.

Beverley came home from the University of Montana every summer and developed into an experienced beekeeper, but soon she graduated from college and left home. Her first professional job? Working for an Apiary magazine in Iowa, writing about bees.

*With bee season coming, what will we do without her?*

We needed lots of bees for pollinating the alfalfa seed and were left with their huge quantities of honey to process.

One blue-sky, sunshiny day in late spring, a remarkable thing happened. A professional beekeeper from Glendive, Mr. McKibbon, came along and saved us. He spotted bee city from the highway that wound down out of the hills above the ranch.

He found Dad and expressed a desire to buy our bees. Halfway into his pitch, Dad stopped him cold.

"They're all yours. Every one. You want 'em—you take 'em. I'm giving them to you free—if you'll agree to leave them here."

"Well yes, I'll need the alfalfa."

"If you want to bring in more of your hives, my fields can use them." Dad added, "At the end of each season, just give us what honey we can use for ourselves."

The beekeeper couldn't believe his good luck. He brought in many of his own hives and set up bee yards at the edges of our alfalfa fields. Bees can pollinate more alfalfa plants and gather more nectar if they live close to the source.

Our hives stayed where they were and, with his added, bees continued to pollinate our alfalfa blossoms as never before. The new owner of our bees processed humungous barrels of honey, endured the stickiness, did all the work—and made sure we had all the honey we could eat and give to our friends and then some.

Now, that was one good deal.

## Rattlesnake Risks

"Girls, come quick! I want you to see this!" Dad called in the back door.

FRANCIE

We ran to where he stood in the middle of the yard, irrigating shovel in hand.

"Come quietly," he cautioned as we looked beyond him in horror.

There a coiled rattlesnake buzzed and only three feet away our favorite black-and-white tomcat faced him, hissing, hair standing on end.

"He's hypnotized the cat," Dad said.

"Don't look in his eyes, he might hypnotize us, too," Beverley warned.

"They've hypnotized each other I think," Mom said softly. "Neither can stop."

We'd heard it from schoolmates that both cats and snakes can hypnotize, but didn't believe it. *Is this what we were seeing?*

The snake and the cat stared at each other, neither breaking its gaze. The snake rattled steadily, flicking his black tongue in and out. The cat stood locked in position, back arched and hair and tail straight up, like a Halloween cat.

"Can't you stop them, Elmer?" Mom didn't move.

I felt I couldn't move either. *Were we being hypnotized? Would the snake attack?*

But it wasn't much of a question.

Dad let us watch awhile, then *Whapp!*

He made short work of that snake with his shovel. As soon as he hit it, the tomcat broke its stare and raced for the barn.

*Rattlesnakes endanger grazing livestock, pets and people, especially children.*

When a rattlesnake strikes, it lunges its head forward, body a spring, mouth open and fangs ready. The fangs are hollow and through those holes it shoots poisonous venom to inoculate the unwary.

As we girls grew older, our folks encouraged us to kill every rattlesnake we saw, since they were a danger to people and livestock alike. The deepest fear is that a rattler will bite a person, especially a child—whose small blood volume makes the amount of injected venom even more deadly.

A dog, excited by a coiled rattler and highly vulnerable to being bit, won't leave it alone. He barks, runs back and forth dodging strikes, alternating retreat and attack.

More commonly, a cow or horse gets struck by a rattler in the grass while grazing, making it painful to eat. She goes off her feed, grows thin and produces less milk than needed for her hungry calf or colt. A saddle horse may end up with a lump jaw, or never seem quite right again. Some animals die from rattlesnake bites, their faces and throats swelled, unable to breathe.

So we killed every rattlesnake we could. We knew if we didn't it might bite a child or a favorite pet another day.

"Always stop when you hear a rattlesnake. Don't move until you

**We killed every rattlesnake we could, knowing they might bite a child or favorite pet someday.**

are sure where it is," Mom told us. "And remember, there may be two and you could step on the other one while getting away from the first."

When town kids visited and we took them hiking, we worried because they often walked carelessly and sometimes ran through tall weeds.

No way could we do that.

*Don't run in long grass or weeds!* The warning voice always played in our heads and brought caution to our steps.

We listened and watched. A rattlesnake warns by rattling before striking, but how could our visitors hear the buzzing or see a snake in time while running? We tried to warn them, but it seemed like they loved to run down our steep gravel hills. We did, too, but knew how and where.

"Stop! Watch for rattlesnakes!" one of us would call. But often they couldn't hear us through the noise of rolling rocks and crunching gravel. Luckily, none were ever bit.

One day when we were about eight and ten, Jeanie and I almost stepped on a coiled rattler in plain sight on our graveled driveway. We didn't hear its warning because we were singing and giggling, being silly, whirling our arms around, tussling and leaning on each other, headed out to get the milk cows.

Suddenly, nearly between our feet we both saw what we hadn't heard—a coiled rattlesnake, angry triangular head reared back, neck kinked, ready to strike.

*Yikes!*

We stopped, each with one foot in mid-air, leaning on each other—and backed up hastily. An instant later we would have stepped on him and surely been struck.

"When you hear a rattlesnake you'll know it," Mom said. "A lot of things sound like a rattler. Insects buzzing. Sometimes a dry weed blowing against a fence. But when it is a rattler, you just know."

Living on the edge of rocky badlands, we had an abundance of rattlers. They denned together for winter in ready-made crevices, rocky outcroppings and small holes in gumbo buttes. We watched carefully while climbing, never reaching to pull ourselves up without first checking if a snake was there waiting to strike.

The snake usually heard us coming before we knew he was there and disappeared, or we were brought up short, hearing his vicious rattle that silenced all other sounds. He whipped his dusty coils into a tight circle—kind of a pad of circles, his flat head aiming for our faces and his rattle pointed up, whizzing the air like the buzz of a

A man working on the stack is ever vulnerable to rattlesnakes thrown up in the hay.

*Suddenly, by our feet writhed an angry rattlesnake, coiled, head rared back, neck kinked, ready to strike.*

thousand bees.

Rattlesnakes crawled into prairie dog holes and the prairie dogs ran out another exit, except babies too young to run. These little ones were a delectable meal for the rattlers who sometimes lived and denned in prairie dog towns.

Their only natural enemies hawks, eagles and bold coyotes, rattlesnakes were our most abundant reptile, and to ranchers, an unnecessary part of God's kingdom.

Sometimes we killed five or six rattlers in an afternoon's ride through the badlands. Anne remembers killing seven rattlesnakes in one day's ride in the rough country of upper Kelly Creek.

We tried to be careful and fortunately we never got bit.

Our method of killing was primitive—we hurled the biggest rocks we could find at the coiled rattlesnake. I was never very good at throwing, so pelted them with lots of rocks till they were thoroughly smashed. We had to be careful while gathering up rocks, as when the snakes first came out of their dens, we might find two or three lying together in the sun along a rocky slope for warmth. They are especially dangerous then since, in a torpor, they might not rattle.

Once when there were no rocks, I pulled off a cowboy boot and threw it repeatedly to kill a snake. When later at home, I told Dad, he reacted fast.

"What! Take off your boot. Right now!"

Inspecting it, he scolded. "Don't ever do that again! I knew a guy got poisoned by a fang stuck in his boot. It broke off, worked its way in and scratched his leg several days later."

With our jackknives we cut off the rattles on each rattlesnake we killed, as did all the cowboys and ranchers we knew, claiming them as our trophies. First, of course, we made sure the snake was dead, fully aware of the danger of it slithering around silently without rattles. Apparently everyone else did, too. In all our years observing rattlesnakes, we never saw a live one with its rattles gone.

With a dull blade, it took a vigorous sawing back and forth of the knife to sever rattles from the tough tail just behind the black button. A bit messy, too—the rattles we dropped in our jar on the window sill often included a bit of ragged flesh.

Then one day Jeanie invented a classier method. Holding the rattle daintily, but firmly, in one hand with thumb and forefinger, and the dead snake's tail in the other, she twisted sharply in opposite directions. The rattle broke off cleanly.

After that, we girls used Jeanie's tweaking method.

"See any rattlesnakes today?" Dad asked each time we came in from riding. Then he wanted to see the rattles to figure how old the snake was. If we didn't kill it he wanted to know why.

*One day Jeanie invented a classier method to remove rattles— by tweaking them off.*

"That's one snake left to bite you. Or bite your horse," he warned.

We tried our best, but sometimes a rattler slithered away into a crevice when we bombarded it with stones.

Dad wanted us to know our adversaries better, so one day he cut a rattler's head apart to show us the fangs. Peeling the jaw bone and flesh away, deep inside the head, we saw the yellowish venom encased in a translucent oval sac, and the visibly hollow fangs, looking like two hypodermic needles, curved to kill. When he cut the sac open, thick dull-yellow mucous dripped out.

ANNE

After this lesson, we were impressed—and very wary.

City visitors to our home seemed a little appalled at our casual treatment of rattles. Most times there would be one lying on the end of the table and several on a window sill. This was true of all ranch homes, not just ours.

One afternoon when Mom entertained her Ladies Aid group from church, a lady had mini-hysterics over a couple of rattlesnake rattles lying around. After that, Mom insisted we deposit our trophies in a little jar on the window sill.

We gave them away willingly when, sometimes, visitors wanted to keep a rattle or two. We knew we'd soon add more to the trophy jar.

Rattlesnakes were sometimes a topic of discussion when ranchers got together. Out of a shirt pocket they brought out a most recent acquisition and debated the snake's age based on number of segments. Having six or eight interlocking rattle segments, or buttons, was common and we found as many as fifteen on one rattle. On others the end was broken off.

Every time a snake sheds its skin, a new segment of the fingernail-like material grows on the rattle's end. We thought this happened once a year, in August, but apparently they sometimes shed more often, especially during a growth spurt. The rattle segments interlock loosely, so shaking produces the loud rattling sound. We tried but could never shake them rapidly enough to mimic the real thing.

Uncle Will told us about one fall when he came upon a large number of rattlesnakes up a certain draw while herding sheep in rugged country near Jordan. They hung around rock crevasses, crawled on sandy surfaces and warmed themselves in the sun. He thought he could kill them all if each day he walked a little farther into the draw and killed a dozen or more.

FRANCIE

"Evidently they had a den and were looking to hole up there for winter. Taking their time, because it was a nice sunny fall and they

*After killing a rattler with rocks, we tweaked off the rattles by grasping the black button and twisting sharply.*

**Nightmares of rattlers slithering over his body, coiling and striking— woke Uncle Will every night.**

warmed themselves on the rocks," he said.

So every day he walked higher up the draw, killing rattlers on every side. But the farther up he got, the more they seemed to surround him. The whole draw and sidehills came alive with rattlesnakes.

Uncle Will began dreaming of rattlesnakes at night. In his nightmares they crawled over the sheep, his dog—and himself—slithering, coiling and striking. He got little sleep.

"I never got to the end of that draw. Every day I killed more rattlers, but still they were all around." He shook his head as if to shake off a memory still too fresh and horrifying.

"I couldn't stand it. Finally I got so spooked I had to move the sheep clean out of there."

Such stories reminded us of our obligation to get rid of rattlesnakes, and we tried our best to reduce the number lying in wait for us, our pets and our livestock.

Jeanie once came upon many rattlesnakes lying on the highway as she drove her Rainbow Club friends from Miles City on a hayride in the back of our truck.

"We went swimming in a reservoir near the Pine Hills and, after picnicking around a campfire, I was bringing them back to the ranch. This was in the fall and we were having warm days but cool nights. When we got onto the highway, rattlesnakes had crawled up there warming themselves on the blacktop.

"Each time we came to one, I stopped to kill it. I'd run over it, then back up and do it again, making sure it was dead. In the light of the headlights, I grabbed the tail, tweaked off the rattles, threw the snake in the barrow pit and got back in the truck. A little farther along there'd be another one.

"The girls squealed every time I killed a snake and grabbed the tail to tweak off the rattles. They thought I was brave to do that. But they weren't at all interested when I explained why—that I killed those snakes to save someone getting bit someday.

"I killed eleven snakes between where we got on the highway and home."

Rattlesnakes ate the eggs of our lovely meadowlarks, fluffy killdeer chicks, pheasant babies and other ground-nesting birds, prairie dogs, gophers and mice. So did bullsnakes, but they were considered beneficial and we never killed them, since they ate mice around the barns.

One time, though, Anne and I caught a big bullsnake in a particularly heinous crime. Walking along the fence row next to an alfalfa field, we were startled when a hen pheasant ran out underfoot and stopped a short distance away. Dropping her wing, she stood

## Singing rattles

As a young man, Dad spent many lonely hours alone across eastern Montana hunting and trapping, freighting eighteen-horse jerk-line teams for Maggie Allen, riding for cattle between line camps and isolated ranches from Jordan to Glasgow, and running wild horses out of the Missouri River Breaks.

He told us that when cowboys there got together they passed the evenings telling stories and playing pranks on each other. One of his favorites concerned a 'tenderfoot' to their ranks from one of the big ranching outfits.

"He crawled into his bedroll bone tired," Dad said. "After he began snoring, the rest of us around the campfire devised a plan. We emptied our pockets of rattlesnake rattles and strung them together on strings.

"Then we whirled our strings, like this. Like you spin a button, pulling it in and out so it sings." He held up his hands, moving them rhythmically closer and farther apart.

"Those rattles sang all right. You never heard such rattling!" he chuckled.

"That cowboy woke up, lying perfectly still. Only his wide open eyes moved as they rolled from side to side trying to figure out where the rattlesnakes were. He couldn't see us as we were outside the light of the campfire.

"We kept whirling the strings, until we could see the man was good and scared. Then we snuck off to our bedrolls for the night.

"At daylight next morning, we crawled out of our bedrolls. That cowboy's eyes were wide and looking all around as he lay perfectly still.

*(continued)*

> **Rattles** *(cont'd)*
> Seeing us he spoke for the first time.
> "'Boys,' he said, 'walk mighty careful. We built our fire in the middle of a rattlesnake nest. They're all around us. I stayed awake all night and didn't dare move.'
> "None of us said a word or cracked a smile," Dad chuckled, "I bet he's still telling about the night he slept on top of a rattlesnake den."

*To our horror, the bullsnake held a yeeping yellow chick in his mouth—and four lumps bulged his body.*

anxiously, trying to entice us to follow.

"Must have a nest," said Anne.

We parted the alfalfa plants and, appalled, saw why the mother pheasant was so distressed.

To our horror a huge bullsnake had invaded her nest. In his mouth he held a yeeping yellow chick. Five or six fuzzy yellow chicks, with black stripes, shrank back in the nest, too young and scared to run away. Four lumps along the length of the snake's body indicated what could only be the rest of his dinner.

Furious, and determined to save the helpless chick if possible, I grabbed the bullsnake by the tail and snapped him hard. The chick popped out of his mouth and ran yeeping into the alfalfa.

I snapped him again for good measure and to my astonishment, out popped another pheasant chick—alive. Like the first, it ran yeeping away. A third chick fell out next. Wet and in disarray, it staggered off, barely able to walk. The fourth chick fell out, dead.

"Get out of here! And don't come back." I hurled the bullsnake as far as I could throw and the surprised creature whipped off in another direction.

Three baby pheasants saved.

At first glance, the bullsnake looks much like a rattler, with similar markings, but is usually larger. If moving, sometimes the only way we could tell for sure, was that the tail of a bullsnake tapers off into a sharp point, while the rattler tail ends bluntly—a stub tail with rattles. Even when driving past a snake stretched out on the highway, this difference is clearly evident.

Also, our prairie rattler whips defensively into a coil when disturbed, ready to fight, while bullsnakes slither away. But not always. The bullsnake can make himself terribly frightening, hissing and making a ferocious buzz, sounding somewhat like a rattler.

When a bullsnake bites, as Anne found out while working on her saddle in an unused barn at the Brandentaylor place, it leaves two tiny fang marks like a rattlesnake, which can be painful but not poisonous.

An old cowboy saying is that 'a rattlesnake doesn't die until sundown.'

Jeanie and I investigated that idea one day. While riding up on the open range looking for a lost cow, we killed a rattlesnake. And then nearby, another one. Jeanie tweaked off the rattles.

Both were most certainly dead, heads and backs thoroughly bloodied and smashed with rocks. Still ….

"What if they really *don't* die before sunset?" I asked. "Maybe they'll crawl away."

"They can't crawl away if we hang them on a barb on the fence, can they?"

" Okay, let's do that."

" Then we'll ride back this way coming home and see if they're still here."

Two or three hours later we brought our cow past that barbed wire fence looking for our snakes.

Both were gone.

"That doesn't prove they crawled away," Jeanie said. "Maybe a coyote came by and ate them. Or eagles or hawks swooped down and carried them away to their nests.

"Yeah. I've seen hawks flying with a snake that way. But…"

" Well, anyway, I've got the rattles as proof," Jeanie said with a grin, patting her shirt pocket. "Even if there's a rattler lurking around without any rattles."

After all, neither the rattles in hand nor the predator theory proved our case. *Impossible, of course, that they revived and slithered off the fence. But what if they did and are now crawling around silently?*

After that I made doubly sure every snake I killed was dead, dead, dead.

## Tractor Driving Beats Milking

Eyeing Beverley where she cultivated corn three fields away, I turned my tractor down the next round, mowing hay.

*Will she stop at the end of that row and head for home?*

One of us had to go in to help Jeanie with the milking. It was almost five o'clock, milking time—the summer sun still high in the sky.

Beverley didn't want to go. Neither did I. No doubt she was also eyeing me, wondering if I'd give in soon. Each of us trying to outlast the other.

She was so responsible and dependable that I thought I could out stay her again. Besides, it was her turn.

Certainly driving tractor was more enjoyable than milking cows. Milking, one of those endless jobs that took forever and just had to be repeated again the next day—twice. Even then, separating and washing up awaited.

Each time I came down that side of the field, I looked across the alkali bog where nothing grew, that Dad was trying to bring back to life, debating what she'd do next. Was she going in?

*No. She made the turn at end of the field and started down another set of rows.*

*After that I made doubly sure every snake I killed was dead, dead, dead.*

*Beverley drove tractor at twelve and became so expert Dad needed her all summer, even after the war.*

She wanted to stay out and finish that field.

*But no way can she finish.* Four rows of corn at a time demanded close concentration and slow, careful driving.

Cultivating required us to till precisely, taking out all weeds between the rows and up close to the young plants. If we were off just a few inches to left or right we could take out four rows of corn, leaving four rows of weeds in their place.

I hadn't yet learned how to do it well, but Beverley was good at it.

Probably she took out some corn rows at first, but soon mastered the technique. It helped that Dad didn't scold us when things went wrong, but pointed us toward doing a better job next time. Sometimes he laughed in a sympathetic way. Not that he didn't have a temper, but if he lost it, he usually told off a cow on the fight or uncooperative machinery instead of us.

*I should go in. Beverley's not going to go. I'll make one more round first. Maybe two…*

Driving tractor was a mostly pleasant and peaceful job, in the midst of our beautiful green fields framed by big shady cottonwoods, breathing in the fresh Montana air with a light breeze and the invigorating smell of freshly turned earth or new mown hay.

Still, it could have been boring, sitting there on the tractor hour after hour—if I hadn't escaped into a storytelling world all my own. Daydreaming, I wrote long stories in my head, continued from one day to the next. Compelling tales of adventure, romance and intrigue, they entertained me while driving up and down the field.

Maybe Beverley did the same, except I knew she wrote down her stories—because she hid them around her room. Periodically when she was gone, Jeanie and I searched her room and read the latest story. Invariably, though, her stories stopped abruptly after solving an exciting crisis and then disappeared for good without any ending, much to our disappointment.

*At twelve, Beverley began raking hay, the easiest tractor job. She soon moved on to mowing and finally cultivating corn, the most precise.*

Beverley began driving tractor at age twelve, beginning with the easiest job—raking hay. Swiftly she advanced to mowing, bucking hay, disking, harrowing, spraying and whatever else was needed, even cultivating corn.

During the war when help was short, she took the place of a hired man and by the time the war was over, she was so expert on the tractor Dad couldn't get through summer without her.

I followed in her footsteps—running the tractor with side-delivery rake by the time I was twelve. At the end of a field I had fun making a quick pivot, setting the brake of one of the big rear wheels so it stood still while the other turned the tractor 180 degrees down the next row. Spectacular, with the long rake on the back. I probably wasn't supposed to do that, but Dad never said anything.

Not everyone agreed twelve was old enough to drive tractor.

One day that first summer, a neighbor Teddy Wasso came to help. A short, stocky, uncompromising Bulgarian, he was a hard worker, tough but good-hearted.

When I drove into the field with the rake, he scowled.

"She's too young for that," he told Dad.

"Francie? Oh, she'll do fine. She's raked hay before."

"No, I won't have it! I'm not going to stay here and watch her driving that tractor. Either she goes or I go!"

Dad needed him that day, so he sent me home. I felt embarrassed, but happy to have an extra day for my 4-H room improvement project. Besides I was still healing up from my kidney surgery following the horse accident and knew Mr. Wasso was concerned about that.

Not all ranch girls worked outside like we did. If they had brothers, the boys worked side by side with the dad, while the girls got stuck helping mom with the drudgery work of cooking, cleaning and laundry—jobs that needed to be repeated endlessly.

*Lucky us!* We counted ourselves fortunate to be free of such stereotypes. Working outside was often more interesting, we thought and, without brothers, all four of us worked both outside and inside the house. Because of her weed allergies, Anne did better with the cattle than in the field. Beverley simply worked through her allergies and toughed it out.

Driving our tractors could never be as simple as turning the key and taking off for the field.

First, we greased-up. Beverley insisted this be done right. Hunting up all the grease zerks—hid in unlikely places and covered with oily black dirt—was a game for Anne and me when too young to drive. We pumped in the heavy dark lubricant with the grease gun till it oozed out the edges and then helped fuel up.

Beverley taught me how to start our John Deere tractors. We gripped and spun the big fly wheel into fast rotations until the engine took off, keeping our fingers free of entanglement, while at the same time quickly adjusting the throttle and other controls to keep it running.

Not easy—for me it was a real challenge. And maybe dangerous.

Once started, I did everything possible to avoid killing the engine until I came back in from the field. I didn't want to go through all that again.

Alfalfa grew thick and lush in our irrigated fields. Dad believed

*Beverley and Francie attach the sprayer to tractor and head to the fields to spray grasshoppers.*

*The power take-off shaft spun an unshielded, double-action steel knuckle to grab a careless person.*

in fertilizer and it showed in our heavy alfalfa crop each year. Once he took the tail-end of a phosphate sack and broadcast the leftover super phosphate in the shape of a giant question mark across an alfalfa field near the highway. His own experiment in the value of fertilizer.

That big question mark showed up six inches higher than the rest of the alfalfa plants for many years, even though the entire field was fertilized annually.

*Francie watches for hidden pheasant nests, mowing the alfalfa that grows heavy and lush in our irrigated fields.*

Often the alfalfa grew too thick for our equipment and the heavy foliage plugged up the sickle and blade end. Periodically we had to stop and pull it out—taking care to first shut off the power take-off leading from tractor to mower.

The power take-off, an unshielded double-action, fast-whirling steel knuckle and its spinning shaft, could suddenly grab and entangle a finger or pants leg of the careless person, whipping it around and around. Dad warned us of farmers killed this way.

The sickle, too, could be deadly, scissoring rapidly back and forth. We took great pains to shut it off before cleaning out obstructions.

That sickle also endangered small animals and pheasants hiding in the thick green foliage. Unfortunately, pheasants nest in June at the same time we began mowing hay.

Beverley and I made one circle around a new field taking a six-foot cut. If a pheasant hen flew out it meant her nest was close. We'd stop, find the nest, count eggs and mow carefully around it on the next round, leaving a triangle of alfalfa uncut. Sometimes I walked through and chased hens out ahead of mowing. If I destroyed a nest, at least the mother would lay more eggs.

We needed to protect those beautiful iridescent red and gold Chinese pheasants and cute little chicks that hatched after three weeks of brooding. We all watched for them when we mowed.

If the eggs had already hatched, bouncing yellow black-striped chicks skittered off in all directions, as many as a dozen or more. By the next round when we came by—pretending not to see her, avoiding eye contact—the mother had them all tucked in the nest again under her protective wings. She watched us anxiously.

Sadly, a couple of times I did mow over a nest and, worse, once cut the legs off a pheasant hen, killing her.

I looked across the alkali patch at Beverley again, mowing down that side, reluctant to leave my storytelling to go milk cows, guessing I'd have to go.

*She's not going in, is she?*

**Whew! Now I could stay in the field until she and Jeanie finished milking.**

*Yes!*

Beverley stopped at the end of the corn row, lifted the cultivator shares out of ground, and there she went, bouncing across the rough field road toward home. Over the 'Putt-putt-putt' of the tractor I could hear her singing.

I waved and she waved back.

Now with a clear conscience I could stay in the field until she and Jeanie finished milking.

*Whew! Off the hook for another day.*

### Rat Trap Harvest

Each fall we filled the granary with grain—wheat, oats and barley and ground feed. The granary was built tight, so there were few mice; those that intruded were quickly caught by one of our barn cats.

And certainly there were no rats—except…

One day Dad found the large droppings of a rat, and rats can destroy lots of grain. Pack rats are many times bigger than a mouse, with a bushy tail—too big for our cats.

That rat, or rats, has gotta go…

"Beverley, I'll pay you for each rat you catch in the granary," Dad offered.

He purchased a large rat trap similar to a mouse trap, only many times bigger. It must have been eight or ten inches long, and with its heavy metal spring, could have caught something much larger than a rat.

*Ouch…*

The granary was a large rectangle building with four big rooms we called bins, a central hallway, and doors to each of the bins off the hall. Planks slid into the open doorways kept the grain in at the right level, and still allowed us to get to it without spilling. As the grain was used, we'd pull out a plank—a simple way to dip a full bucket without any grain spilling out the open doorway.

Night came, and Beverley went into the granary and set her trap.

Carefully, she stepped into the first bin where the grain was only several feet deep. Pulling the wire spring back, she cautiously hooked the metal trip into the spring. The bait plate she had already loaded with a tasty bit of cheese. Hard to entice a rat to try her bait when it was surrounded by bushels and bushels of thrashed wheat, but she hoped the unusual smell of cheese among all the grain might do it.

The next morning was Sunday, and Beverley, who was in high

*"Beverley, I'll pay you for each rat you catch in the granary."*

school at the time, had to drive into town very early.

"I'll milk for you tonight if you'll take care of my rat trap tomorrow morning," Bev told Jeanie. "If I catch a rat you've got to remove it and then reset the trap ready to go again."

Jeanie looked a little dubious, but the offer of milking decided it.

Saturday night Bev set her trap for the first time, but she reckoned without the hired man. The hired man at the time was a thin little man, a Mexican, by the name of Jessie.

Like most of our hired men, Jesse liked to go to town on Saturday night and celebrate in the bars.

Jessie had an awful craving for drink. He had gone without all week, so when Saturday night came, he thumbed a ride into town and headed straight to Miles City's bars. There he found his tequila or whatever brand of alcohol he loved.

Fact is, he found quite a bit of tequila.

Sometime in the wee hours of Sunday morning, he caught a ride back to the ranch, and by now drowsiness was setting in. Not wanting to disturb the household by chance of waking us up at that hour, he decided to sleep in the granary. It would be warm in there, and very comfortable lying on a soft bed of grain.

Jeanie, in due time Sunday morning, went to check Bev's trap as promised. Silently she peered into the first room where Beverley said she set her trap.

Sprawled on the wheat was—a man.

*A man? Asleep in the granary?*

She sneaked nearer; sure enough, there was a man, and upon closer look, she discovered it was the hired man Jessie.

*Where was the rat trap?*

Jeanie quietly looked around the bin.

No rat trap.

As she brushed the side of the bin door, a board leaning there fell right across Jessie's shoe. The sudden falling board disturbed Jessie from his drunken stupor.

As he turned, there was a loud snap.

"Yee-ow-ww!" Jessie let out an awful yowl, and rubbed his hand down across his rear end.

Clinging to his hip, Jeanie saw the rat trap. It had Jessie caught firmly in its spring jaws. Jeanie ran.

Biggest rat Beverley ever caught.

> *Jeanie saw the rat trap clinging to his hip, Jessie caught firmly in its spring jaws.*

# Blowing the Spring

"Damn!" Dad's seldom used expletive split the air. He stared in dismay at the water trickling into the watering tank, then walked in quick strides to the house.

"Marie" he called to Mom, "I've just *got* to do it. That stock tank's been running all night and it's barely full." Then turning on the kitchen sink water he said, "Look! There's very little water here either."

"Oh, Elmer," Mom sadly responded, "I wish you didn't have to use dynamite!"

But as they discussed it, there seemed to be no other way. The spring filled in every few weeks and although Dad cleaned it out bringing surging water, it plugged up again and the water went deep underground. The only remedy? Blow a hole wide and deep enough to contain its large flow.

Last winter, the tremendous gush filled the tanks for watering our livestock plus our calves in three feed lots. Even when the ground froze, this deep spring provided more water than we needed for the livestock and our house water. Usually at this time the springs and ground water ebb, yet with all the cattle at home it was crucial to have a dependable water supply.

"I'll go to town and get some dynamite," Dad said. "If I don't fix that spring before winter sets in we won't have water for the livestock. I've got to blow that spring now."

We all wanted to ride to town and watch while he bought the dynamite, the closest thing we could get to fireworks during wartime. But he didn't want anyone in the car with him if something went wrong. We could see Mom's concern about the risks, so we worried too.

When Dad returned, he showed us the long, yellowish dynamite sticks, sealed over with wax. Their short string like fuses came out one end. He had an extra roll of fuses to tie to the end of each stick. He said this longer fuse length allowed him to get a safe distance from the dynamite before lighting.

We couldn't believe our good fortune when he agreed to let us girls go with him and watch the blowing of the spring. Earlier that summer we watched him clean out our spring, carefully digging out the mud and roots, then covering it over with metal rails and boards sheeted with tin. Nothing could get in from above. But occasionally the sides caved in or tree roots grew into the water pipe, and Dad would have to go up and clean it out again. There seemed to be unlimited amounts of water—until there came another cave in.

Mom's warning of "Girls be careful!" was a bit chilling. We

Two corrals opened onto this main water tank, with water piped to a third pen below. Often in winter all three pens filled with calves or lambs.

*When Dad returned, he showed us the long, yellow dynamite sticks, sealed over with wax.*

Western Ranch Life in a Forgotten Era

climbed in the back of the truck for a fresh air ride up the hill to the spring.

We loved to explore here. Crowned with rimrocks of sandstone, these hills endlessly fascinated us. From these high hills, the towering cottonwoods looked like little bushes shading the miniature ranch buildings below. The green fields stretched forever. The Yellowstone River curved its way alongside, reflecting silver in the sun almost to Miles City. Its dangerous, fast-flowing waters hidden by greenery and lost to reflection in the distance.

Coming from the east, Highway 10 was a winding road dropping down out of the higher hills into the lower Yellowstone Valley, our valley. To people traveling that road across the high drylands on a hot summer day, our ranch must have looked like a verdant oasis, or a safe refuge in winter.

Who knows what fascinating bits of history these rims had seen in the past several hundred years? It didn't take much to visualize Indians running buffalo over the sandstone cliffs or early pioneers winding their horse-drawn wagons through gaps between towering rocks. All this within a half mile of our house. No wonder as kids, we spent much of our free time scouring these hills.

A large round-top hill circled with rims ended this rocky chain. Here the rimrock boulders were so close together we couldn't climb them. Nooks and crannies among these rocks made excellent hiding places for rattlesnakes that lived there, waiting to catch an unwary mouse—or my fingers reaching for a hand hold.

Separating this hill from the next was a narrow coulee winding from the prairie dog flat below to the top of the rims.

It was here among the sandstone rocks that we spent many hours as kids. Our dependable spring surfaced here in a grove of cedars, chokecherries, box elder trees and wild roses.

As winter turned to summer, magpies built their round stick nests in these thickets of trees. Each year they added another pile of sticks until the tree seemed overbalanced. We spent many hours watching them bring food to their young and teaching them to fly. From watching them my older sisters were inspired to catch a baby magpie, our beloved Kaw-Kaw, and bring him home to live with us. Kaw-Kaw lived in an upstairs room all that summer begging food tidbits, learning to fly and teasing us with his raucous cry. We thought he might learn to speak, too. But one day he glided out the window, tumbled into a tailspin and our dog Jack, ever on the alert, seized him and broke his back.

We called this area of the ranch Alabama Bill's. When our parents first bought the ranch, it came with residents Bill O'Conner and his wife SugarBabe. I suppose she had another name that we never

*We visualized Indian hunters running buffalo over the cliffs and pioneers winding their wagons through the rocks.*

knew—Bill called her SugarBabe and so did we. They were African-Americans who came West to avoid Southern race prejudice. Their house was once a homesteader's shack along with seven acres that they purchased from a disillusioned owner.

As the years went by, they built on rooms and porches, until it became a rambling low building almost leaning against the rimrocks, looking somewhat like an extension of the rocks themselves. Springs came out below the sandstones, so they had a plentiful source of water. They even dug a small garden, so this was probably a great way to live during the depression years of the late twenties and thirties. And the view was worth a million.

Highway 10 was one of only two U.S. highways that ran east and west across the state then, both paved. It ran below their house through our ranch and was an easy way for SugarBabe and Bill to get to Miles City if they needed to buy something—and if they had the money. In the thirties money was tight for everyone. Even though it was the main highway across Montana, from Chicago to the west coast, in those days with gas rationing there was very little traffic, mostly local people. Everyone knew this couple and stopped to give them a ride.

SugarBabe was a large woman, tall and about the heaviest woman we knew, but strong. She also was cranky and constantly scolded her husband when she could catch him. Alabama Bill, who was a small thin man, seemed to avoid her whenever possible. He liked Dad, who in turn enjoyed Bill's happy outlook.

They had an old gray horse and a milk cow that our parents let them pasture on our place. Bill used the horse to 'cultivate' their little garden each spring. He hooked his horse to an old set of bed springs, sweet talked SugarBabe into sitting in the middle of it for weight, and pulled it across the garden. Surprisingly, it did a reasonable job of pulling up weeds and breaking up the soil for planting.

One time Dad came across them working their garden, and he asked Alabama Bill if that wasn't awfully hard on SugarBabe bouncing on those rough bed springs. He suggested that Bill and his wife might change places.

"Sho," answered Bill, "But I's got to lead de hawse." Then he came close to Dad's ear and whispered, "SugarBabe, she weigh so much, she dig a good furrow!"

As the depression lifted, Bill found work in a car dealership in Miles City, and he and SugarBabe stayed overnight in town more and more. Finally they abandoned their home on the rimrocks and sold their few acres to Mom and Dad. It was a fun and spooky place for us to take friends and kids who came with their folks on business. If they were interesting and nice, we let them explore Alabama

*Francie climbs in the rocks and a twisted cedar tree in the hills near our enclosed, ever-flowing spring.*

Bill's tumbledown house. If they were smarty types, we made up spooky tales, and dared them to go in.

The sandstone rimrocks at Alabama Bill's appealed to our adventurous nature. Summer days can be very hot and the cool fresh water springs, surrounded by trees and tall brambles often had a breeze that blew down the rimrocks above and made a cool haven. Birds flitted among the cedar, chokecherry and buffalo berry trees. Every so often a bright yellow warbler or goldfinch flashed out between bushes and the rat-a-tat-tat of a busy woodpecker echoed among the rocks.

That summer day we kept a wary eye on Dad's progress as he dug holes and prepared to blow the spring. We loved exploring in these rocks.

I watched Beverley, Jeanie and Francie carving into a sandstone boulder. It was fun carving sandstone as it quickly etched away. We'd find a harder rock with a point for carving, such as granite. Once in awhile we found a chipped stone which we imagined was sharpened and used for a knife by Indians long ago. There were occasional arrowheads and lots of moss agates.

Others carved in these rocks before us. One of our neighbors, 'Grandpa' Hill, said the original military wagon road, the Buford trail, ran right through here beside our ranch spring and through the break in the rimrocks. According to him, this old road was the only wagon road for years, convenient to water the horses at our spring as they pulled up the hill going east. The first water for several miles for those crossing the dry land going west.

Some of these travelers carved names and dates into the soft rock. There was even a large face carved on a sandstone point. Many times I wished I could have been there with those silent rocks watching as the early people came and sheltered among these massive sandstones.

By now my sisters had finished carving and were exploring the rocks and occasionally exclaiming over colors in agates they found. I caught sight of two golden eagles soaring high above the rims. Forgetting about hiding from the dynamite, I watched the eagles, following them along the hill to see if they had a nest, wondering what it would be like to be so high above the ground. They were effortlessly circling, not even moving their wings.

Then I heard Dad yell, "Girls are you safely away? I'm ready to blow!"

Too late! I looked frantically for a place to hide. Just to my left I saw a small, dark cave under the rims. I ducked in. The floor was finely ground sandstone, the roof so low I crawled on my knees.

Then as my eyes adjusted to the dark, I noticed a dark shadow

> *Grandpa Hill said soldiers on the Buford military trail watered their horses here at our spring.*

against one side of the cave. It first looked like another rock. To my horror the shadow moved. Or did it? Yes, it moved again.

Yikes! A porcupine lumbered toward me, its nose twitching and long back quills standing straight up in a ruff—its tail readied to swat. In a flash I remembered a dog some town friends brought out when they visited. How painful when the owners pulled quills out of its muzzle, and how pitifully the dog whined and cried. Was that to be my fate? I was afraid of the dynamite, but more afraid of the porcupine.

Quickly I decided to chance the rocks and mud flung up from the dynamite, and crawled out, abandoning the cave to the porcupine.

Out in the open—Boom!

Mud, small rocks and water hurled skyward. The sound deafened my ears on that quiet hillside and the concussion blew by in a wave. There was nowhere to hide as the blobs of muddy debris flung themselves toward me. I ducked quickly into some scratchy buck brush, thankful no mud or rocks hit me and the globs of mud snagged on the bushes above. Since I often came home from the hills covered with scratches I knew no one would notice what the bush did to me, and I wouldn't have to tell that I had been distracted by the eagles instead of safely hiding.

"Okay girls, you can come out now."

Dad gave the all clear and I ran down to see what the dynamite had done. He seemed pleased with the results, and waded into the mucky mud with a shovel to clean out the small debris. He spent a couple of hours building a bigger holding tank, while I ran with my sisters among the sandstone rimrocks.

This gravity flow spring remained an amazing advantage our ranch had over other ranches and farms, where well drilling was necessary to get water. Pipes, six feet underground to avoid freezing, brought our water down the hill. Our spring continued year after year to give us all the fresh water we could use for the livestock, house, lawn and Mom's lilacs and profuse yellow roses.

When Dad finished cleaning and enlarging the holding tank, we climbed in the back of the truck for a quick ride home. Once there, we jumped out of the truck and ran to the big watering tank. Sure enough, water gushed out the faucet and spilled over the end of the full tank.

We had enjoyed a full afternoon—exciting to be close to the dynamite and watch the mud fly, lots of sandstone boulders and rimrocks explored, and no one asked about my many brush scratches. I decided to keep quiet about my adventure with the porcupine. And best of all, I had no porcupine quills to dig out.

The spring from the hill furnished all the water needed for our house and main water tank, even in winter when all our cattle and horses drank from it.

*Dad yelled, "Girls are you out of range? I'm ready to blow!" I wasn't.*

# 11 Neighbors

## Matchmaking Horseplay

The extreme shyness of our bachelor neighbors and their seeming enjoyment of our pranks, challenged us girls to invent new and complex twists to enliven their days.

ANNE

The Winninghams were our closest neighbors and often shared machinery as well as farming information with Dad. The two young men, Elmer and Louie, in their mid-twenties, lived with their mother and uncle, after the dad died of cancer. Their mother hadn't been to Miles City for ten years. She only left home twice in my memory: once when the Yellowstone River flooded their house; the second time when we girls prevailed on her to attend our rural school Christmas program.

Mrs. Winningham was a small grandmotherly type, an inveterate talker, and fed us cookies when we visited, while Dad talked ranching with the men. She must have been a lonely woman the way she opened up when neighbors stopped by, but never did she go to town. She shopped mostly from the Sears, Roebuck and Montgomery Ward catalogs, while the men did the grocery shopping.

*She must have been lonely, the way she visited when neighbors stopped by, but she never went to town.*

The Winningham young men were very nice and it seemed a shame they weren't getting married—bringing wives and children into the community for us girls to enjoy. In fact, they didn't seem to pursue any young women, as they were very shy, especially Elmer.

We invited unmarried women to dinner when they worked with Dad and would be eating with us. Unfortunately, they just blushed and slipped out to the machine shed as soon as possible.

One day, teasing them, we told Elmer and Louie they just weren't 'sweet' enough.

This gave us an idea. So for a month or more, when they came over, we placed three candy suckers on the seat of their pickup (including one for Uncle Charlie). Each time we could arrange it, the suckers grew bigger and bigger, until we were making peanut brittle candy in large frying pans. The last candy sucker we made in a washtub and added three dowels for handles.

When they still didn't find sweethearts, Jeanie came up with another idea. One day while cleaning out a closet and tossing clothes into a pile, she noticed several landed in such a way that they looked exactly like a woman. The blouse fell as if tucked into the skirt, with a pair of shoes just below.

"Aha!" she said. "What do you suppose we can do with this?"

The closet cleaning immediately stopped. Francie and I saw the possibilities of making this into a beautiful woman, but exactly what did Jeanie have in mind? She didn't keep us waiting long.

"Wouldn't it be funny," she said, "to make a gorgeous lady and put her in the Winninghams' pickup the next time they come over?"

Francie and I burst into giggles. We imagined them getting into their pickup and finding a strange maiden waiting.

We found more discards and went to work sewing the garments together. Francie stuffed rags inside the blouse and skirt in all the strategic places. What to do for a head? She sewed together a pair of old flesh-colored panties stuffed with rags, and painted on eyes, eyebrows, nose and mouth. Next we found yellowish-brown yarn and straggled that around the face, topping it off with an old flowered hat sent by an aunt.

A marvelous-looking lady indeed!

We watched expectantly the next few days until the Winninghams' pickup appeared in our yard. Elmer was inside, the shyest of the Winninghams. He didn't know what was in store for him! We giggled with glee—the time had finally arrived.

He parked the pickup as usual between the house and shop and went down to see Dad. Perfect. There, no one would see as we carried out our plan. Our ragdoll was a beauty, though a little floppy, but Jeanie was able to prop her in the middle of the front seat. One arm lay enticingly over the back of the driver's seat, her head drooping a little to one side, with the hat set at a charming angle. Her hair did look a little straggly, but not everyone carried a comb.

We ran back to the house and watched.

Elmer approached his pickup.

"He's coming!" shouted Francie to the rest of us.

We all dashed to an upstairs window to see the fun.

He stopped just a few steps from the pickup, turned around and went back into the shop.

"He saw her!" Francie exclaimed. "He's too scared to go meet her!"

We all giggled.

After a long wait he appeared again, this time walking slowly and

*Francie and I burst into giggles. We imagined them finding a strange maiden in their pickup.*

tentatively toward his pickup. What would he do next?

We crowded around the window, straining to see. Elmer slowly pulled open the door and gently tipped his hat. He stepped back a moment, then jumped in the pickup and roared the motor.

As he began driving away the oddest thing happened. He stared right at the window where we were hiding and surprised us with a big wink.

## Yellowstone Ice Jam

Jeanie and I woke early that mid-March day to the sound of cattle bellowing—a deep, desperate, haunting sound, unlike anything we'd ever heard before, even at weaning time.

We came into the kitchen, rubbing the sleep from our eyes.

"What is it? What's happening?" we asked.

"Ice jam! The ice jammed at Winninghams. The river is flooding their barns and pastures." Mom said. "All those poor cows…"

She paused in her task of making a big supply of sandwiches and coffee.

"Dad went to help before daylight when we woke up and heard the cattle. Winninghams' house is flooded, too. It's cut off from all roads. They can't get out."

"Where's Beverley?"

Mom pointed out the window. There we saw Beverley already half-way across the fields hitting a fast pace down toward the railroad tracks a mile away.

Jeanie and I grabbed some breakfast and took off running. We struck off through the fields with Beverley way in the lead and the frantic bawling of cattle propelling our feet.

We came up on top of the first set of railroad tracks, the Northern Pacific, then the second set, the Milwaukee tracks, and suddenly—there before us—was the river.

It lapped nearly against the rails, far from its normal riverbed beyond the big bare cottonwoods on the other side of Winningham's house and ranch buildings where it belonged. Water reached the first floor windows of their story-and-a-half house.

The ice apparently had jammed against the Milwaukee railroad bridge, downstream a mile or so. We knew that ice jams clogging the Yellowstone River could extend for miles, forcing the river to shift

> We woke to the sound of cattle bellowing—a deep, desperate, haunting sound, unlike anything we'd ever heard before, even at weaning time.

*The ice jammed at the curve just above the Milwaukee railroad bridge, forcing the rushing Yellowstone River waters out of their channel.*

outside its main channel, inundating low-lying lands and homes.

Most years, when the ice broke up it went out harmlessly and suddenly—after a few days of warning crackling and popping. The force of rising water cleared the channel quickly as the ice forced other huge chunks out on both banks with loud smashing, grinding sounds. Ice chunks fifteen feet deep fused into castles and cliffs and lay frozen on the shore for days or weeks, until they thawed completely.

But periodically the Yellowstone River turns into a monster. It has a destructive history of ice jams in the Miles City area, where it hits the Tongue River.

Unusually cold winters leave their mark by freezing thicker ice, forming heavier icebergs, Even more potentially disastrous, winters of deep snow bring vastly more water downstream in the first March thaw. Too much water surges everywhere while the still-frozen ground remains unable to absorb it.

It compounds the difficulties when the ice goes out upstream before it breaks up down below. Then the broken ice sweeps down, piles up across the frozen river and lodges tight against bends and bridges. The river channel chokes with ice and there's nowhere for the water to go but bursting from its banks.

Later, when a big ice jam finally breaks loose, it surges on downriver, sometimes causing new jams and major flooding all along the way.

Water can rise fast in unexpected places, within minutes flooding places formerly believed safe, as it had at the Winninghams during the night.

Now the full force of the Yellowstone separated us from their ranch buildings. The whole rushing main channel of muddy river water was between Winninghams' house and the railroads.

Huge ice chunks swirled and crashed against each other, grinding ominously.

Beyond, we could see an icy wall and hear the boom and crash of ice—where the icebergs smashed and rode up on each other to form a solid dam. Spring floodwaters had nowhere to go but onto the fields and pastures, isolating house and barn like a water-soaked island over there.

We joined Beverley, Dad, and a handful of men running along the tracks, shouting to each other. Hazel Johnson was there, too, with her dad Warner Johnson who was section boss for the Milwaukee Railroad.

Three or four Hereford cows stood with us on the railroad tracks, wet, bedraggled, and all of them bleeding from barbed wire cuts, with dark red blood streaming down their white chests.

*The Yellowstone River can turn into a monster. It has a destructive history of ice jams around Miles City, where the Tongue River flows in.*

Before us the waves sloshed against the gravel, carrying swirling sheets of ice and, appallingly, bellowing white-faced cattle beseeching us for help, fighting to keep their heads above water as the current took them.

Because they were fenced into a small pasture against the railroad fence and the water surged above the fence, there was no way they could climb out. Two tight strands of new barbed wire topped the woven wire fence.

Some cows swept downstream to the next cross fence and were held there, emitting deep bellows of despair. Others tried to scramble onto floating chunks of ice. A few circled close, lowing desperately for help from the people gathered on the railroad tracks.

Dad roped a cow that swam close and another man secured the rope over her nose. Several men then worked to drag and hike her heavy body over the fence to the safety of the tracks while Dad braced and held the rope taut.

Another group of three or four men had their rope around a second cow and she thrashed through the water, too, bawling plaintively.

Sadly, they weren't making much headway. The water churned about six inches over the top of the fence and the men couldn't pull the cows, heavy with calf, across the sharp barbed wire. Both cows fought frantically to get up and over to dry land. Both were bleeding as they struggled against the barbs. Other men tried to take down the fence. But no one had a tool to cut the wires.

*From the railroad tracks, Mary Hill and Jeanie view the flood of ice and debris that swept through Winninghams' fields, isolating their home and drowning helpless cattle and wildlife.*

Mom arrived with her sandwiches and coffee. Our neighbors, soaking wet, gathered around with somber faces and quiet voices.

We girls tried to see everything, while staying out of the way.

Farther out in the river current other cows swept by, whites of their eyes flashing, moaning and rolling their eyes in terror.

We learned that the cows in this pasture were the thirty-some prize registered Herefords that Elmer Winningham had bought for his brother Louie, with money sent home from the Army each month. It was about 1943, during World War II, and Louie, our local hero, fought Nazis on the European front, while Elmer stayed home to farm on a draft deferment.

In another pasture, far out of reach, the family cattle swam valiantly while giving hysterical voice to their desperation. We could see only the tops of fence posts where many were hung up against a cross fence.

Just then the county sheriff drove up towing a rowboat on a trailer.

> *Farther out in the river other cows swept by, whites of their eyes flashing, moaning and rolling their eyes in terror.*

Dad hurried over to help unload. He had worked with the sheriff during the last Miles City flood, checking flooded houses on 'the island,' the low-lying side of town where the worst floods hit. During his years living on the Missouri he had crossed the river many times in a rowboat and even swam his beef herd across at Glasgow to ship on the railroad.

The ranchers launched the boat against the top wires of the fence and Dad jumped in and took the oars.

"Who's coming to help?" he called.

The others hesitated. No one much wanted to get into that rocking boat with waves and ice splashing at its sides. Probably few if any of the ranchers could swim.

"Now, Elmer, we need to rescue the Winninghams," said the sheriff.

"Sure. But first we need to save some of these cows."

As they held it against the shore, a high school girl, Warner Johnson's daughter stepped up, smiled her beautiful smile, and climbed in the boat. About seventeen, Hazel was the prettiest girl we knew, with a smile radiant as a movie star. She was always nice to us young kids, too.

Another man jumped in and they shipped over the fence and out into the fast current. They brought in the first cow from midstream, but again came up against the fence.

As they handed off the rope to others, grabbing another, the sheriff came over.

"You've gotta go on over and get those people," he said earnestly. "One of those big bergs hits the house and it's liable to go!"

But the boat headed out to bring in another cow. None of the ranchers wanted to give up on the cattle.

"They're okay over there. They want us to save their cows!" Dad shouted over his shoulder.

By the next trip the men had cut the top two barb wires of the fence and forced down the woven wire. Things went better after that. No more bleeding cows. And the men on the tracks worked swiftly to pull in the cows towed to shore by the rowboat crew.

Now they finally had an efficient way of saving them, the ranchers seemed united in their unwillingness to leave the drowning cattle. The sheriff marched up and down, increasingly concerned for the Winninghams' safety.

We girls were concerned, too, but like the ranchers we felt sick at heart over the desperate cattle.

We could see the water was now halfway up their first-floor windows, and occasionally saw someone waving from an upstairs window under the peak of their house.

*"You've gotta go get those people," the sheriff said earnestly. "One of those big icebergs hits the house and it's liable to go!"*

"The thick trunks of all those cottonwood trees help protect the house and deflect the ice bergs from hitting it," Mom murmured to us.

Soon a bedraggled bunch of twelve or fifteen cows stood silently on the tracks, heads down, as if too tired or distressed to make a move.

The water stopped rising. Only a few cattle still swam out in the river. We heard a few plaintive cries, then an ominous silence. The surviving cattle that could be saved had been rescued. We didn't want to think about the others.

Only sloshing sounds of waves hit against the railroad gravel.

Finally Dad set off alone to rescue the Winninghams, saving plenty of room for them in the boat. But he didn't think they'd come. And they didn't.

*When they saw him coming, they crawled out the upstairs window and sat on the porch roof above the flood.*

When they saw him rowing the boat their way, they all crawled out the upstairs window and sat on the porch roof above the flood. There were four of them: Elmer—the bachelor son—his elderly mother and father and his uncle Charlie.

"They don't want to get into that flimsy little boat—with that whole rushing channel of muddy Yellowstone to get across!" Jeanie whispered to me.

I agreed. It didn't look very safe in the boat.

Dad talked to them a while, then turned the boat back toward the railroad tracks where we waited.

"They think the water is going down," he told us. "They want to check their barns and see what cattle or horses might have survived. They think they'll be able to get out soon."

Later Mrs. Winningham packed a bag and came up to stay with us a few days while her menfolk cleaned away the silt and debris from their house.

We girls spent all morning watching the cattle rescue. The neighbors saved maybe a third of the cows. The others went on down the river, somehow scrambling or lunging themselves over cross fences, or drowned against fences.

Finally, the water receded somewhat. The current turned sluggish.

That afternoon, fortified by another of Mom's sandwiches, Jeanie and I walked down the rails as far as we could go toward the Milwaukee bridge. By this time the water had fallen, but the railroad tracks near the lower part of our slough remained deep in water and debris.

We walked the rails, balancing ourselves on the slippery steel. Jeanie was really good at this and could keep her balance for long distances without stepping down. For me, it was up and down, slipping off, half-stepping between the black wooden ties and back up on the rails.

For once, we didn't need to watch for trains. Up ahead the tracks of both railroads sank out of sight in our flooded slough. A little farther on, poking up between stands of big cottonwood trees and brush, broken rails stood crazily on end, heavy ties still attached at peculiar angles.

Wherever we looked there were interesting and unfamiliar sights. Icebergs swirled and lodged between the big bare cottonwood trees, some carrying dark, mysterious snags.

A bewildered skunk clung to a sheet of ice that swept by, so close we could almost reach out and touch him. Bare-branched trees and brush surged ahead with the occasional carcass of a dead calf or deer caught in the branches. We tried not to see the dead cows that floated out in the current or lodged against the railroad embankment.

Then, up ahead, the little skunk's iceberg smashed up against others and stopped, building a temporary bridge to shore. He lost no time humping his way over the ice chunks toward land.

We watched as he scrambled out and disappeared into the dry marsh weeds, glad that he was spared.

*A bewildered skunk clung to a sheet of ice that swept by, so close we could touch him.*

## Red Ribboned Skunk

"Here girls, look what I found," Dad called to us as he came through the kitchen door. He had a galvanized metal bucket in his hand and a smile on his face.

We all rushed over and peered into his bucket. There in the bottom, were two baby skunks, the white of their stripes glistening against shiny black fur. They wiggled their noses as they stared at us without fear. They were very young and just fit in our two cupped hands.

"Do we get to keep them?" Jeanie asked, astonished, knowing what a problem skunks were on our ranch.

Not only did skunks leave unpleasant odors but they also created havoc in Mom's hen house. She raised several hundred chickens and sold eggs to grocery stores in Miles City.

A skunk in the hen house disturbs all the hens—they get excited, squawk and flap about. Mom said it caused blood spots in the eggs, which then couldn't be sold. Not to mention that skunks ate all the eggs they could and broke the rest.

Mom did not like skunks.

"Where's the mother?" ever practical, Francie asked.

Dad said they didn't have a mother—and upon reflection, I suppose by then they didn't. Probably he happened upon their mother and afterward discovered her two babies.

In this case, our baby skunks were such cute little fellows, even Mom didn't mind them. They loved to be cuddled and liked to snuggle in our clothing. Before Dad brought them to us, he removed their scent glands. A little ether poured on a rag, held to their noses, then the quick flip with a knife, and the little skunks didn't feel a thing.

They slept almost all day and when they awoke, they wanted to play. They acted just like little kittens. We let them run loose in the house. At night they slept in a cardboard box, curled up next to each other on top of soft rags.

We fed them table scraps, raw eggs and a bowl of milk. When their milk bowl was empty and they were still thirsty, they stomped their feet to show their dissatisfaction and demanded more. They also stamped their feet at us if we startled them.

But they never turned their rear ends toward us or acted like they wanted to spray and, of course, they couldn't spray if they wanted.

With the ready food supply, the little skunks grew rapidly.

Then one day they disappeared.

Strange. *Did they escape the house and run away?* When we got older we realized they were just at the age to explore the hen house, and Mom and Dad must have decided it would be best if they just disappeared.

The fun of having baby skunks around stayed with us. Several years later, Jeanie got an opportunity to try out an idea. She found a baby skunk in the barn. The mother had left it for a short time and Jeanie couldn't resist scooping up the baby.

"What do you suppose we can do with this?" she asked.

*One day the little skunks disappeared. Strange. Did they escape the house and run away?*

Francie and I eyed the baby. After all, this was a wild skunk that still had scent glands. Would it spray her? But Jeanie was confident in her ability to handle baby skunks.

"Baby skunks won't squirt if you're gentle with them," she said.

Francie and I looked at each other. We weren't about to get that close. Let Jeanie find out if baby skunks could spray.

About that time a much-loved gray-haired bachelor, Ben Thal, the neighbors' hired man, drove into the yard. Not only was he a kindly man, but also a very good mechanic. He could help Dad fix and drive any piece of machinery we had.

Ben had news for us. Today was his birthday and he brought an invitation from the neighbors to come to their house for his birthday supper.

"I know what we can give him for his birthday!" Jeanie said with a gleam in her eye. "Ben doesn't have a pet and Art Hills' don't have any chickens, so it's a perfect home for our new Skunky."

Mom looked askance at Jeanie's idea, then she smiled, "My, won't he be surprised!"

We looked at each other—would Mom really let us do that?

"He needs to have a present," Mom said, "and he really likes holding your skunk. Soon Skunky will be too big to keep in the house. And you know what he'll do in the hen house if he gets loose."

Jeanie took the little skunk over to Mom's sewing box where she pulled out a bright red ribbon. We nestled the baby skunk into a small square box, cut a few air holes, and tied the red ribbon around its neck. Next, we gently closed the lid over him, wrapped the box in gift paper being careful to leave air spaces, and topped it with a red ribbon bow.

"There! A beautiful present!" Jeanie exclaimed. "Ben said 'Skunks will get in your chicken house.' But he'll know what to do with this baby skunk."

Even Dad had a grin on his face when we piled into the car. Jeanie held our beautifully wrapped gift carefully on her lap.

When the adults retired to the living room to visit, we ran out to the car to get our present. We couldn't help giggling as we placed the box on Ben's lap. He no doubt felt it wiggling, saw the air holes, and knew this would not be the usual present of candy or cookies.

Ben was a good sport and pretended to be pleased with Skunky.

Later we wondered what he did with his pet. It was altogether too possible that Ben turned the skunk loose and it returned to our barn and Mom's hen house.

If so, the joke was on us.

One evening, as we were getting ready to go to the rodeo in Miles City and ran out to shut the chicken house door, we found a surprise.

The hens were all on the roosts. The deep, golden straw strewn around the floor smelled sweet and a feeling of deep, dark peace had settled in. A dozen wooden nests lined one wall about three feet from the floor, high enough to keep the nests above the reach of skunks.

Suddenly we saw it—the florescent white stripes of a huge skunk curled up in a nest. The eggs, gathered earlier, had left a cozy, eggish smell, attracting our visitor. A skunk has three-quarter-inch anal scent glands on both sides of the rectum. When startled he lifts his tail as a defense mechanism and the gland openings, usually

*"I know what we can give him for his birthday!"*

A skunk heading for Mom's hen house sets off a family alarm, causing us to take immediate extreme action.

retracted, protrude and squirt nasty fluid out ten to fifteen feet.

Skunks have no qualms about a person getting close. No problem for them, they just let you have it. *Pee-yew!*

Mom said we had to get rid of the skunk before we left for the rodeo.

We took the twenty-two, but knew if we shot him in the nest, the smell would last forever and the hens would hate us and quit laying. Mom thought we might haze him quietly out the door—she was always hopeful about things like that.

We had a better idea.

We'd read a cowboy story that said a skunk can't squirt with its tail down. The writer explained how, if you find a skunk in the hen house after dark, you can blind it with a powerful flashlight and push it into a cardboard box with a good lid. Have a helper shine the light in the skunk's eyes while grabbing it with one hand—tucking the tail down—and grip its back with the other hand. Then drag the skunk quickly out of the nest, dump it in the box, still holding the tail under its body while the helper slams down the lid before the tail goes back up. Finally, snatch out your hands and any nasty fluid stays inside the box.

Jeanie, Francie and I discussed the plan. We'd take the box a safe distance away and run, leaving the skunk to find his own way out of the box.

Beverley wasn't having any of it. But Francie said if Jeanie would grab the skunk, she'd blind it with the flashlight and put the lid on the box.

*Francie and Jeanie shrieked when the skunk flipped up its tail and let them have it. Both guns!*

Curious, I hung back by the chicken house door. If this worked, I intended to see it.

Beverley was nowhere in sight. Whenever she dressed up to appear in Miles City, whether at a rodeo or church, Bev looked very poised and beautiful. She didn't want to be embarrassed, nor did she ever want us to contribute to her embarrassment. Our lovely older sister.

Mom said, "You're not going to try that!"

But it seemed like we ought to try—it was almost rodeo time. If it worked, we'd have a new skill.

Francie and Jeanie sidled up to the nest. Jeanie grabbed the skunk, pressed its tail down, jerked it out of the nest and jammed it at the box.

But the skunk exploded out of her hands with more power, strength and weight than we could have imagined.

Francie and Jeanie shrieked and ran while the skunk flipped up its tail and let them have it. Both guns!

Francie got away first with the flashlight and Jeanie stumbled after her in the dark, the smell making her retch. We got outside the chicken house gasping and choking.

None of the contents of the scent glands got on Francie, but Jeanie's brand new western shirt was ruined. It was stained and fouled with yellowish skunk liquid and smelled terrible.

Jeanie ran for the house.

"No, wait!" Mom called.

Mom helped her wash in a pan of warm water in the shop and brought clean clothes. They scrubbed with strong soap, while Francie and I watched from a distance.

Mom hung the pretty western shirt on the fence to let it air out and determined that most of the smell all along was in the shirt.

After what seemed ages, we were okayed to go on our way to the rodeo.

We headed for the packed bleachers, with Beverley several rows away, pretending not to know us. A certain aroma still hung around Jeanie. Heads turned before and after she passed.

"Anne, you have to sit on one side of Jeanie," Francie whispered. "And I'll sit on the other side."

As we squeezed into our row, a woman in the crowd behind us said to her companion, "Do you smell skunk?"

Jeanie turned quickly and with a twinkle in her eye retorted, "Maybe there's a dead skunk under the bleachers."

The rest of the evening we carefully blended with the crowds. But from time to time the wind stilled and Francie and I worried that the odor settled in too close around us.

Unfazed, Jeanie would turn and ask innocently, "Do you smell skunk?"

*We headed for the packed bleachers. A certain aroma hung around Jeanie and Beverley pretended not to know us.*

## Neighborhood Fun

"I'll Take You Home Again, Kathleen," Dad sang in his pleasing baritone, with Beverley at the piano.

I think I was next, soloing with my half-sized violin. Fortunately, I was too small to know how bad it sounded and the audience applauded generously.

This was a gathering of our entire community visiting at our house for one of the winter card parties. Money was tight in those austere war years, with people just emerging from the depression of the 1930s. Everyone raised gardens and preserved food for winter. Any income was needed for restocking farms and ranches. There was no extra money and certainly none for entertainment.

*Money was tight in those war years, with people just emerging from the 1930s depression.*

Yet, the people in our community made their own fun. During cold winters they came together for card parties about once a month, either at someone's home or Tusler, our rural schoolhouse.

Years later, neighbors said these were some of the best times of their lives.

It was such fun! First, there was the excitement of cleaning our house, dusting off our bookcases and getting all the coal dust out of the corners—we used coal for heating, hauled from an underground mine in the Pine Hills and it left its residue.

My job as the youngest, was to clean the piano, and I used wonderfully smelling furniture polish to rub in until the piano gleamed. The ivory piano keys we cleaned with ammonia, which gave off piercing fumes, but that job was given to one of my sisters, who carefully kept the ammonia off the polished wood.

I arranged the sheet music on the piano music stand and set up those with cover photos of singers with beautiful smiles.

Mom and my older sisters made sandwiches of canned meat ground with pickles and seasonings with homemade mayonnaise on fresh bread. Guests also brought food for the midnight lunch. Dad brought up full coal scuttles from the basement coal room to keep the house warm during the cold night.

Our house had two big living rooms separated by double French doors, which normally we kept closed to save on heat and coal. We opened these rooms and set up as many as five card tables, some brought by guests.

People started arriving after they'd finished supper and their evening chores.

The older men played cribbage at one end of our long kitchen table. People at card tables played whist or pinochle. These were progressive games and after each set the winning couple moved forward a table and all changed partners. At midnight, when lunch was served, the top winning man and woman, as well as low man and woman received small inexpensive gifts provided by the host.

Kids did not take part in adult card games until we became skilled. Even then, we preferred to make our own fun. Sometimes we played card games—Old Maid or Rummy—or board games and checkers.

Mostly, we liked to go to an adjoining bedroom and hear ghost stories.

"Let's tell ghost stories!" one of our friends would urge and we'd all look at Beverley and Jeanie.

Jeanie made up interesting and scary ghost stories with a mischievous bent. When she got into the tale, her eyes glowed with the mystery of what was coming. Beverley told suspenseful fairy tales,

*Anne learned both guitar and violin at an early age.*

including her listeners as characters in the fun. The rest of us hung on every word. Little ones were secretly glad when the ghost stories ran out and they could return to safety in the next room where their parents played cards.

Part way through the evening there'd be a break and people entertained. That's when people might sing solos; others told jokes and short stories or recited a poem. In those days before TV, many willingly shared their talents and humor. The audience was supportive and no one expected perfection.

Sometimes Beverley played piano for group singing or organized entertainment. The crowd liked her 'Truth or Consequences.' She'd call on people who enjoyed performing to stand up and answer questions. If they guessed wrong, their consequence was performing some silly and amusing stunt she had planned. Everyone laughed uproariously at these antics.

These parties gave our family and the neighbors relief from long winters and the work of feeding cattle, breaking open icy water holes and hauling coal. Winter could be long and cold in eastern Montana and we found many ways to enjoy it.

One hard, but fun, job was putting up ice. Fun, because it brought neighbors together. Since many people had no electricity until several years after WWII, we prepared for the sizzling summer days by storing our ice house full of blocks of ice from the river.

In mid-winter when the Yellowstone River froze to a suitable depth, and it was safe to be out on the river, word passed among the neighbors and a day selected. The Yellowstone could be dangerous when thinly frozen, so the colder the better.

Dad hitched up our work team Slim and Jumbo with the long wagon that he put runners on for feeding hay in winter. This made a good sled to pull on the river, hauling ice. Other ranchers brought their teams and sleds and long saws for cutting the ice.

They sawed open a square hole in a safe part of the river where deep ice, maybe eighteen inches thick, covered the moving water. Then they spent the day sawing blocks of ice as large as two feet square, loaded the hay wagons and hauled them home.

Mom filled her big twenty-cup enameled pot with coffee and heated it on a fire built at the river's edge. People brought sandwiches for lunch and at noon this work day became a social outing with its share of jesting, laughter and friendly conversation.

Everyone filled their ice house within a few days. Ours was dug a couple of feet into the ground with walls of tightly chinked cottonwood logs and a heavy log and sod roof, built just for this purpose. We set the ice blocks tightly together with sawdust packed between each layer and thickly on top. The natural insulation of

*We often picnicked and hiked with the Hill kids, Dorothy, Edwin and Mary. On this day we drove miles with horses and wagon for a load of coal to heat their home. Ed dynamited a coal seam out of the creek bank to release chunks of coal.*

*Our neighbors worked together to cut and put up ice from the Yellowstone River.*

## Skating

We skated many warm sunny Sunday afternoons, but even more special was skating on clear moonlight nights. Sometimes friends and neighbors met us for an after-dark skating party.

When the temperature drops, the air loses moisture and can be stunningly clear and the stars brilliant. Skating on a big sheet of ice, with the only light from a pale moon reflecting on the snow and ice, turns the surroundings into a wonderland. With the only sound the swish-swish of ice blades, even a jackrabbit bouncing from his sagebrush hiding place on the bank makes no sound.

We'd skate until numb with cold, then warm by the fire with a cup of cocoa. Stars overhead sparkled with ice crystals in the air and shone all the brighter far from town or ranch lights.

*When it was our turn for 4-H, they often walked five or six miles to our house, and then walked home.*

sawdust, logs and sod slowed the melting of ice, which kept frozen until mid-September.

We didn't get electricity on the ranch until copper wire became available several years after World War II, so ice was a special luxury during hot summer months. It meant we could keep milk and perishables in the ice box and make ice cream every Sunday we wanted. Grass and waving sunflowers grew on the thick earthen roof of the ice house in summertime, making further insulation.

Women in our community formed the Tusler Homemakers Club and met monthly, organized through the Extension office to help rural women learn—and teach—the latest homemaking skills. The local County Home Agent and women she trained presented self-help programs ranging from growing gardens, canning and feeding families to the latest health recommendations. It was also a place to catch up on local news, share recipes, child care tips and the recent activities of husbands.

The Home Agent appreciated Mom's knowledge and encouraged her to share her expertise.

One time Mom's lesson came with a dire warning.

"My pressure cooker has large screws with wing nuts to hold the lid down tightly," she explained, showing them. She owned one of the first pressure cookers in our community.

"But I'm so afraid the pressure will build up inside the cooker and blow the lid off. I tell my children to help me watch the gauge, but not get too close. I've got to tell you that even our little dog Spotty watches that gauge. The steam pressure builds up, up, up and when it reaches 30 pounds, the needle moves into the red zone.

"Once the needle goes into that dread danger zone, there is a possibility it might explode. I've heard of pressure cookers blowing up and people getting seriously hurt by the flying lid, hot steam or broken glass. I've heard of pressure cooker lids being blown right through the ceiling.

"If the pressure gauge ever gets into the red zone I tell my kids to run out of the room fast. So when they see the needle creeping up into the red zone, they shout, and I run and pull the pressure cooker to the back of our cook stove away from the hot fire.

"Spotty, who sleeps behind the door, leaps up as soon as he sees the girls' eyes getting big, looks at the gauge and dashes out the screen door, pushing it open himself."

Her friends laughed as she finished the story, but in Mom's vivid descriptions, they recognized the need for caution.

At age ten we girls joined the Blue Ribbon 4-H Club, also organized through our County Extension Service, which helped young people 'learn by doing' in the latest home, family and agricultural skills, as well as develop their leadership potential.

Our leaders and the other members lived in the Kircher community nearer town. When it was our turn for the meeting they often walked in a group the five or six miles to our house, and afterward walked home again. We members carried projects of all kinds, received many awards and developed speaking and leadership skills that helped us throughout the rest of our lives.

As part of our 4-H activities, we exhibited a variety of projects at our county fair. Beverley, usually one of the top exhibitors, won considerable money. Jeanie won Montana's top award for bread baking, but since it was during the war, the state Extension Service was unable to send her on a deserved trip with winners from other states. Francie lived and worked with families in Switzerland six months on the International 4-H Youth Exchange program of peace and world understanding. Anne won the regional states demonstration award in Portland, Oregon. With her presentation on ironing a shirt in an efficiency-tested way, using the fewest possible motions in the shortest time, she appeared on fledgling television stations around Montana.

*Riding horseback out of the Pine Hills, neighbors show us the ice caves hidden in a hole half-way up a steep gumbo bluff.*

Occasionally, we attended country dances where local musicians provided the music, and someone passed the hat at midnight to pay them while everyone took a food break. Entire families came, the youngest soon falling asleep on a pile of coats.

This was a time of literally no spare money, no TV, even very little radio because we saved on batteries. Yet neighbors in our community and others enjoyed themselves in many ways. They made their own entertainment and forged lasting bonds.

That's why people still say, "Those were the good old days—the best times of our lives!"

# Brink Ranch Legacy

Western Ranch Life in a Forgotten Era

# Brink Ranch Legacy

## A Good Place for Cattle Ranching

From the time the first cattlemen came into this valley, our ranch was recognized as a prime location for ranch headquarters. This was cattle country and several prominent ranchers established their headquarters here in succession from the 1880s—including Walter Jordan, Henry Tusler and J. J. Shambaugh.

Was one of these men the bridegroom who built the large modern house for his Southern bride? Local tradition points to Shambaugh. But some details don't quite add up.

Situated at the lower end of the Yellowstone Valley where steep bluffs circle in toward the river, the ranch gives good protection from cold wind and blizzards. Vast unfenced grasslands lay higher up on plateaus and draws. Until homesteaders claimed the land, the enterprising cattleman ran his stock on free open range.

Level bottom lands below the bluffs offer rich soil and produce quantities of hay and grain for stockmen to feed their cattle and horses in winter.

Located only ten miles from Miles City, the ranch provided a ready supply of beef for townspeople and the military base at Fort Keogh. An easy day's drive meant cattle could be butchered as needed. In addition, thousands of steers shipped out from the railroad stockyards at Tusler and Miles City every fall to eastern markets.

Most highly valued of all in drought-prone eastern Montana, the ranch has three excellent sources of water. First, pure sparkling drinking water flows out of the hills from a large spring that runs all year. Piped down to the buildings, it provided an abundant supply of fresh water for house, barns and corrals. In one pasture an artesian well flowed continually.

Second, an irrigation system begun in 1885 brings water from Tongue River for hay and cropland. The main ditch winds above the fields, providing an assured water supply during the growing season. Along the ditch bank, towering cottonwood trees grew up, shading homes all the way.

Third, along the northwestern side of the ranch runs the Yellowstone River. In open range days it provided water for livestock grazing in that area, but was inaccessible to us because of the two railroads between.

*Early cattlemen recognized this as a good location for ranch headquarters.*

The cattlemen who established their headquarters here benefitted from all this, living on the ranch from the early 1880s until the economic recession that followed World War I.

Walter B. Jordan purchased what was to become the Brink ranch headquarters from the Northern Pacific Railroad as soon as available and, in the custom of large cattle companies, ran his cattle over a large area of free range. He branded *J-Bar-L*, with a *Sythe* horse brand. As the Post Trader at Fort Buford, near the junction of the Missouri and Yellowstone rivers, he arrived from Iowa in 1872. When Fort Keogh and Miles Town sprang up with boomtown opportunities, Jordan became a cattleman and bought half of railroad section twenty-nine for ranch headquarters. This was part of the enormous land grant given the Northern Pacific as an inducement for opening up the vast lands of the northern Louisiana Purchase. The grant encompassed alternate sections—like the black squares on a checkerboard—extending out forty miles on each side of the railroad tracks, a total of eighty miles.

"A grand empire," one Senator complained on the floor of the U.S. Congress.

The other sections—the red checkerboard squares—eventually opened to homesteading. Early homesteaders could claim a quarter section or 160 acres; later this acreage was increased with various stipulations. So our ranch lands adjoining that original half-section were homesteaded by different owners, then purchased one at a time as the homesteaders left, likely by Tusler or Shambaugh. Later we added a half-section of school land with more springs and good rangeland.

Partnering with his brother-in-law, Joseph Leighton, Jordan launched a number of successful business ventures and helped establish the original irrigation system in the valley. For a time he moved supplies across what became our ranch lands, on the Buford Trail, with bull teams and his famed Diamond R Freighting Company.

Henry Tusler, already a well-known cattleman when he bought the ranch from Jordan in 1897, partnered in several large companies, running thousands of cattle in the area. Tusler came to Montana around 1880 as superintendent of the Concord Cattle Company of Concord, NH. In the summer of 1883 he helped bring in three cattle herds of twenty-five hundred each, locating them on both sides of the Yellowstone River, as well as on the Tongue and Powder rivers. He branded *Two-Lazy-Js*.

Although he died not long after moving here, he left his legacy in the nearby Tusler School, post office and railroad station, named for him.

When Henry Tusler died the land passed from his estate to J. B. Poe in 1904. A search of library and county records fails to reveal more information on this owner. Poe owned the ranch for seven years.

J. J. (John) Shambaugh was the next cattleman to live here. Old-timers told us he built the ranch house and that he was the hapless bridegroom. But there are puzzling aspects.

We first heard the story of the Texas bride from Mrs. Winningham, an elderly neighbor, soon after we came to the ranch.

"Around the turn of the century, J. J. Shambaugh, who owned your ranch, went to Texas on a cattle buying trip. There he found a bride," she told us.

"He came home and built the largest, fanciest two-story ranch house in this part of the Yellowstone Valley for her. It even had running water and a bathroom way back then, the first in any of the ranch homes.

"Yes, that was an elegant house. Especially on the south half built for the owner's family. When completed, he proudly brought his bride to live here. But she didn't like it. Refused to live here and went back to Texas." She sighed, "Shambaugh eventually followed her there."

Mom gave us more details on why the bride departed as she learned them from various old-timers. "In Montana, ranch owners and their families were expected to eat with the hired help, of course. Sadly for Shambaugh, his Southern bride did not approve

*Western Ranch Life in a Forgotten Era*

of this. As one old Pine Hills cowboy told me, 'the Texas girl was raised in more genteel surroundings, where people dressed up for dinner. She couldn't accept our rough frontier way of life.'"

Mom chuckled, as no doubt the old cowboy had, but her eyes sharpened. "Here in Montana, hired people have always been treated like equals. The people who worked here valued their dignity and self-respect and it probably showed." Then her eyes softened, "Things might have differed a great deal in the south where the new bride grew up. She didn't want to eat in this plain room and certainly not with the hired men. And she didn't want her husband to eat here while she ate alone in their dining room."

If we were sitting around the supper table when Mom told the story, she might point out the kitchen woodwork that she had painted a pleasant off-white.

"See how plain and square the woodwork is around the windows and doors in here. It's all cut from local cottonwood trees. And you can see the wooden pegs used for nails. Now look in the living room."

We looked through the open door at the more ornate wood trim and saw she was right.

"Two carved blocks crown the top of every door and window, and there are designer moldings throughout," Mom said. "Mrs. Winningham said that lumber was shipped in by rail and unloaded at the Tusler railway station."

"And the stairways," we chimed in, laughing. "Look at Francie's puny stairway!"

Those steep stairs turned sharply in a half-spiral to get to her large bedroom, but she could run up them fast.

Beverley, Jeanie and I took the long way around to our bedrooms up the wide front stairs with their broad landing and foyers at top and bottom, all of the woodwork beautifully carved. The wide front door built for the owners and their guests—which we rarely used—opened to this staircase and led upstairs to three sizeable bedrooms. I often slid down the long banister.

Naturally, our family made efficient use of the entire house without distinction.

Abstract records list the ranch house as built in

## Our Ranch House

Our ten-room home was designed with a dual purpose in mind: an elegant section for the owner's family to the south, on the right, and a plainer section to the north for hired workers.

On the family side the front door faced onto the wide, wrap-around porch. On the north was a door to the barns, shop and likely a bunkhouse. Our main entrance was on the other side, to the east, where we stepped directly from porch into the kitchen. Verandas ran the length of the house on both sides.

Going inside the front entry on the right, a foyer opened into a wide staircase with a landing and carved banister. There, living and dining rooms were separated by double French doors. A third room at that end on first floor—Mom and Dad's bedroom—could also have been a library, music or sewing room, since there were three good-sized bedrooms above on second floor. All the woodwork in this part of the house was beautifully carved with two cornice blocks

A question exists as to who built this house. Reportedly it was Shambaugh, but not all records verify this.

1905, six years before Shambaugh owned the place. It could have been built much earlier, as the records from that time are approximate, says Dick Mitchell, owner of Security Abstract and Title Company in Miles City.

The county deed book documents that Shambaugh bought the land in 1911. But a difficulty in relying on deed records is that they give only the final date when paid in full and the deed transferred to its new owner. For example, it is possible Shambaugh lived there and built the house soon after 1904, then paid off his mortgage in 1911.

He is the man who, according to tradition, built the large ranch house—with one elegant half designed for the wealthy owners and the other half square and practical for ranch and household help. However, the record fails to affirm that Shambaugh was this man.

Another piece of the puzzle that doesn't quite fit: 'Little Henry' Tusler and his two sisters paid us a visit in the mid-1940s. I was about ten and the Tusler siblings, children of 'Big Henry,' perhaps in their fifties or sixties.

"This was my bedroom," said one of the sisters, entering our upstairs storeroom.

But how could she have lived in this house if it was built by Shambaugh after the Tusler family moved away? Certainly, he made a local impact. When we came many neighbors called our place the Shambaugh ranch.

He sold the ranch in the 1920s to George Bennington. From that time until we moved to the ranch in 1939, the place was owned by land agencies and real estate companies that apparently gave little care to the land or were unable to keep it up because of worldwide depression.

FRANCIE

Our family ranched here for thirty-three years, longer perhaps by twenty years than any previous owner. Dad and Mom built up the rundown ranch—drained fields gone to alkali, rebuilt sagging fences and corrals, improved ranch buildings. And many times over the years, they painted, repaired and redecorated the house, both inside and out, till it shone and rivaled its original beauty. They also purchased more land—half of the school section and the Slater place southwest of us, when they

at the top of every door and window. Even baseboards had their own ornate pattern. Mrs. Winningham told us the lumber was shipped in by rail and unloaded at the Tusler railway station near our school.

On the workers end of the house, woodwork was plain and square, sawn from local cottonwood trees and nailed with slim wooden pegs. On first floor was the large kitchen where the hired people ate, the separator or milk room, bathroom and pantry. The kitchen had birdseye maple wainscoting, covered with layers of paint, paneling the lower third of its plastered walls.

Seven doors opened from the kitchen, a fact that baffled Mom when she tried to arrange efficient work space. She always hoped to build out the kitchen to the west, in planning a more efficient U-shape design. Two kitchen doors led outside, east and west, with the other five doors opening into the original dining and living rooms, the pantry, stairway and bathroom. A narrow spiral staircase led from the kitchen up to a small hallway and two large dormitory-style bedrooms.

To us, the finer distinctions of the dual sections of the house didn't matter because, of course, we used all the rooms. Each of the six bedrooms had a roomy walk-in closet.

According to Butch Krutzfeldt, current owner, oral history indicates the builders of this house used some of the wood from the Yellowstone steamboat, which sank nearby in 1879. If so, we think it's possible this might have been the birdseye maple wainscoting in the kitchen, since it was decorative, yet matched none of the other woodwork.

The full basement, of limestone blocks two feet thick, was separated into three large rooms. It also had two stairways, one to the outside. Completed around the turn of the century, the house was solidly built, with never any structural problems in all the years we lived there. Mom kept it painted a pleasing cream color, with white trim.

Someone once told us, "Seeing your house from the road, I always think it's the kind of house where you have chicken dinner every Sunday." Often, we did.

came up for sale. About three or four miles farther up the irrigation ditch from us, the Slater place had a low log house and a set of rundown corrals and sheds, mostly overgrown with big weeds. Mom and Dad let Bill Hickok, an older bachelor we girls called Wild Bill, live there. Dad fixed up the corrals so Bill could keep a horse or two. Located on Jones Creek, this was an irrigated place with some pasture that grew hay and grain. Sometimes we grazed our cattle on the fields after harvest in the fall.

In the early 1970s, after Mom died of cancer, Dad retired and moved for eighteen years to the Seychelles Islands in the Indian Ocean. He sold our ranch to Lyle Pettus of Miles City, a land agent. Pettus sold off some hilly parcels of the land for what he deemed scenic suburban living sites.

The puzzling questions remain: Was it really Shambaugh who built the ranch house? Was he the mysterious bridegroom? He came later to the scene, several years after the house was listed as built, and he stayed quite a long time—into the 1920s. Or at least he owned the place till then and might have leased it out.

Jordan, the first owner of the ranch, although known as wealthy, came too early and as a partner of his wife's brother, Joseph Leighton, a well-known name in Miles City. Not the bridegroom.

As for the Tuslers, if Henry Tusler built the house, he would have been an older man with children and the bride story doesn't fit him either.

Could there have been an unidentified owner, who perhaps built the house, followed his new bride back to Texas and defaulted on his purchase contract? Because of the stories told by old timers, this seems unlikely.

The timing is right for Poe as the builder. Is it possible he was somehow related to the Tuslers? Did he share in the Tusler estate rather than purchasing the land deeded to him in 1904? If so, could not the Tusler girls have continued to live there after their father's death? Then, since he only owned the place seven years, could he have been the spurned bridegroom who left for Texas? The record is silent on this point.

Nevertheless, we have to go with J. J. Shambaugh. Perhaps he actually built the house after 1911. Perhaps he built it earlier under a mortgage. Perhaps the Tusler daughter had it wrong or we misunderstood her. And maybe, as Mom seemed to think, he left his bride in Texas and continued ranching without her.

Our mother believed Shambaugh was the bridegroom—and she was seldom mistaken. The stories the old-timers told made sense to her, even in light of the Tusler visit. And after all, we arrived within thirty-five years of the event, so they likely had firsthand knowledge.

Still, the mystery remains.

*Our parents removed the long, narrow, dilapidated porch and replaced it with a broad compact one with built-in seating. We used this side of the house, facing east, for our front door.*

# Ranch Ownership

The main headquarters of the Brink ranch were built in the north half of Section 29, Northern Pacific Railroad land located ten miles east of Miles City. *(Land description: Township 9N, Range 48E, Section 29, North Half.)* The information below refers primarily to this original half section (Art Hill owned the south half). Additional land was purchased through many years; ultimately, the Brink ranch property came to include parts of these sections: 16, 17, 20, 29 and 30, and the Slater place. The dates below indicate when final deeds were recorded; however, the buyer was often in possession of the property earlier, in some cases much earlier.

**1881**—**Northern Pacific Railroad** received land grants from the U.S. Land Office. The NP completed tracks through Miles City in 1881. In Montana Territory the NP land grant was doubled to include alternate sections 40 miles out on each side of railroad tracks, a strip extending 80 miles wide across southern Montana. The NP completed tracks through Miles City in 1881.

**1891**—**Walter B. Jordan** deeded from NP railroad, NW1/4 of Section 29.

**1895**—**Henry Tusler** from Walter B. Jordan, NW1/4 of Section 29.

**1901**—**Henry Tusler** from NP Railway Company, NE1/4 of Section 29.

**1904**—**J. B. Poe** from Tusler Estate, NE1/4 of Sect. 29.

**1911**—**J. J. Shambaugh** from J. B. Poe. N1/2 of Sect. 29.

**1926**—**George Bennington** received from J. J. Shambaugh.

**1926**—**National Real Estate Organization** (a corporation) from George Bennington.

**1929**—**Missouri State Life Insurance Co.** foreclosed on National Real Estate Organization. Ownership transferred to MSLI, which held the mortgage.

**1933**—**General American Life Insurance Co.** received from Missouri State Life Insurance Co.

**1944**—**C. D. and Dorthea Wolfe,** land agents, from General American Life Insurance Co.

**1939**—**Elmer and Marie Brink** first leased, then purchased the ranch, from C. D. and Dorthea Wolfe. Made improvements while buying the ranch.

**1950 (about)**—**Elmer and Marie Brink** purchased from School District part of Section 16, School Section land.

**1974**—**Lyle Pettus**, land dealer, purchased from Elmer Brink.

**1984**—**Delano M. Hardy and Marie A. Hardy.**

**1989**—**Leonard M. Follmer and Helen M. Follmer.**

**1991**—**Joseph A. Peila, Jr. and Catherine M. Peila.**

**2004**—**Montana Producers, LLC, William J. Krutzfeldt and Julie A. Krutzfeldt.**

*Sources: William J. Krutzfeldt; Custer County Clerk and Recorder; Dick Mitchell, Security Abstract and Title Company; Miles City Public Library.*

# Buffalo Rapids destroys the Yellowstone

For forty-seven years the steamer Yellowstone plied the Yellowstone River and other waters of the upper Missouri. The first steamship to reach Fort Union at the mouth of the Yellowstone, in 1832, it tied up many times over the years at the Miles City pier, loaded with soldiers, pioneers, military supplies and building materials. Then one fateful day—on June 4, 1879—it hit the dangerous rocks of the Buffalo Rapids below our bluffs and broke apart.

"The river was running high and in some manner the pilot swerved the boat off the deepest section of the channel in going through the Buffalo rapids and the old ship was wrecked, being badly pounded on the rocks," reported the Glasgow Courier in December 1923.

The old ship lay where it was wrecked along the river bank for several years. Then it was sold, hauled to Miles City and fashioned into a two-story building. Some of the lumber was reportedly used in our ranch house, according to current owner Wm. J. (Butch) Krutzfeldt. Near the site our dad found a ship's large anchor.

The Yellowstone was built for the fur trade by Kenneth McKenzie of the American Fur Company. A side wheeler, it was 130 feet long, carried 144 tons of freight, had a draft of sixty-six inches. In its second year of 1833 the Yellowstone took on 7,000 buffalo robes at Fort Union—valued at three dollars each at the fort. During the high water of the 1877 June rise "as many as six boats were tied up at the [Miles City] pier ... all loaded with supplies and building material."

Hostile Indian attacks were a risk. Others were snags, low water, shifting channels and buffalo migrations that sometimes blocked passage for days.

## Major Trails and Roads across our Ranch

The road that wound past the Face and crossed the plateau above was what our neighbor Bill Hill called the Buford Trail, an old military freight and stage road.

ANNE

The U.S. Army, building Ft. Keogh just after the disastrous 1876 Battle of the Little Bighorn, used this overland road to bring in supplies and move troops. When the river was high, supplies were shipped up the Yellowstone, but it was navigable only in times of high water.

In places the Buford Trail is still evident in old wagon tracks, gravel grades along a hillside and dry wash crossings. The old cement tank in Spring Coulee was perhaps constructed by Ft. Keogh soldiers as a watering stop for the military trail. It connected Ft. Keogh with two other military forts: Ft. Buford at the mouth of the Yellowstone River and Ft. Lincoln on the Missouri River at Bismarck.

This trail followed the river except when the terrain became too rough at the lower part of our ranch. The end of the valley was just beyond, with its steep cliffs breaking off into the Buffalo Rapids of the Yellowstone. There the Buford Trail turned a right angle and climbed steeply out of the valley, up the grassy draw near our spring, past the Face and between the point of rimrocks and to the flat above. To this day, the paved highways stay this distance from the river.

Still, the Buford Trail took the long way around from Ft. Keogh. It went northeast to Ft. Buford and then southeast to Ft. Lincoln. A shorter route was needed, so the Ft. Abraham Lincoln trail was laid out overland. It used the same roadbed east of our ranch and another forty miles, then cut east toward Ft. Lincoln at Bismarck, a distance of three hundred miles.

Initially this trip to Bismarck took ninety-six hours by buckboard at a fast clip with two or three teams of horses. With changes of horses, the time was cut to seventy-three hours and at times sixty hours.

Across our land, this new trail merged with the Buford Trail, marked by a line of cedar posts to which a piece of canvas was attached. (Francie remembers seeing some of these cedar posts, each as if standing alone on the open range. The cattle used them for scratching posts, and she thought that was their purpose, but apparently not. The cedar was burnished to a beautiful sheen by range cattle rubbing off their insect and winter hair itch. One or two still waved a bit of dirty canvas rag nailed at the top.)

About every sixteen miles, stations were set up with buildings, corrals, horses, supplies and atten-

*We came to believe the Face marked the entrance to the Buford Trail for stage drivers and weary soldiers. Carved three or four times larger than a human face and three inches deep on a sandstone point of rock just east of our fresh water spring, it has been disfigured by vandals.*

dants. After our ranch was established, its buildings and corrals became a handy stop for freighters and the stagecoach.

The name 'Buford Trail' receded into history in our locality, replaced by the better known Ft. Lincoln Trail. The new route to Ft. Lincoln was first used in 1879 to haul mail.

This freight and stage road—the combined Buford and Lincoln Trails at our end—was used until after the turn of the century when cars came into wider use. Horses could pull wagons where early cars with their narrow tires got stuck. Just above the big spring near the Face was one of these places. Sand and gravel lodged there in deep deposits and a motorized vehicle going uphill needed to keep going fast to get through at all. Even then, it could sink too deep and have to be dug and pushed back down to a grassy stretch. Many times we girls spent an hour or two shoveling and pushing our old cut-down truck—'the Bug'—as we tried to reach the open range or school section areas of our ranch by this old freight road.

This same route of the Buford and Lincoln Trails, winding up onto the high plateau from the level valley below, was followed, roughly, by each of the succeeding four interstate roads that crossed our land.

New and better roads for autos were needed. The Yellowstone Trail, first interstate road across the northern tier of states, became the answer. It also cut through the middle of our ranch, as did old U.S. Highway 10, new U.S. Highway 10 and Interstate 94.

Local people built the Yellowstone Trail in a grassroots movement. Neighbors came together on 'Trail Day,' a day of both work and celebration. Cowboys, ranch hands and family from our ranch undoubtedly joined others with their best work horses, plows and dirt scraping equipment, while women brought food on that festive work day.

They dug out barrow pits on each side of the road for runoff, and built up the road with dirt and gravel, smoothing and packing it into a hard crown. This enabled the dirt road to dry quickly after rain or snowmelt so vehicles didn't leave such deep ruts. Sometimes, as in our driveway, which was part of the Yellowstone Trail, the roadside was built up with rocks to prevent washout when the dry creek bed ran bank full from heavy rain or flash flood. They added culverts where needed to save the road.

Local people chose the route—the most advantageous, shortest way that was least disruptive of farmlands and property—and often with an eye to increasing tourism. It only had to meet with the YT road at the county line.

Coming east from Miles City, the route followed the railroad on the old Buford and Lincoln freight roads to the middle of our fields, then turned in a big loop near the section line and headed for the hills and flat above.

But at this point, instead of going up the steeper draw of the Buford Trail through the rimrocks at the Face—too steep for cars—the new route turned to the right and led east directly past our house and barns. This made a good approach to the ranch buildings that we used for our driveway until new Highway 10 was built on the other side.

The Yellowstone Trail continued on beyond the ranch buildings with its large house, running beside the corrals and the big white barn that held horses and hay. Until it burned, this barn was a well-known Yellowstone Trail landmark. It was believed the fire was caused either by a tipped lantern or a cowboy dropping ashes from a forbidden cigarette. It burned totally to the ground and, when our family moved here in 1939, only charred timbers bucked into a pile marked the spot.

When I was four or five, with no big sisters to tag along after when they were in school, Mom gave me an empty coffee can, and the ranch dog Jack and I spent many hours picking up square nails from that barn—nails that could cause a flat tire in farm vehicles. I had a great sense of satisfaction knowing I was preventing flat tires. And it was fun—like hunting Easter eggs. My guardian Jack kept me safe from rattlesnakes and I thought he would pick up nails too, if he could.

From the white barn, the Yellowstone Trail

climbed gently along the south bank of a big dry wash into the pasture we called the Whitmeyer place, east of us. From there it wound its way up onto the high plateau beyond. The climb was more gradual, longer, and allowed for faster and smoother travel than had it gone past the Face.

Once up on top, the Yellowstone Trail crossed a broad flat and cut through some rugged country toward the head of Spring Creek before winding down again toward the railroad at Dixon and Deep Creeks and on through the next irrigated valley at Shirley. Thus, the Yellowstone Trail generally followed the old Buford freight trail, and cut a new route when that proved too rough.

In all, five important roads, between states and interstate, cut across our ranch, each of them seeking the easiest and fastest route up onto the high plateau from the fertile irrigated farmlands of our valley. This inevitably depended on road-building technology and financing at the time. All five roads found their way up between the Face of the Buford Trail and the east edge of our land.

First came the military road of 1876, the Buford and Lincoln Trails, followed by the Yellowstone Trail of 1913, renamed U.S. Highway 10 in 1926 and paved by 1930. This highway parted from the original at the irrigation ditch half-way through our ranch, crossing on a bridge and carving a new grade around the big round-top hill across from our house.

In 1952, new Highway 10 was built on the opposite side of our buildings from the older roads, followed there by Interstate-94 in 1976. Mom said this took too much of our land—cutting a wide sweep right through the middle of our home pasture—and rendered the other side useless with no water or access for livestock. Sadly, too, the majestic rock-crowned buttes we called Battleship Rock and Bathtub Rock, landmarks for miles around, that we delighted in climbing and exploring, were broken up and used for roadbed and fill.

Thus our ranch saw extended travel across the centuries, and still does. On the Yellowstone River, by Native Americans and early explorers, by Captain William Clark of the Lewis and Clark Expedition, and by the Far West steamboat as it brought home casualties from the Custer Battle. Overland came ancient Indian trails, the historic Ft. Buford and Ft. Lincoln military trails, two railroads, the Yellowstone Trail and three paved U.S. Highways.

Lots of traffic concentrated in a small area. No wonder the Face surveyed it all with a scowl.

---

## Five Roads across the Brink Ranch

**1877. The Buford Trail,** a military wagon road likely following an old Indian trail, moved supplies, mail and soldiers between Ft. Keogh and Ft. Buford. In 1879 the Lincoln Trail followed the same route up into our hills and about forty more miles to O'Fallon Creek, where it turned east to Ft. Abraham Lincoln in Bismarck.

**1912-1915. The Yellowstone Trail** became the first transcontinental road across the northern tier of states. It was built for auto traffic by local people, in a grassroots movement that began only about a hundred miles east of the Montana border.

**1926. U.S. Highway 10** built on nearly the same route as the Yellowstone Trail, with improved grade and drainage and cutting across corners. By 1930 this highway was paved through our ranch.

**1952. New U.S. Highway 10** swept in a straighter course than old Highway 10, cutting on the other side of our house, built with large earth-moving equipment then available.

**1976. Interstate 94** built through our ranch roughly parallel to new Hwy 10, but in an even more direct course, as part of the Eisenhower road system.

## The Lincoln Trail

*Written by Mrs. Gulnare Lutts of the Beaver Valley Homemakers Club, Fallon County.*

On July 1, 1879 at 6:00 a.m. the first official mail carrier left Fort Keogh [on the Fort Abraham Lincoln Trail]. Through wind, sun, rain, hail or blizzard the stages rushed, at first once a week then two, then three times a week. The first trips took 96 hours but the time was shortened to 73 hours and occasionally 60 hours.

The ... buckboard, was drawn by four to six horses or mules, depending on road conditions, driven at a gallop. To prevent fire or snow blocks, a guard area was burned off each fall, about 100 feet on each side of the trail. Due to weathering, the ruts often cut 12 to 15 inches deep, so new parallel tracks were made.

There were many hardships besides the weather and loneliness. Indians were a constant threat and northeast of the present town of Plevna is a marker commemorating the death of a driver known as Fritz ...

Around these gumbo buttes and across these ridges and valleys, the old trail wound its way ... government mail stage, covered wagons, soldiers, people searching for homes, wealth, or adventure with horse, ox teams, and mules plunged and plodded along this undulating trail.

In 1878, one freight train of 95 wagons, each drawn by four to six horses or mules, and each loaded with civilian goods of all kinds made up the largest train to make this trip. All were constantly watched and harassed by the Indians, whose lands and way of life were, by trick and treaty, being forever forced from them ...

A few grassy ruts may be seen on the ridge to the southwest.

## The Yellowstone Trail

*First interstate road across the northern tier of states*

In 1912, the Yellowstone Trail Association was started as a grassroots movement by people organized in Lemmon, South Dakota, who yearned for more passable roads for their autos. Soon joined by others, they proposed this trail as the 'shortest route between the Twin Cities and Yellowstone Park.' By this time, people were enjoying their new touring cars and wanted to travel. However, wagon roads were full of rocks and ruts and, worse, long stretches of deep gumbo mud in rainy and thawing weather. There were few bridges.

During this time, the NP Railroad spent huge sums advertising Yellowstone Park and the large hotel they had built at Gardner, at the northern entrance to the Park. Grand excursions promoted by the railway included stage coach tours through the Park: "Visit these natural wonders and arrive on the comfort of the Northern Pacific Railroad."

Trail organizers met with leaders of each county in every state along the way. In their plan, county volunteers, ranchers and farmers determined the local route and built the road across their county with local funds and labor. This successful idea birthed the long Yellowstone Trail, eventually joining the east coast at Plymouth Rock to Puget Sound on the west coast. In Miles City local leaders met in 1913 and started the trail soon after.

Yellow signs and where available sandstone obelisks painted yellow marked the Yellowstone Trail. Black arrows warned of sudden right-angle turns, as the trail was planned along section lines and avoided cutting across property lines. Land owners, who helped build the road themselves, didn't much want a public road cutting through their fields. Later, federally financed highways along the Yellowstone Trail confiscated land and cut across the corners.

Section lines mattered less in Montana range country than through settled farm lands of the Dakotas. Too, the rugged terrain made straight roads impossible. Easier routes meant working with the lay of the land. Where possible, the Yellowstone Trail followed the railroad and the Yellowstone River. But at the river channel's swing points, where the swift current cut deeply into high cliffs, the easier route swung farther south across

*(continued)*

## Yellowstone Trail (cont'd)

less rugged terrain. Meandering up a grassy draw rather than climbing the crest of a gumbo butte, or crossing a washout creek higher up instead of near the river, were choices selected by volunteers working east of Miles City.

The Yellowstone Trail generated great enthusiasm. At Livingston, a road from the west met the road from east, while one branch headed to Gardiner and the park.

Interestingly, few traces of the original Yellowstone Trail exist today. The pioneers chose the route well—much of the original is buried by subsequent highways. Only here and there a gravel grade, known mostly to local historians and old-timers, veers off to seek alternate terrain.

A grassroots movement so successful it almost disappeared from sight, the Yellowstone Trail is being rediscovered, remapped and its route again honored with the familiar yellow signs.

## Ft. Keogh and Miles Town

Fort Keogh was built at the junction of two rivers—the Tongue and the Yellowstone—soon after the Battle of the Little Bighorn in 1876. Before this, only a few fur traders and mountain men, other than Native people, travelled the area. Many of them came west by canoe or small boat up the Yellowstone; others followed ancient Indian trails.

Miles City grew up near Fort Keogh in 1877 as a drinking and trading post for soldiers stationed there. Originally called Miles Town, it was surrounded by rugged country and lush valley prairie grasses.

Miles City became the destination of many cattle drives from Texas to the grasslands of the northern plains. Soon big cattle herds trailed north out of Texas to fatten on the high-quality, high-protein grass. Ranchers boasted that fully grown steers grew larger on northern ranges—so did horses and even cowboys! Big ranches and cattle companies spread out and claimed the free open lands.

Railroads came early, too. The Northern Pacific arrived in Miles City in 1881 with a huge acreage of lands to sell. Its deal with the government, in building hundreds of miles of track through vacant lands, included a land grant of alternate sections extending out forty miles on each side of the railroad, like a checkerboard. This totaled an actual NP ownership of forty square miles for each mile of track laid. Ranchers and farmers in the valley began buying up these railroad lands, as did Walter Jordan, who first purchased the land for our ranch headquarters from the NP.

Alternate sections opened to homesteading, free in quarter-sections or 160-acre tracts. Initially, this meant four homesteads to the section. Jordan and other ranchers picked these up as they became available.

## The T & Y Ditch

Our irrigation ditch was one of the earliest in Montana. The Miles City Irrigation and Ditch Company was incorporated in December 1885, by Walter B. Jordan, first owner of our ranch, and two partners. By 1911 it was renamed the Tongue River and Yellowstone River Irrigation District—called the T&Y Ditch—and combined with another local company.

The T&Y drained out of an early diversion dam on the Tongue River, built in 1885, twelve miles up from where the Tongue runs into the Yellowstone at Miles City. Later the large Tongue River Dam and Reservoir was built more than a hundred miles farther south on the Wyoming border.

The main ditch skirts the hills for all that distance, bringing water through the relatively narrow valley of the Tongue and then for another ten miles through the broad bottomland fields of the lower Yellowstone valley.

It serves about 300 families with 9,400 irrigated acres. The hundred-plus-miles of canals in the irrigation system bring water to major crops such as alfalfa, corn, wheat and sugar beets, and to vegetable gardens and orchards.

Ours were the last fields on the ditch, the place where high gumbo bluffs close in against the river. Then as now, the main ditch circled our lower pasture against the bluffs, sending any leftover water into a swampy area between the railroad tracks and, eventually, into the Yellowstone River.

Henry Tusler, another early owner of our ranch, filed a deed for water rights in the early 1880s.

*Travelers had three choices going west: they came up the Yellowstone River, in background, or after 1881 traveled on the adjacent railroad, or followed trails and roads down through our ranch. The roads descend these hills from the flat above. Winter feeding of our cattle below the springs shows Yellowstone River and two railroads in background.*

## Celebrating Travelers, Events and Routes through this Land

A rich historical heritage throbs across this land, pulsating against the hiker's feet when climbing toward the hills above—hoofbeats of Indian ponies, stampeding Texas longhorns, cavalry horses and the hastening steps of fleeing outlaws, native women leading dog travois, homesteaders tilling their land.

Along the riverbank, one hears the dip and splash of paddles as Captain William Clark and his party canoe down the Yellowstone River. The paddlewheel *Far West* steams past carrying wounded from the Battle of the Little Bighorn and first news of the disastrous battle.

A World War II troop train rumbles by; in the windows sit rows of young men bound for the front, platoons of uniforms in silhouette, caps aimed straight ahead. A freight train follows: a hundred cars loaded with the sobering machinations of war—jeeps, tanks and mounted machine guns chained to open flatbeds, all painted a dull, dark khaki.

On old Highway 10 a long-distance trucker shifts down, and again down lower, climbing around the long hill from valley to plateau above. On a yet newer highway, drivers of big cars with gleaming tail fins, the Greyhound bus and muddy pickups speed by, and a hitchhiker heads for home.

These events occurred on our ranch and the Yellowstone River bordering it, over centuries. All these people traveled across our land or beside it on the river.

It happened here because, as the last place in the valley, our ranch located where it narrows and the river swings in close against steeply eroded bluffs. Valley trails had to turn up at this point. And they did, as road builders over the years sought the easiest course between fertile valley lands and high bluffs. Interestingly, they never deviated far from the early roads, themselves likely based on ancient trails, even as they shifted from one draw to the next.

This was and still is the natural route of entry and exit for the southern half of Montana, a gateway for travelers up and down the Yellowstone River. The easiest way through the valley, avoiding river crossings, cuts right across our land on the south side of the river.

Thus most travelers who have come west as far as Miles City, visited Yellowstone Park or crossed the continent at this latitude, almost certainly traversed the land that became the Brink ranch, or they rode the rails or river at its edge. This land and that of our neighbors in the Yellowstone valley share a rich history that embraces Native American buffalo hunts, cattle kingdoms, homesteading, depression, wartime apprehension and agricultural prosperity.

*Western Ranch Life in a Forgotten Era*

# Historical Timeline

- **Eons of geological time.** For millions of years vast inland seas cover this land. Volcanoes, earthquakes and glaciers violently shape and shake the terrain. The Yellowstone River and its tributaries eventually break through, charting a course to the Gulf of Mexico. Events like these leave their mark in distinctive geologic formations—rugged bluffs and badlands, rock-crested buttes, plateaus and prairies, dinosaur fossil beds, moss agates, petrified wood and abundant coal and oil deposits.

- **Pre-history. This is Indian country.** A land where many tribes live and come to hunt and trade, yet often encounter culture clashes, war and retreat. Wildlife abound—elk, buffalo, bighorn sheep, antelope, deer, grizzly bear, mountain lions and wolves. Prime grazing range, it is crisscrossed by nomadic tribes following migrating herds of buffalo for hundreds, perhaps thousands, of years as still evidenced by clusters of stone tepee rings partly buried in the earth throughout the region. Horses, arriving in the 1740s, bring major change for nomadic people through increased mobility, speed and effectiveness.

- **1742. The La Vérendrye brothers explore up** the 'Roche Jaune' or Yellow Rocks River in April, writing that they saw mountains, possibly the Sheep Mountains or Big Horns. Hardy mountain men—British, Spanish, French and French Canadian—follow into the Yellowstone valley, trapping, hunting and trading with native people.

- **1803. The Louisiana Territory is purchased from France for $15 million.** New lands include the Yellowstone and Missouri River drainage basins.

- **1804-1806. Lewis and Clark launch the Corps of Discovery up the Missouri River from St. Louis to the Pacific Ocean. On their return,** Captain William Clark explores the Yellowstone valley, declaring, *"The whole face of the country was covered with herds of buffaloe, elk and antelopes."*

On July 30, 1806, Clark's small party floats in dugout canoes and round Indian-style bullboats down the Yellowstone past what later became the Brink ranch. They camp the night before at the junction of Tongue River and in the evening at the mouth of Powder River, making forty-five river miles that day. Clark writes of the dry creeks in this stretch having *"the appearanc of dischargeing emence torrents of water ... and emencely of mud also"* at times. He reports *"great numbers of Buffalow on the banks,"* large quantities of wood *"as is common in the low bottoms of the Rochejhone."* Beaver are plentiful and they caught three *"small and fat"* catfish and a soft shelled turtle.

At the point where the Yellowstone swings against our high bluffs they encounter the hazardous Buffalo Rapids—a section with large rocks in the river. Clark names them 'the Buffalo Schoals' because of a buffalo caught there. He writes that they *"passed a Succession of those Shoals for 6 miles the lower of which was quit(e) across the river and appeared to have a decent of about 3 feet. Here we were Compeled to let the Canoes down by hand for fear of their Strikeing a rock under water and Splitting. This is by far the wost place which I have Seen on this river from the Rocky mountains to this place."* The next day Clark recorded seeing a grizzly, *"the largest I ever saw."*

Later, after Clark rejoins Lewis at the mouth of the Yellowstone, they meet trapping and trading parties coming upriver. Two of their men petition to join them, including John Colter.

- **1807. John Colter and Manuel Lisa's party travel up the Yellowstone in search of furs.** Colter spends the winter trapping in the upper Yellowstone. He is the first white man to see the Yellowstone Park geysers, soon dubbed 'Colter's Hell,' in disbelief of his astounding stories.

- **1828. Fort Union, first trading post on the Yellowstone,** is built downstream at the river's junction with the Missouri. Ft. Union opens up the Yellowstone valley to a rich fur trade of beaver and other highly sought furs. Additional posts move fur trade farther up to the mouth of Tongue River and Rosebud Creek. Later the military post Fort Buford joins Ft. Union at the mouth of the Yellowstone.

● **1832. The first steamship,** *The Yellowstone,* reaches Ft. Union on the Yellowstone River. From this time on, steamboats carry goods farther and farther upriver and take out a fortune in beaver furs, buffalo hides and tongues, salted down in barrels. The peak season for traffic on the river is 1867.

● **1862. Gold.** Gold seekers find what they want in Montana mountains in 1862 and an invasion of this part of the West is on.

● **1862. The Homestead Act becomes law,** granting 160 acres to heads of household over age 21, provided they build a house, dig a well, plow a certain acreage and live there five years. Modified many times, the law grows complex and not infrequently subject to fraud. Homesteads can be commuted after a fourteen month residence by paying $1.25 per acre. In each area homesteading must await completion of land surveys.

The Enlarged Homestead Act of 1909, increasing size to 320 and then 640 acres, brings a new wave of settlers, hopeful of a miracle in dryland farming, to the plateaus and high lands along the Yellowstone. Of all states in the U.S., Montana records by far the most homesteads with a total of 151,600, most granted after 1900. Wheat farming becomes popular until a long drought and market drop after World War I ruins many farmers. This 'homestead bust' forces many to abandon the state.

● **1864. Montana Territory is designated** upon the division of Dakota Territory, acquired in the Louisiana Purchase.

● **1864-1877. Hostilities deepen between whites and Native Americans.** For over a century most Indian tribes welcomed traders. But tensions increase through a long tragic history of broken treaties, settlement strains, deceit, theft, murder, rivalries and fights fueled by liquor, buffalo slaughter by whites, attacks on settlers and travelers, and retaliation by the U.S. Army. Fears of a full-scale Indian war spread across the northern plains.

● **1876. (June 25) General Custer and 7th Cavalry defeated in the Battle of the Little Big Horn.** Attempting to force hostile Indian bands back onto reservations, Custer's troops ride from Ft. Lincoln—likely riding through our ranch as they travel near the Yellowstone River—and then turn up Rosebud Creek. Learning that a large Indian encampment lies just ahead and fearing escape, Custer decides not to wait for reinforcements. He divides his force, planning a three point attack. He leads about 250 men down a long draw into an ambush of thousands of Cheyenne and Sioux warriors. When the dust clears, Custer and all the men with him lie dead.

The paddlewheel steamer *Far West,* already stationed at the mouth of the Big Horn River, churns its way up that small river as far as possible. Over several days it loads fifty-four wounded soldiers from Reno's command and pushes off for Bismarck, arriving there more than a week later with first news of the disastrous battle. Indian runners inform the missionary Thomas Riggs of the outcome and Custer's death within twenty-four hours—he's 500 miles away near Pierre, SD.

● **1876. Fort Keogh, a military post** named for Brevet Major Keogh, who died with Custer, is established under command of General Nelson Miles near where the mouth of the Tongue River flows into the Yellowstone. Later Ft. Keogh becomes a USDA agriculture experiment station.

● **1877. The Buford Trail, is built as a military wagon road to** move supplies and men between Ft. Keogh and Ft. Buford, likely following an old Indian trail through our ranch. This faster, more dependable overland route supplants most steamboat traffic since the river is navigable only in times of deep water. In 1879 the Lincoln Trail takes the same route up into our hills from the valley to Ft. Abraham Lincoln in Bismarck.

● **1877. Milestown springs up near Ft. Keogh** as a trading and drinking post for soldiers. Four years later renamed Miles City.

● **1879. Steamboat,** *The Yellowstone,* **sinks at Buffalo Rapids,** near the Milwaukee Railroad bridge, June 6. This was first steamboat to reach the mouth of the Yellowstone River, in 1832.

● **1879. The first big herd of Texas longhorn steers trails into this area.** Earlier, smaller herds come as beef-on-the-hoof for military forts, Indian agencies and mining camps. Now with the country safe from hostile tribes and nearly emptied of buffalo, the fertile grasslands beckon. Wealthy cattle barons bring in tens of thousands of cattle and turn them loose to graze on free range. They discover that on the high-protein northern grasses small Texas steers 'grow out two-hundred pounds heavier' than in the south, and are surprised to find that not only can cows survive northern winters on range grass, but produce strong healthy calves in the spring. Some of these cattlemen, Walter B Jordan, Henry Tusler and J.J. Shambaugh are early owners of our ranch.

In the boom years of the 1880s, numerous cattle companies headquarter around Miles City, which gains a reputation as the 'Cow Capital of the West.'

● **1881. Northern Pacific Railroad reaches Miles City.** The railroad builds between the Yellowstone River and land that would become the Brink ranch. America's first northern transcontinental railway, the NP reaches the west coast in 1883.

As an incentive for building across the unpopulated west, Congress awards the NP alternate sections of land, doubled in Montana Territory to forty miles on each side of the tracks. This gives the railroad actual ownership of forty sections for *each mile of track*, with additional sections farther out if grant lands are taken. Thus, the NP benefits greatly from land sales, as well as increased traffic of passengers, goods and farm products shipped.

Half of one of these railroad sections becomes the ranch headquarters.

● **1882-1883. Buffalo make their last stand here and within a distance of 150 miles to the east.** When nearly extinct, about 1880, the last 100,000 of an estimated sixty to seventy-five million bison migrate to the Miles City area. Here they split into two herds. Half travel north under heavy hunting pressure and are soon killed off. The other half move east into the relative safety of the Standing Rock Sioux reservation, which at that time extended nearly to the Montana-North Dakota border. Thus, it was not white hide hunters, as popularly believed, but Teton Lakota hunters in traditional hunts on this reservation who, over the next three or four years, killed the last 50,000 free-ranging buffalo. By October 1883 the last buffalo were gone. Riding the open range above our ranch we sometimes found their skulls, horns and dark-stained bones in sandy creek banks, perhaps from this final buffalo migration.

● **1884. Tongue River Indian Reservation is established for the Northern Cheyenne** up the Tongue from Miles City, where many have settled, with headquarters at Lame Deer. Later renamed Northern Cheyenne Indian Reservation.

● **1889. Montana gains statehood**.

● **1908. The Milwaukee, a second transcontinental railroad,** lays track between the NP and the Yellowstone River, after building a bridge across the river just below our high bluffs. Without a land grant, the Milwaukee finances its own track laying. Special immigrant trains bring thousands of land seekers to Miles City and every boom town up and down the valley on both railroads.

● **1910. Several years of drought** thin out settlers on desperately dry lands. Most of the property sells to ranchers.

● **1913-1915. The Yellowstone Trail is built** for auto traffic by local people in a well-coordinated grassroots movement, the first transcontinental road across the northern U.S.

● **1916. Grazing homesteads of 640 acres** are permitted on excess lands not previously homesteaded and proved up. These homesteads require grazing classification with no possibility of irrigation. No mineral rights allowed.

● **1917-1918. World War I.** The U.S. declares war April 6, 1917. War ends Nov. 11, 1918. Young men fight in the trenches overseas and people at home buy war bonds, make surgical dressings and raise money for the Red Cross. During war, the Spanish flu hits hard, killing soldiers and civilians world-wide and devastating Montana communities.

● **1926. U.S. Highway 10 builds on basically the same route as the Yellowstone Trail**, cutting across section corners, unlike the YT, which mostly followed land survey lines. By 1930 this highway is paved through our ranch.

● **1929. The stock market crashes, bringing on the Great Depression of the 1930s.** For eastern Montana however, the depression lasts some twenty years. The state suffers from drought and low farm prices during most of the 1920s, as well as the 1930s.

Through all these years, hard times deepen. Dust storms darken the skies. Grasshoppers invade. People feed Russian thistles to cattle. Extreme weather in 1936 breaks records for hottest, driest and coldest. Ill-conceived federal programs such as condemning, killing and burying cattle and sheep in efforts to raise prices meet with failure and despair.

Rich Missouri River bottomlands are confiscated for Fort Peck Dam, including our parents' first homestead. The huge Ft. Peck reservoir promises much, but fails to deliver benefits to Montanans. Roosevelt's New Deal programs help many survive hard times. Like many others, our family survives by sheer determination, without federal aid.

● **1941. World War II begins.** U.S. enters the war Dec. 7, when Japan bombs Pearl Harbor; war ends with Japan's surrender Aug. 14, 1945. The draft takes most young men at age eighteen. Banners begin appearing in windows, a blue star for each son or daughter serving, and sadly, a gold star for each life given. Wartime restrictions bring scarcity of goods: rationing of sugar, tires and gasoline. Citizens join to mobilize resources. We plant Victory gardens, buy war bonds, recycle metals and cooking fat for military use, organize civil defense teams, practice blackouts and raise funds for the Salvation Army and Red Cross. A famous poster declares 'Uncle Sam Wants You!' with Uncle himself pointing a scrawny finger at each of us. Women enter the workforce in west coast factories that shift abruptly from building cars and tractors to ships, munitions and implements of war.

● **1952. New U.S. Highway 10 builds through our ranch** cutting high on the other side of our house from the older roads, followed there by I-94 in 1976.

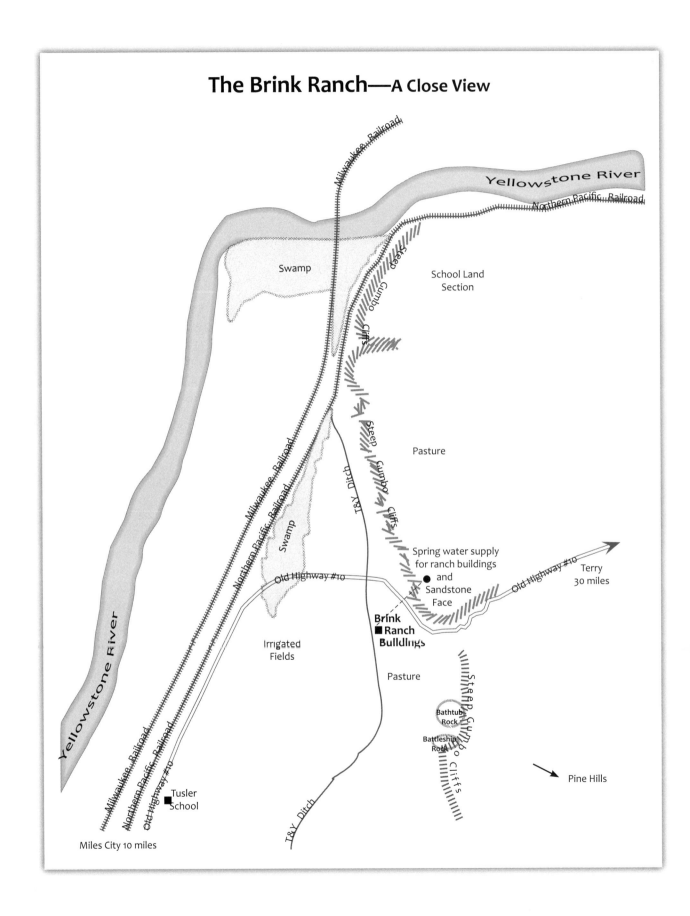

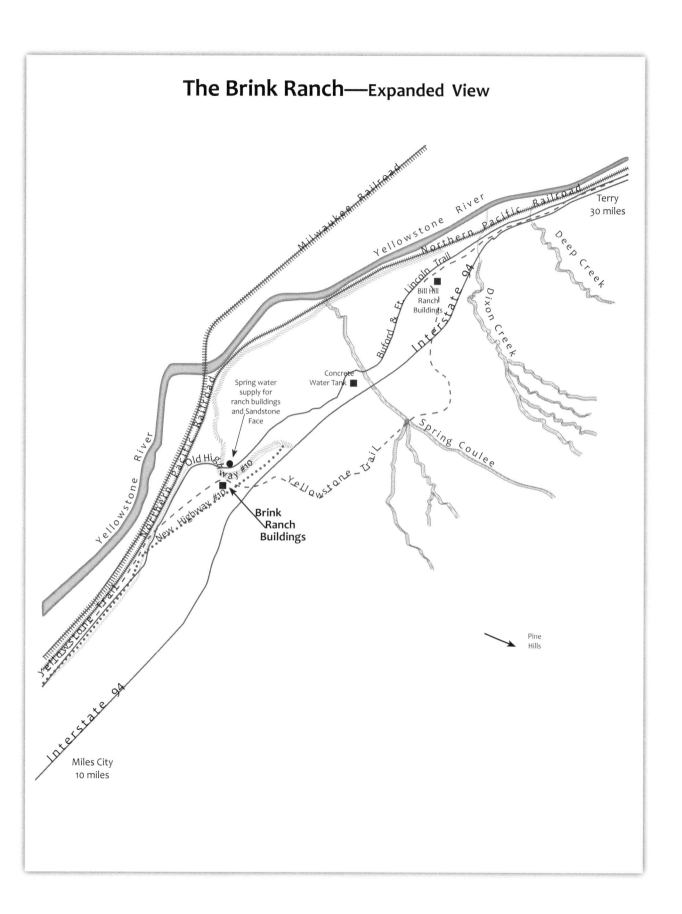

# Tribute

Our mother and father both came west by 1909—Mom as a child of five in a covered wagon and Dad, barely thirteen, alone, following the track of a cowboy hero from his hometown into the Missouri River Breaks near Jordan.

They knew wild west settlement days first-hand and through tales told by old-timers—lawmen, Texas trail drivers, cattle kingdoms and days of free range. With always an interest in community improvement, they focused on family and the practical issues of survival in tough times.

Born around the turn of the century, they were married in Glasgow in 1928. They made their first home in the rich Missouri River bottomlands with miles of unfenced cattle range in the Missouri Breaks, but were displaced a couple of years later by Fort Peck Dam and its rising waters.

Adeline Marie Barrett Brink (1901-1965) is descended from a long line of educators whose earliest relative in the U.S. was involved in French-Indian wars in 1751. Her mother Louisa Mosier was a teacher and her father a Quaker and horse breeder who ran his cattle herds and horses on the Rosebud Indian Reservation. They homesteaded near Shade Hill in South Dakota, later moved to Wenatchee, Washington, and soon returned to Montana to ranch in Garfield County and near Lewistown, where Mom went to high school.

Mom graduated from the University of

Montana in 1924 and did postgraduate work at the University of California at Berkeley and several Montana colleges. In a teaching career that spanned 40 years, she coached the girls' basketball team in Winnett in 1926 and 1927, taught mathematics and languages in Lewistown and Winnett high schools and later was a teacher at Jordan, Mildred, Ashland, Ismay, and a number of rural schools in southeastern Montana.

With a ranch wife's versatility, she coped equally with the 'hospital band' of sheep, a bloating heifer, grain harvest—or a classroom full of children. When her own children were babies she insisted that they have fresh milk—despite the fact that corralling a wild range cow was nearly as big a production as milking it. Dad finally bought her a goat.

Always active in education circles, Mom worked in Miles City chapters of the American Association of University Women and the Daughters of the American Revolution.

Through the years she inspired countless Montana youngsters—including a number of ranch hands—to widen their horizons by seeking college degrees. She spent a lifetime in living Montana history, collecting it and passing it on to Montana school children.

Elmer Theodore Brink (1896-1990), was born in St. Hilaire, Minnesota to Otto W. and Wilhelmina (Minnie) Brink. Otto, a blacksmith who carried the iron for his trade on his back from the nearest railroad town, built wagons and carriages, and later expanded his business into farming and race horses. He came at age twenty in 1882 and Minnie in 1880, both from Sweden.

Our father rode a racehorse circuit and later rode saddle broncs in rodeos. During the early 1900s he drove freight wagons with a jerk line team of up to twenty-four horses from Miles City to Jordan and north to Glasgow.

Dad managed numerous cattle drives and was a World War I veteran. He helped organize the Custer County Farm Bureau, and was an active member of the American Legion, Toastmasters, Last Man's Club, NRA rifle club, and along with Mom, belonged to the First Methodist Church and enthusiastically supported 4-H. Upon retiring from the ranch he and Mom moved to Decker and managed the Tongue River Dam, which furnished water for the Tongue and Yellowstone irrigation project, while she taught nearby.

After Mom's death, Dad became a world traveler, spending time in Australia, New Zealand, India, Africa, and Indonesia, going around the world several times both by ship and air. He built a house in the Seychelles Islands in the Indian Ocean where he made his home for seventeen years. In this he pursued a dream planted by a returned veteran who said he'd jumped ship there and found 'the best fishing in the world.'

Both Mom and Dad are buried in the Custer County Cemetery, Miles City.

We are grateful to our parents for the energy and zest they imparted to us each day.

# Beverley Brink Morales Badhorse

(1929-2003)

Beverley's sense of adventure took her on two career paths. She spent the first half of her career in the sophisticated world of metropolitan newspapers. Then, knowing their desperate needs, she moved to Lame Deer, Montana, and took on the problems of Native Americans and reservation life. Her enthusiasm and dedication to principle and hard work brought major achievements in both.

At age one, Beverley rode horseback every day to school with her mother who taught a country school on the Missouri River. She developed a love of learning, skipped several grades, and graduated from Custer County High School in Miles City. While there, she challenged the unfairness of a rule that only boys could join the science club and, with the backing of several teachers, became the first girl admitted.

She earned her journalism degree from the University of Montana at age 19 and later a masters degree from the University of Wyoming.

As women's editor at two major newspapers, Ft Lauderdale's Sun-Sentinel and the Dayton Daily News in Ohio, Beverley hit the glass ceiling—the highest level large newspapers allowed women. Yet she won over 60 journalism awards, virtually every award given at that time, including the prestigious University of Missouri national competition for Best Women's Pages in 1960. Unusual for women journalists of the time, she had three children while working full-time.

Beverley also wrote for the Mexico News in Mexico City, Whitefish Pilot in Montana, the Council Bluffs Nonpareil in Iowa, and frequent special features for the Billings Gazette and Great Falls Tribune.

When she left the world of city newspapers for the Northern Cheyenne reservation, Beverley founded one of the first independent newspapers in Indian country. She called it A'tome—*Listen*.

Moving on to much-needed work with Indian tribal colleges, Beverley devoted herself to the development of tribal resources and health improvement, often in conditions of great poverty. She helped tribal leaders envision solutions, obtained funding and ensured follow-through on specified goals. In Montana she worked many years for Dull Knife College of the Northern Cheyenne and Ft. Belknap College, also assisting tribal colleges in North and South Dakota, Wyoming and the Denver Indian Health Center. Along the way she wrote the books *Wyoming Land of Echoing Canyons* and *Fruits of Destiny: Biography of a Clipper Ship Sailor*.

*Bev interviewing Burt Lancaster in Mexico City.*

Beverley scored a major coup in obtaining a television station for Dull Knife College.

On the day the federal offer arrived on her desk, she applied for a television station for Dull Knife, the only one allowed per state. It meant all other state colleges and universities would pay licensing fees to the selected college, a much needed amount. Because of her quick and dynamic proposal, Dull Knife was immediately selected.

However, unaware it had been awarded, larger Montana colleges locked in a dispute over which should apply for the TV station. To resolve the question, the Board of Regents invited all colleges to Helena to select one. Beverley's was too small to be considered but she was invited to the meeting.

When she arrived, a little late, pausing in the doorway to locate a seat, the whole assembly turned and faced her. It was just announced that the licensing was no longer available—that she had won it. Before she could sit down, she was summoned to the governor's office.

The governor leaned back in his chair, frowned and sized her up.

"Beverley, how'd you do it?" It was a moment of triumph for Beverley.

In her final years Beverley worked in Fairbanks, Alaska, to be near her son Hector, as Senior Grant Writer for the Tanana Chiefs Conference. She obtained millions of dollars in funding for Athabascan tribes and was instrumental in the vision and building of their large new Tribal Hall, a source of much pride, built with logs in the old way. She made strong friendships and through her caring support was adopted into the Alaska Native community.

Truly charismatic and always generous, Beverley loved her work, the people she worked with and those served by her efforts. She lived in the moment, keeping a focus on her vision. Beverley was preceded in death by her daughter Marie, son Daniel, husbands Hector Morales and Jack Badhorse. She is survived by son Hector Morales and his wife Mary of North Pole, Alaska.

*Beverley, at right, encouraging her assistant in Ft. Belknap College office.*

# The Authors

## Anne Brink Sallgren Krickel

Anne is an international award winning harp jazz and pop performer. She entertains on a wide variety of musical instruments—the dobro, banjo and antique instruments such as the dulcimer, zither and psalmodican. She teaches music and private harp lessons as well as harp workshops in Montana and Georgia.

She received a BA degree in medical technology at Carroll College in Helena, Montana, with internship in Portland, Oregon, and post graduate courses in the fields of music and health including special courses at the Centers for Disease Control and Prevention in Atlanta. She worked as Bacteriologist at the Montana State Health Microbiology Laboratory for ten years. Anne taught Microbiology at Carroll College and bacteriology workshops for training laboratory technologists in hospitals throughout Montana.

With her husband, John, she taught many adult education classes on a variety of subjects—her favorite, wilderness backpacking. Anne continues her love of nature and the outdoors in the Rocky Mountains.

Her strong sense of volunteerism has led to her service on boards of many community non-profit organizations. She is a recipient of the Dean Day Smith Award for community service. Anne is the mother of three daughters, has four grandchildren, spends winters in Portal, Georgia and lives in Phillipsburg, Montana.

## Jeanie Brink Thiessen

An elementary teacher for forty years, Jeanie taught in Montana, New Mexico and California. Her degree in education is from the University of Northern Colorado, and her master's from California Polytechnic Institute.

Of her many years teaching, Jeanie says her most memorable was the year she taught three ranch children in a designated Isolated School north of Miles City. She lived alone far from any ranches. Her one-room school had no water supply, no indoor plumbing and no electricity. She carried coal, built fires for heat and had no car, but kept her saddle horse in a shed in the school yard.

During her teaching career, Jeanie (Betty Jean) received numerous awards and distinctions, including acceptance into Phi Kappa Phi and Delta Kappa Gamma Society International, a professional honorary of women educators.

Always an innovator, Jeanie used her own money to buy computers for her ten-year-old students around 1979. Hers was one of the first classrooms to use computers, and these children quickly became proficient, also designing their own web page.

Jeanie was named *Woman of Distinction* of her Senatorial District for her pivotal role in shaping the economic, educational and social fabric in Hacienda Heights. On weekends she continued her outdoor activities as volunteer in the California Ski Patrol. Jeanie has two sons, three grandchildren and lives with her husband Mac in Hacienda Heights, California.

## Francie Brink Berg

The author of thirteen books, Francie has written in the fields of western history and health for over forty years. She graduated from Montana State University, Bozeman, in family consumer science, and holds a master's degree in family social science and anthropology from the University of Minnesota. Born during the depression at home on their ranch in the Missouri River Breaks, she has worked as a county extension agent in Scobey, Montana, and taught high school and college.

Her books on western history include: the *Old West Series—North Dakota Land of Changing Seasons, South Dakota Land of Shining Gold,* and *Wyoming Land of Echoing Canyons; Ethnic Heritage in North Dakota;* and the *Last Great Buffalo Hunts: Traditional Hunts in 1880-1883 by Teton Lakota.*

As founder and sixteen-year editor and publisher of *Healthy Weight Journal* and author of *Women Afraid to Eat, Children Afraid to Eat,* and *Underage and Overweight, What Every Family needs to Know (www.healthyweight.net),* Francie reported the scientific research on weight issues and eating disorders to health professionals and consumers worldwide. Through her work with wellness, she has presented seminars at international and national conferences, and been a guest on national television, including Oprah, Lezza and Inside Edition. Francie continues to enjoy riding and keeps both horses and beef cattle in her pastures. She has four children, nine grandchildren and lives in Hettinger, North Dakota.

# Readers say...

**THESE HEARTFELT STORIES CRY OUT** to be shared with everyone—especially if you love Montana. Humor, hardships, ingenuity and family strength—a must read for anyone who has grown up on a ranch. These stories capture the fun, joy, trials and tribulations of growing up on a ranch in Montana. Should be required reading for all Montana history classes—an assignment students will enjoy.

—Ardis J. Rice, Librarian
Lewis and Clark County Library, Helena

**MONTANA STIRRUPS, SAGE AND SHENANIGANS CAPTURES** a fascinating time in Western American history, that interesting bridge between the hardscrabble life of Montana ranchers before and after modern mechanization took hold on the plains. This is a time when horses were all-important, and the tales of ranch ife, as told by three sisters, reveal that in delightful fashion.

Anne observing that "when our lead cow turned tail to the wind, we knew we were in trouble," draws the reader directly into a winter blizzard, when "we were so stiff with cold, and the bottoms of our feet stung with a thousand needles when we eased out of the stirrups and hit the ground." You know these gals were as gritty as the best of the men, who claim most attention in the story of the West.

This is the dramatic story of one ranch family, told by three sisters who lived and breathed what today is an American Epic. There is no greater authority than those who straddled a horse to round up cattle or ran a stick-shift truck through the muck and snow.

The fabulous Brink sisters will educate, enlighten and entertain you, all the while revealing the grandeur that is ranch life in Montana.

—David Borlaug, President
Lewis & Clark Fort Mandan Foundation

**THESE VIVID AND ACTION-PACKED STORIES** will help us old-timers tell our grandchildren and great-grandchildren of our lives back then. This book reminds me of the patriotism of our family and community during World War II, the closeness to the land, the stewardship we felt and the satisfaction of hard work. In my lifetime I've seen dramatic changes, as chronicled here. These authors tell it like it was in colorful, fast paced stories. Historically accurate, this book is a pleasure for all who appreciate the spirit of the west.

—Tim Babcock
Governor of Montana (1962-1969)

**THE MEMORIES** in *Montana Stirrups, Sage and Shenanigans* are a treasure. When they strike a harmonic chord in the memories of the reader it is a double gift.

—Rev. Warren S. Craig
First Presbyterian Church, Anaconda

**IT'S GREAT TO HAVE THE HISTORIC** significance of this ranch documented. Without these authors' research and memoirs the details would be lost to Montana history. Their former ranch was headquarters for one of Montana's early ranches, which ran thousands of cattle on both sides of the Yellowstone River in the 1880s, and is crossed by historic trails. It is now possible to find, through this ranch, remnants of the 1877 Ft. Buford Trail and the 1913 Yellowstone Trail—and to actually walk in the ruts of these early trails.

—Louise Galvin
Lewis and Clark County Rancher
President LP-Bar, Inc.

**WITH PLENTY OF ADVENTURE, HUMOR, DRAMA AND THE DRIVE TO SUCCEED** that were the foundation of old west stories, *Montana Stirrups, Sage, and Shenanigans* is a fine pick for those who seek more of these stories of legend. Enhanced with many historic black and white photographs.

—Midwest Book Review
Reviewer's Choice

**HOLD ONTO YOUR HATS AND GRAB THE REINS;** these three Montana-raised women are taking you for an authentic ride into ranch life as they lived it. Reading their lively narrative, you'll know what it's really like to rescue drowning cattle from Yellowstone ice jams, butcher roadkill to help Rez friends, move cattle in blizzards. You'll see the hazards of ranch life from overturned tractors to snakebite to wheatfield fires. You'll welcome stranded travelers, hunt deer and elk, and enjoy the bounty of a ranch garden, and make doughnuts and divinity.

—Linda M. Hasselstrom
**Author of *No Place Like Home: Notes from a Western Life***

**THE BOOK BRINGS BACK** long-forgotten memories—the way we did things back then. Very different from ranching now.

—Patty Spears Helm
Helm Hereford Ranch, Miles City

**THE HUMOR, SPUNK AND SPIRIT** of these sisters shines through the pages of their joint memoir of ranch life. Their personal stories describe an era when living ten miles from town brought a degree of isolation and self-sufficiency almost unfathomable by today's standards. The bond between these sisters seems forged by the demands of ranch life, the extremes of Montana's climate, and their own inventive natures.

—Donna Healy, Feature Writer
*Billings Gazette*

# MONTANA STIRRUPS, SAGE AND SHENANIGANS
## WESTERN RANCH LIFE IN A FORGOTTEN ERA

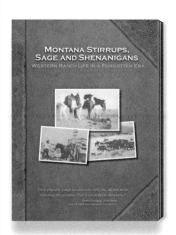

An old west epic with a modern touch, this book is filled with personal stories of ranching, wildlife and western humor. Includes over 260 historic photographs from the Brink family collection. The authors—Francie Brink Berg, Anne Brink Sallgren Krickel and Jeanie Brink Thiessen—grew up on a historic cattle ranch in eastern Montana on the Yellowstone River.

Flying Diamond Books
402 South 14th Street
Hettinger, ND 58639

For more information, reviews and how to order
visit the website, email (info@montanastirrupsandsage.com)
or call 1-605-347-1806

**www.montanastirrupsandsage.com**